GENEALOGIES OF INTERACTING PARTICLE SYSTEMS

LECTURE NOTES SERIES
Institute for Mathematical Sciences, National University of Singapore

Series Editors: Chitat Chong and Adrian Röllin
Institute for Mathematical Sciences
National University of Singapore

ISSN: 1793-0758

Published

Vol. 38 *Genealogies of Interacting Particle Systems*
edited by Matthias Birkner, Rongfeng Sun & Jan M Swart

Vol. 37 *Mathematics of Shapes and Applications*
edited by Sergey Kushnarev, Anqi Qiu & Laurent Younes

Vol. 36 *The Geometry, Topology and Physics of Moduli Spaces of Higgs Bundles*
edited by Richard Wentworth & Graeme Wilkin

Vol. 35 *Combinatorial and Toric Homotopy: Introductory Lectures*
edited by Alastair Darby, Jelena Grbić, Zhi Lü & Jie Wu

Vol. 34 *White Noise Analysis and Quantum Information*
edited by Luigi Accardi, Louis H Y Chen, Takeyuki Hida, Masanori Ohya, Si Si & Noboru Watanabe

Vol. 33 *Sets and Computations*
edited by Sy-David Friedman, Dilip Raghavan & Yue Yang

Vol. 32 *Mathemusical Conversations: Mathematics and Computation in Music Performance and Composition*
edited by Jordan B L Smith, Elaine Chew & Gérard Assayag

Vol. 31 *Geometric Analysis Around Scalar Curvatures*
edited by Fei Han, Xingwang Xu & Weiping Zhang

Vol. 30 *Modular Representation Theory of Finite and p-Adic Groups*
edited by Wee Teck Gan & Kai Meng Tan

Vol. 29 *Forcing, Iterated Ultrapowers, and Turing Degrees*
edited by Chitat Chong, Qi Feng, Theodore A Slaman, W Hugh Woodin & Yue Yang

Vol. 28 *Slicing the Turth: On the Computable and Reverse Mathematics of Combinatorial Principle*s
by Denis R Hirschfeldt, edited by Chitat Chong, Qi Feng, Theodore A Slaman, W Hugh Woodin & Yue Yang

*For the complete list of titles in this series, please go to
https://www.worldscientific.com/series/LNIMSNUS

Lecture Notes Series, Institute for Mathematical Sciences,
National University of Singapore Vol. 38

GENEALOGIES OF INTERACTING PARTICLE SYSTEMS

Editors

Matthias Birkner
Johannes Gutenberg-Universität Mainz, Germany

Rongfeng Sun
National University of Singapore, Singapore

Jan M Swart
The Czech Academy of Sciences, Institute of Information Theory
and Automation, Czech Republic

World Scientific

NEW JERSEY • LONDON • SINGAPORE • BEIJING • SHANGHAI • HONG KONG • TAIPEI • CHENNAI • TOKYO

Published by

World Scientific Publishing Co. Pte. Ltd.
5 Toh Tuck Link, Singapore 596224
USA office: 27 Warren Street, Suite 401-402, Hackensack, NJ 07601
UK office: 57 Shelton Street, Covent Garden, London WC2H 9HE

British Library Cataloguing-in-Publication Data
A catalogue record for this book is available from the British Library.

Lecture Notes Series, Institute for Mathematical Sciences,
National University of Singapore — Vol. 38
GENEALOGIES OF INTERACTING PARTICLE SYSTEMS

ISBN 978-981-120-608-5

For any available supplementary material, please visit
https://www.worldscientific.com/worldscibooks/10.1142/11439#t=suppl

Contents

Foreword

The Institute for Mathematical Sciences (IMS) organizes thematic programs of longer duration as well as shorter workshops and public lectures. The themes are selected from among areas at the forefront of current research in the mathematical sciences and their applications. Each volume of the *IMS Lecture Notes Series* is a compendium of papers based on lectures and tutorials delivered at the IMS. The aim is to make original papers and expository articles on a subject of current interest available to the international research community. These volumes also serve as a record of activities that took place at the IMS. We hope that through regular publication of these Lecture Notes the Institute will achieve, in part, its objective of reaching out to the community of scholars in the promotion of research in the mathematical sciences.

Chitat Chong
Adrian Röllin
Series Editors

Preface

The present volume grew out of the one-month program "Genealogies of Interacting Particle Systems", held at the Institute for Mathematical Sciences, National University of Singapore, from July 17 – Aug 18, 2017. Interacting particle systems are infinite systems of interacting Markov chains. Since their introduction in the 1970s they have found many applications in statistical physics and population biology. In biologically motivated systems, it is often natural to determine the state of individuals living at a given time by tracing their ancestors, or potential ancestors, backwards in time. The collection of all (potential) ancestors with their relationships forms a random network that is called the genealogy of the interacting particle system. There is a close connection to real genealogies of biological populations, but the general idea is useful in other settings as well, for example, for systems motivated by theoretical physics.

Central to the program were a one-week workshop as well as a week of "Learning Sessions", an experimental format where participants presented topics outside of their own direct research area. As described in a small article in the *IMS Bulletin* (Volume 46(7) (2017), 16-17), these sessions proved fruitful in spreading knowledge and promoting interaction. They also yielded five lecture notes that serve as an introduction to new and developing fields of research and that are published in the present volume. In addition, the present volume also contains five original contributions by participants of the program, which include both survey articles and original research.

We would like to thank the IMS for generous financial support and the IMS staff for their dedication and expertise. Organising a meeting is a real pleasure when you are backed up by a professional team that you can always rely on! Last but not least, we would like to thank all participants, many of whom put in a lot of effort to make the program a success.

February 2019

Matthias Birkner
Johannes Gutenberg-Universität Mainz

Rongfeng Sun
National University of Singapore

Jan M. Swart
The Czech Academy of Sciences,
Institute of Information Theory and Automation

Volume Editors

Part I

Lecture Notes from Learning Sessions

Coexistence and Duality in Competing Species Models

Yu-Ting Chen
Department of Mathematics and Statistics
University of Victoria
Victoria, British Columbia
Canada V8P 5C2
chenyuting@uvic.ca

Matthias Hammer
Institut für Mathematik
Technische Universität Berlin
Straße des 17. Juni 136, 10623 Berlin, Germany
hammer@math.tu-berlin.de

We discuss some stochastic spatial generalizations of the Lotka–Volterra model for competing species. The generalizations take the forms of spin systems on general discrete sets and interacting diffusions on integer lattices. Methods for proving coexistence in these generalizations and some related open questions are discussed. We use duality as the central point of view. It relates coexistence of the models to survival of their dual processes

Contents

1. Introduction

Competition of species is a basic phenomenon studied in ecological dynamics. The mathematical analysis goes back to the foundational model by Lotka and Volterra in [19, 30]. Total sizes of two populations of different species, or types, are the main objects of study. The populations are assumed well-mixed, and their sizes are in the continuum. The dynamics for these total sizes follow a two-dimensional extension of the logistic differential equation. Now restricted resources and competition among individuals in the populations suppress the exponential growths. See the monograph by Hofbauer and Sigmund [12] for an excellent introduction to this classical model.

In the biological literature, choosing new ingredients to extend the Lotka–Volterra model has been a subject of research during the last few decades. One popular method is the use of stochasticity and spatial structure. By stochasticity, types of individuals in the whole population are randomly determined. In the presence of spatial structure, it is no longer the case that everyone can interact with everyone else. Populations are arranged according to Euclidean spaces, integer lattices, or more general geometric structures. In particular, the Euclidean space and integer lattice in two dimensions are biologically relevant. The corresponding tori are useful for mathematical analysis and computer experiments.

Spatial structure and stochasticity can be combined to create new properties. Spatial structure naturally points to the use of partial differential equations, and in the presence of stochasticity, of stochastic partial differential equations. See [13, 20] for some examples. Besides, stochastic models on discrete spatial structures arise as popular candidates to incorporate both space and stochasticity. They can be connected to stochastic partial differential equations via scaling limits [5, 7], so that useful tools from stochastic analysis can be applied. On the other hand, these discrete models can circumvent the serious issue of well-posedness of stochastic partial differential equations beyond one spatial dimension.

This article is an introduction to stochastic spatial generalizations of the Lotka–Volterra model. For definiteness, our discussions are centered around two well-known generalizations, one due to Neuhauser and Pacala [22] and the other due to Blath, Etheridge and Meredith [1]. We explain in Section 2 how these models are natural extensions of the Lotka–Volterra model. These extensions incorporate stochasticity in the forms of the voter model and the stepping stone model.

The models in [1, 22] can be briefly described as follows in terms of the spatial structures and the dynamics. First, the model by Neuhauser and Pacala is a dynamical model (started at time 0) for individuals in two populations of different types. These individuals occupy different sites of a discrete space, and so the model is at the microscopic level. (In terms of the terminology in [18, Chapter III], the model is a spin system.) In contrast, the model by Blath et al. [1] considers continuous population sizes as in the Lotka–Volterra model, but here the two populations are structured into 'villages' occupying vertices of $\mathbb{Z}^d$. Given weights for the strengths of interactions among the villages, a countable system of interacting diffusion processes defines the dynamics of the sizes of the populations.

As in the Lotka–Volterra model, the major interest in these generalizations is centered around the question of characterizing large-time behavior. In these lecture notes, we explain the related properties with the use of duality. For the Neuhauser–Pacala model, our discussion considers the associated spin system (η_t) in terms of its dual process $(\widehat{\eta}_t)$. These two processes are related by equations taking the following form for suitable bivariate dual functions H:

$$\mathbb{E}[H(\eta_t, \widehat{\eta}_0)] = \mathbb{E}[H(\eta_0, \widehat{\eta}_t)], \tag{1.1}$$

when η_0 and $\widehat{\eta}_0$ are both constant. Hence, the question of characterizing the large-time behavior of (η_t) can be converted to the same question for $(\widehat{\eta}_t)$, and the hope is that the dual process is easier to study. In the case of [22], duality can even take a finer form, namely Harris's graphical duality, which is sometimes also called pathwise duality. Here, (η_t) and $(\widehat{\eta}_t)$ can be coupled on the same probability space. See [9, 14, 28]. For the case of interacting diffusions in [1], we discuss large-time behavior of the model using duality in the form (1.1). Finally, we refer the reader to the lecture notes in the present series by Sturm, Swart and Völlering [29] and to the papers by Jansen and Kurt [14, 15] for further discussions of duality.

Organization of this article. This article is organized as follows. In Section 2, we first explain some features of the Lotka–Volterra model that motivate the generalizations in [1, 22]. Section 3 is devoted to a discussion of the (symmetric) Neuhauser–Pacala model. We first explain the duality from various points of view. Then we explain its usefulness for solving the stationarity of the model. Section 4 continues the spirit of Section 3 for a discussion of the Blath–Etheridge–Meredith model.

Acknowledgments. This article was written for the learning session 'Competing species models' organized by the authors and held in August 2017 at the Institute for Mathematical Sciences of the National University of Singapore. Supports from the IMS for both authors and a grant from the Simons Foundation for Y.-T. C. are gratefully acknowledged. The authors also thank an anonymous referee for comments.

2. Death-birth events in the Lotka–Volterra model

In this section, we discuss the Lotka–Volterra model and its connections to the stochastic spatial generalizations in the main papers under consideration [1, 22].

Recall that the competitive Lotka–Volterra ordinary differential equations for two populations of different types are based on the logistic equations. There are interactions within and between the two populations. If we denote by $N_0 = N_0(t)$ and $N_1 = N_1(t)$ the total sizes of the populations of type 0 and 1, respectively, then the system of dynamical equations for N_0 and N_1 is given by

$$\begin{cases} \dot{N_0} = r_0 N_0 \left(1 - \dfrac{N_0 + \alpha_{01} N_1}{K_0}\right), \\ \dot{N_1} = r_1 N_1 \left(1 - \dfrac{N_1 + \alpha_{10} N_0}{K_1}\right). \end{cases} \tag{2.1}$$

Here, r_i is the intrinsic growth rate of the population of type i, K_i is the carrying capacity of the population of type i, and α_{ij} is called the strength of interspecific competition. These are given positive constants. Without the term $(N_0 + \alpha_{01} N_1)/K_0$ to suppress the growth of 0-individuals, they grow exponentially with rate r_0. A similar interpretation applies to the equation for N_1. By (2.1), the total population size $N_0 + N_1$ does not stay constant.

Now we rewrite (2.1) in the following form (with obvious definitions of constants):

$$\begin{cases} \dot{N_0} = N_0(a - bN_0 - cN_1), \\ \dot{N_1} = N_1(a' - b'N_1 - c'N_0), \end{cases} \tag{2.2}$$

and then turn to the densities of 0-individuals and 1-individuals defined by

$$p_0 \stackrel{\text{def}}{=} \frac{N_0}{N_0 + N_1} \quad \text{and} \quad p_1 \stackrel{\text{def}}{=} \frac{N_1}{N_0 + N_1}.$$

By (2.2), we get

$$\dot{p_0} = -\frac{N_0 N_1}{(N_0+N_1)^2}\left(a' + bN_0 + cN_1\right) + \frac{N_0 N_1}{(N_0+N_1)^2}\left(a + b'N_1 + c'N_0\right). \tag{2.3}$$

To simplify the last equation, we assume

$$a = a', \tag{2.4}$$

that is, intrinsic growth rates r_0 and r_1 are identical. Then (2.3) reduces to the following equation:

$$\dot{p_0} = -\frac{N_0 N_1}{(N_0+N_1)^2}\left(bN_0 + cN_1\right) + \frac{N_0 N_1}{(N_0+N_1)^2}\left(b'N_1 + c'N_0\right). \tag{2.5}$$

Now, we change time by using the functions

$$t \longmapsto p_j\left(\int_0^t \frac{1}{(\lambda p_1(s) + p_0(s))b'(N_0+N_1)}ds\right), \quad j = 0, 1. \tag{2.6}$$

Then (2.5) can be transformed into a closed equation, since this time-change amounts to the effect of dividing both sides of (2.5) by

$$(\lambda p_1 + p_0)b'(N_0+N_1).$$

To simplify notation, for $j = 0, 1$, we still denote the corresponding new function in (2.6) by p_j. Then (2.5) implies

$$\dot{p_0} = -p_0\frac{\lambda p_1}{\lambda p_1 + p_0}(p_0 + \alpha_{01}p_1) + p_1\frac{p_0}{\lambda p_1 + p_0}(p_1 + \alpha_{10}p_0). \tag{2.7}$$

The coefficients of (2.5) follow since, under the present assumption in (2.4), the constants a, b, c, a', b', c' in (2.2) satisfy the following equations:

$$\lambda \overset{\text{def}}{=} b/b' = K_1/K_0, \quad c/b = \alpha_{01} \quad \text{and} \quad c'/b' = \alpha_{10}. \tag{2.8}$$

Equation (2.7) suggests natural interpretations similar to the Moran process in population genetics. The first term in (2.7), that is

$$-p_0\frac{\lambda p_1}{\lambda p_1 + p_0}(p_0 + \alpha_{01}p_1) = -p_0 \times (p_0 + \alpha_{01}p_1) \times \frac{\lambda p_1}{\lambda p_1 + p_0}, \tag{2.9}$$

has the interpretation as

$$\begin{aligned} &-(\text{probability to find a 0-individual}) \\ &\quad \times (\text{death rate of a 0-individual}) \\ &\quad \times (\text{birth probability of a 1-individual}). \end{aligned} \tag{2.10}$$

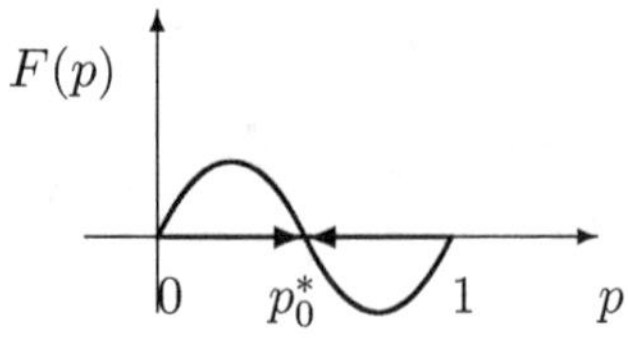

Fig. 1.

Here, taking

$$p_0 + \alpha_{01} p_1$$

as a death rate is consistent with the dynamical equation of N_0 in (2.1) now that $r_0 = r_1$ by our assumption in (2.4). Indeed, $p_0 + \alpha_{01} p_1$ is the same as $N_0 + \alpha_{01} N_1$ up to the (time-dependent) total population size $N_0 + N_1$. Also, by the definition of λ in (2.8),

$$\frac{\lambda p_1}{\lambda p_1 + p_0} = \frac{K_1 N_1}{K_1 N_1 + K_0 N_0}.$$

This ratio can be taken as a birth probability of 1-individuals. The higher the carrying capacity K_1 is, the more likely 1-individuals can replicate. These altogether are responsible for the decrease of p_0, which explains the minus sign in (2.9). The second term in (2.7) can be interpreted similarly with the roles of 1 and 0 exchanged. But now this term is responsible for the increase of 0-individuals in the population.

Equilibrium in the dynamical equation for p_0 in (2.7) is very easy to obtain. Since $p_0 + p_1 = 1$, we get

$$\begin{aligned} \dot{p_0} &= \frac{p_0(1-p_0)}{\lambda(1-p_0)+p_0}\left\{(1-\lambda\alpha_{01}) - p_0\big[(1-\lambda\alpha_{01}) + (\lambda - \alpha_{10})\big]\right\} \\ &= \frac{F(p_0)}{\lambda(1-p_0)+p_0}, \end{aligned} \tag{2.11}$$

where F is a polynomial in p_0. Setting the right-hand side to zero shows that the foregoing equation has a unique stable equilibrium given by

$$p_0^* = \frac{(1-\lambda\alpha_{01})}{(1-\lambda\alpha_{01}) + (\lambda - \alpha_{10})} \tag{2.12}$$

if $0 \leq \alpha_{10} < \lambda$ and $0 \leq \alpha_{01} < 1/\lambda$, where the restrictions on α_{10} and α_{01} ensure that p_0^* falls in $(0,1)$. The stability of p_0^* is plain from the graph of F given in Figure 1.

A stochastic spatial generalization of (2.11) to spin systems is introduced by Neuhauser and Pacala in [22]. Put in a general framework, the model can be described in the following way. Let q be an irreducible transition probability kernel on a nonempty set E and suppose that q has a zero trace:

$$\sum_{x\in E} q(x,x) = 0.$$

For $\sigma \in \{0,1\}$, $x \in E$ and $\eta \in \{0,1\}^E$, we define the local frequencies of σ's by

$$f_\sigma(x,\eta) = \sum_{y\in E} q(x,y)\mathbf{1}_{\{\sigma\}}(\eta(y)).$$

The canonical example here is that q is given by the transition probability kernel of simple random walk on a connected graph. In this case, $f_\sigma(x,\eta)$ reduces to the usual frequency of σ's in the neighborhood of x:

$$f_\sigma(x,\eta) = \frac{1}{\deg(x)} \sum_{y:y\sim x} \mathbf{1}_{\{\sigma\}}(\eta(y)),$$

where $\deg(x)$ denotes the number of neighbors of x and $y \sim x$ means that y is a neighbor of x. The use of these kernels (E,q) is meant to allow perturbations of graph structures in the usual $\mathbb{Z}^2$ or two-dimensional discrete tori. Hence, we can view the spin system to be defined below from a more general perspective.

Now the flip rates of the sought-after generalization are set to be

$$\begin{aligned} &0 \to 1 \text{ with rate } (f_0 + \alpha_{01} f_1)\left(\frac{\lambda f_1}{\lambda f_1 + f_0}\right), \\ &1 \to 0 \text{ with rate } (f_1 + \alpha_{10} f_0)\left(\frac{f_0}{\lambda f_1 + f_0}\right). \end{aligned} \tag{2.13}$$

In this spin system,

$$f_0 + \alpha_{01} f_1 \quad \text{and} \quad f_1 + \alpha_{10} f_0$$

are the death rates and

$$\frac{\lambda f_1}{\lambda f_1 + f_0} \quad \text{and} \quad \frac{f_0}{\lambda f_1 + f_0}$$

are the birth probabilities, as in (2.9). Hence, in the sense of flip rates in [18, page 122–123], the $\{0,1\}^E$-valued Markov process (η_t) under consideration is characterized by the following limits:

$$\begin{aligned}
&\mathbb{P}(\eta_t(x)=1|\eta_0(x)=0)\\
&\quad=[f_0(x,\eta_0)+\alpha_{01}f_1(x,\eta_0)]\left(\frac{\lambda f_1(x,\eta_0)}{\lambda f_1(x,\eta_0)+f_0(x,\eta_0)}\right)+o(t),\quad t\searrow 0+,\\
&\mathbb{P}(\eta_t(x)=0|\eta_0(x)=1)\\
&\quad=[f_1(x,\eta_0)+\alpha_{10}f_0(x,\eta_0)]\left(\frac{f_0(x,\eta_0)}{\lambda f_1(x,\eta_0)+f_0(x,\eta_0)}\right)+o(t),\quad t\searrow 0+,\\
&\mathbb{P}\big(\eta_t(x)\neq\eta_0(x),\eta_t(y)\neq\eta_0(y)\big)=o(t),\quad t\searrow 0+,\quad \forall\, x\neq y.
\end{aligned} \tag{2.14}$$

The flip rates introduced in (2.13) are applicable for general (E,q)'s defined above. In particular, it can be shown by limit theorems of semimartingales (cf. [16]) that the density of 1's in this particle system on a complete graph over N vertices converges to the solution of (2.7) as $N\to\infty$ in the space of càdlàg functions equipped with Skorokhod's J_1-topology.

Our discussions in the rest of this paper for this model will be on the construction of this spin system in the symmetric case with $\alpha_{01}=\alpha_{10}$ and $\lambda=1$ and on the analysis of its equilibrium. Note that if $\alpha_{01}=\alpha_{10}=1$ and $\lambda=1$, then the model reduces to the voter model and is henceforth excluded from the discussion below.

A model introduced by Blath, Etheridge and Meredith [1] generalizes the Lotka–Volterra differential equations in (2.1) from a point of view very similar to those in the biology papers by Bolker and Pacala [2] and Murrell and Law [21]. Here, we only consider the case that the underlying spatial structure is an integer lattice $\mathbb{Z}^d$ for some $d\geq 1$. Population sizes at points in $\mathbb{Z}^d$ are in the continuum and subject to the new features of migration and stochastic growth defined by branching noises. More precisely, $(\eta_t)=(\eta_t(x);x\in\mathbb{Z}^d)$ and $(\eta'_t)=(\eta'_t(x);x\in\mathbb{Z}^d)$ model population sizes of 0-individuals and 1-individuals at all sites $x\in\mathbb{Z}^d$, respectively, are given by the following system of stochastic differential equations: for all $x\in\mathbb{Z}^d$,

$$\begin{aligned}
d\eta_t(x)&=\sum_{y\in\mathbb{Z}^d}m_{xy}\big(\eta_t(y)-\eta_t(x)\big)dt\\
&\quad+\eta_t(x)\left(\alpha-\sum_{y\in\mathbb{Z}^d}\beta_{xy}\eta_t(y)-\sum_{y\in\mathbb{Z}^d}\gamma_{xy}\eta'_t(y)\right)dt\\
&\quad+\sqrt{\sigma\eta_t(x)}dB_t(x),
\end{aligned} \tag{2.15}$$

$$
\begin{aligned}
d\eta_t'(x) = {} & \sum_{y\in\mathbb{Z}^d} m_{xy}'\big(\eta_t'(y) - \eta_t'(x)\big)dt \\
& + \eta_t'(x)\left(\alpha' - \sum_{y\in\mathbb{Z}^d} \beta_{xy}'\eta_t'(y) - \sum_{y\in\mathbb{Z}^d} \gamma_{xy}'\eta_t(y)\right)dt \\
& + \sqrt{\sigma\eta_t'(x)}dB_t'(x),
\end{aligned} \tag{2.16}
$$

where $\{B(x), B'(x); x \in \mathbb{Z}^d\}$ are i.i.d. one-dimensional standard Brownian motions.

The equations in (2.15) and (2.16) now model spatial effects in a different way. For example, the sum

$$
\sum_{y\in\mathbb{Z}^d} m_{xy}\big(\eta_t(y) - \eta_t(x)\big)dt
$$

models migration of individuals by a nonnegative matrix m since

$$
\mathsf{L}f(x) = \sum_{y\in\mathbb{Z}^d} m_{xy}[f(y) - f(x)]
$$

is the generator of a Markov chain moving along sites of $\mathbb{Z}^d$ subject to the q-matrix $\{m_{xy}; x \neq y\}$. Also, the term

$$
\eta_t(x)\left(\alpha - \sum_{y\in\mathbb{Z}^d} \beta_{xy}\eta_t(y) - \sum_{y\in\mathbb{Z}^d} \gamma_{xy}\eta_t'(y)\right)dt
$$

has a natural correspondence to the logistic differential equations in (2.1). It is assumed in addition that these matrices $m, m', \beta, \beta', \gamma, \gamma'$ are 'homogeneous' and have finite ranges in the sense that, for example, m_{xy} depends only on $\|x-y\|_\infty$ and $m_{xy} = 0$ for all $\|x-y\|_\infty \geq L$ for some $L > 0$ independent of x, y.

The objects corresponding to (2.7) are the density processes

$$
p_t(x) \stackrel{\text{def}}{=} \frac{\eta_t(x)}{\eta_t(x) + \eta_t'(x)}
$$

of 0-individuals at all sites x. The derivation in (2.3) can be generalized by Itô's formula and formally conditioning on $\eta_t(x) + \eta_t'(x) \equiv N$ for all x. Then, assuming also that $m_{xy} = m_{xy}'$ and that $\beta_{xy}, \beta_{xy}', \gamma_{xy}, \gamma_{xy}'$ are zero for $x \neq y$ (purely local interactions), one is led to the following system:

$$\begin{aligned} dp_t(x) &= \sum_{y\in\mathbb{Z}^d} m_{xy}\big(p_t(y)-p_t(x)\big)dt \\ &\quad + sp_t(x)\big(1-p_t(x)\big)\big(1-\mu p_t(x)\big)dt \\ &\quad + \sqrt{N^{-1}p_t(x)\big(1-p_t(x)\big)}dW_t(x), \quad x\in\mathbb{Z}^d, \end{aligned} \tag{2.17}$$

where $\{W(x); x \in \mathbb{Z}^d\}$ is a system of independent standard Brownian motions, and s and μ are real parameters which can be expressed explicitly in terms of the parameters of $\{\eta_t\}$ and $\{\eta'_t\}$. We refer the reader to [1, pages 1482–1483] for details of the derivation. In particular, if we consider the symmetric case where η and η' follow the same parameters, then $\mu = 2$ and $s = \beta_{xx} - \gamma_{xx}$. Note that existence and uniqueness (in the strong sense) of a $[0,1]^{\mathbb{Z}^d}$-valued solution of the system of SDEs in (2.17) can be obtained independently from classical results for infinite-dimensional SDEs due to Shiga and Shimizu [26, Theorem 2.1]. In Section 4, the discussion for these interacting diffusions in (2.17) is independent of the system defined by (2.15), and so we use general parameters s and μ from now on.

3. Spin systems on discrete sets

Throughout this section, we focus on the symmetric Neuhauser–Pacala model, that is $\lambda = 1$ and

$$\alpha_{01} = \alpha_{10} = \alpha \in [0,1). \tag{3.1}$$

In this case, there is no bias in birth rates and death rates induced by these parameters, and the flip rates defined in (2.13) can be simplified as follows:

$$\begin{cases} 0 \to 1 \ \text{ with rate } (f_0+\alpha f_1)f_1 = (1-\alpha)f_0f_1+\alpha f_1, \\ 1 \to 0 \ \text{ with rate } (f_1+\alpha f_0)f_0 = (1-\alpha)f_1f_0+\alpha f_0. \end{cases} \tag{3.2}$$

The assumptions that $\lambda = 1$ and $\alpha_{01} = \alpha_{10}$ mean that the populations of types 0 and 1 behave in the same way.

Below we first introduce in Section 3.1 various constructions of this symmetric model. The constructions will progressively lead us to the duality for the model, which is called *parity duality* in this paper for reasons that shall become self-evident. In Section 3.2, we discuss the related basic results and open questions.

3.1. *Constructions and parity duality*

In the following, we first view the symmetric Neuhauser–Pacala model in terms of two sets of flip rates. These sets of flip rates define the symmetric model with a detailed coupling by independent Poisson processes. On a finite set, the construction leads to a matrix representation of the model and easily induces the parity duality. We will explain these steps in detail. By the end of this subsection, we give a sketch of how generalizations of the parity duality can be obtained on infinite sets.

First, the flip rates defined above in (3.2) are decomposed into two fundamental mechanisms given by

$$\begin{aligned} &\text{pairwise annihilation:} \begin{cases} 0 \to 1 & \text{with rate } (1-\alpha)f_0f_1, \\ 1 \to 0 & \text{with rate } (1-\alpha)f_1f_0, \end{cases} \\ &\text{voting:} \begin{cases} 0 \to 1 & \text{with rate } \alpha f_1, \\ 1 \to 0 & \text{with rate } \alpha f_0. \end{cases} \end{aligned} \tag{3.3}$$

As before, we suppress sites and configurations to be updated in this notation. The corresponding $\{0,1\}^E$-valued Markov processes can be characterized analogously as in (2.14).

The first set of flip rates in (3.3) corresponds to pairwise annihilation. The children of two neighbors randomly chosen according to the transition probability kernel q try to invade a focal site, namely the site chosen to be updated, subject to pairwise annihilation. To see this interpretation, first we write the flip rates for site x given population configuration η as follows:

$$\begin{aligned} &(1-\alpha)f_0(x,\eta)f_1(x,\eta) = (1-\alpha)f_1(x,\eta)f_0(x,\eta) \\ &= (1-\alpha)\left(\sum_{y\in E} q(x,y)\eta(y)\right)\left(\sum_{z\in E} q(x,z)[1-\eta(z)]\right). \end{aligned}$$

Hence, a spin flip at x is triggered if and only if a 1-individual y and a 0-individual z are chosen (according to the kernel q).

Let us use this property to explain the implied effect of pairwise annihilation. If the underlying spatial structure is a graph and we let a filled vertex denote a vertex occupied by a 1-individual and an empty vertex denote one occupied by a 0-individual, then we can visualize the complete set of possible transitions as follows:

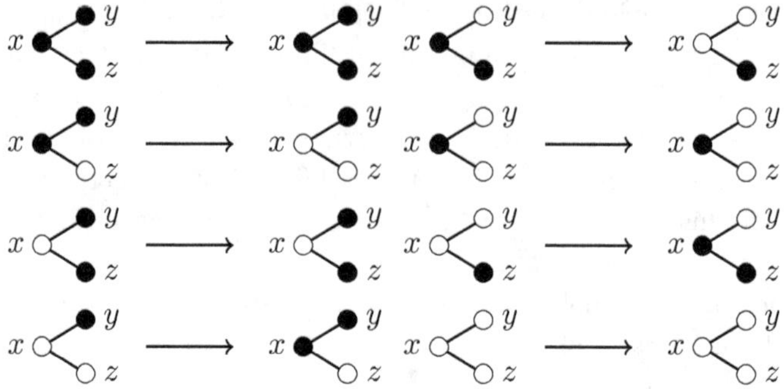

(3.4)

Consider the following two examples from the first row of (3.4). For the transition on the right-hand side, the focal site is x, and the two neighbors at sites y and z, randomly chosen according to the probability $q(x,\cdot)$ without replacement, try to invade site x by their children. The resident 1-individual at x and the invading child of the 1-individual at z annihilate each other, and so the child of the 0-individual at y takes over the site x after the update. A similar interpretation applies to the focal site x for the transition on the left. But this time, two of the three 1-individuals (two from y and z plus the resident 1-individual) annihilate each other, so that there is only one 1-individual left at x (we do not care where this survivor comes from).

The second set of flip rates in (3.3) defines a voting mechanism. Now, the flip rates can be written as

$$\alpha f_1(x,\eta) = \alpha \sum_{y\in E} q(x,y)\eta(y) \quad \text{and} \quad \alpha f_0(x,\eta) = \alpha \sum_{y\in E} q(x,y)[1-\eta(y)]$$

so that there is a spin flip at x if and only if a neighbor of the opposite type is chosen:

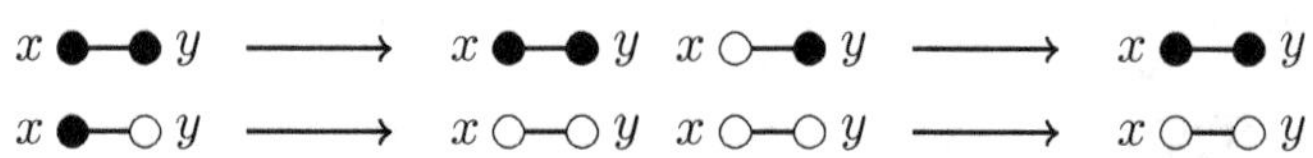

(3.5)

On the left of the second row of (3.5), the focal site chosen for update is again x, and the new type is chosen randomly from one of the neighbors, which is y, so that a 0-individual replaces the resident 1-individual at x.

Now we explain how to use the quotient group $\mathbb{Z}_2 = \mathbb{Z}/2\mathbb{Z}$ to 'linearize' the mechanisms in (3.3). For any $x,y,z \in E$, we first define an $E \times E$ matrix with entries in $\mathbb{Z}_2$ by

$$\{x\} \times \{y,z\} \stackrel{\text{def}}{=} (a_{v,w})_{v,w\in E} \quad \text{with nonzero entries } a_{x,y} = a_{x,z} = 1.$$

Then for the pairwise annihilation mechanism, we consider the standard superposition of Poisson processes and the following Poisson updates by linear transformations: for all $\eta \in \mathbb{Z}_2^E$ and $x, y, z \in E$:

$$\begin{aligned} &\eta \to \eta + \{x\} \times \{y, z\}\eta \\ &\qquad\qquad \text{with rate } r(\{x\} \times \{y, z\}) = (1-\alpha)q(x, y)q(x, z). \end{aligned} \tag{3.6}$$

Here, the configuration after the update can be written as

$$\eta + \{x\} \times \{y, z\}\eta = [\eta - \eta(x)\mathbf{1}_x] + [\eta(x) + \eta(y) + \eta(z)]\mathbf{1}_x.$$

The difference in the first pair of brackets shows that the types outside x are kept fixed and we clear the site x by setting it to be zero. The sum in the second pair of brackets shows precisely pairwise annihilation due to the group structure of $\mathbb{Z}_2$ and always gives us a value in $\{0, 1\}$. Here and in what follows, we continue to use addition modulo 2 whenever $\mathbb{Z}_2$ is used.

For the voting mechanism, we introduce independent Poisson updates by the following linear transformations:

$$\eta \to \eta + \{x\} \times \{x, y\}\eta \quad \text{with rate } r(\{x\} \times \{x, y\}) = \alpha q(x, y). \tag{3.7}$$

Now the configuration after the update can be written as

$$\begin{aligned} \{x\} \times \{x, y\}\eta &= [\eta - \eta(x)\mathbf{1}_x] + [\eta(x) + \eta(x) + \eta(y)]\mathbf{1}_x \\ &= [\eta - \eta(x)\mathbf{1}_x] + \eta(y)\mathbf{1}_x, \end{aligned} \tag{3.8}$$

and so we have removal and adoption of types at x by the first and second terms of (3.8), respectively. We write the matrices defining the updates in (3.6) and (3.7) as $\mathrm{Id} + J$ in the following.

On a finite set E, the above linear construction of the symmetric Neuhauser–Pacala model by Poisson events can be made more explicit by a random-walk type construction. Here, we time-change a discrete-time Markov chain by an independent Poisson process (see [23, Section 2.6]). In detail, with initial condition $\mathbf{1}_A$, we define a $\mathbb{Z}_2^E$-valued process by

$$\eta_t^A \stackrel{\text{def}}{=} (\mathrm{Id} + J_{N_t}) \cdots (\mathrm{Id} + J_1)\mathbf{1}_A, \tag{3.9}$$

where $J_1, J_2, \dots$ are i.i.d. random matrices with entries in $\{0, 1\}$ and law

$$\mathbb{P}(J_1 = J) = \frac{r(J)}{\sum_{J'} r(J')}$$

and (N_t) is an independent Poisson process with rate

$$\mathbb{E}N_1 = \sum_{J'} r(J').$$

By the construction in (3.9), we can imitate the usual time-reversal duality of random walks and easily get a duality of the symmetric Neuhauser–Pacala model:

$$
\begin{aligned}
\langle \mathbf{1}_B, \eta_t^A \rangle &= \mathbf{1}_B^\top (\mathrm{Id} + J_{N_t}) \cdots (\mathrm{Id} + J_1) \mathbf{1}_A && (3.10)\\
&= [(\mathrm{Id} + J_1^\top) \cdots (\mathrm{Id} + J_{N_t}^\top) \mathbf{1}_B]^\top \mathbf{1}_A \\
&= \langle \widehat{\eta}_t^{t,B}, \mathbf{1}_A \rangle \overset{(\mathrm{d})}{=} \langle \widehat{\eta}_t^B, \mathbf{1}_A \rangle, && (3.11)
\end{aligned}
$$

where $\langle a, b\rangle = \sum_{x\in E} a_x b_x$ for any $a, b \in \mathbb{Z}_2^E$. In other words, the inner product $(\xi, \eta) \mapsto \langle \xi, \eta \rangle$ is used as the bivariate dual function and

$$
\widehat{\eta}_t^B \overset{\mathrm{def}}{=} (\mathrm{Id} + J_{N_t}^\top) \cdots (\mathrm{Id} + J_1^\top) \mathbf{1}_B \tag{3.12}
$$

is the dual Markov chain.

To specify the update rule of the dual chain defined in (3.12), it is now more convenient to say that there are only 1-individuals in the population and the sites not occupied by 1-individuals are left empty. (Recall that for the symmetric Neuhauser–Pacala model, we use the interpretation that there are 0-individuals and 1-individuals in the population.)

By transposing the linear transformation in (3.6), we obtain a mechanism featuring branching with pairwise annihilation:

$$
\begin{aligned}
\xi \to \xi + \{y, z\} \times \{x\}\xi \equiv {} & [\xi - \xi(y)\mathbf{1}_y - \xi(z)\mathbf{1}_z] \\
& + [\xi(x) + \xi(y)]\mathbf{1}_y + [\xi(x) + \xi(z)]\mathbf{1}_z.
\end{aligned} \tag{3.13}
$$

If x is chosen as the focal site and is occupied by a 1-individual, it gives birth to two children and they try to invade two sites y and z subject to the pairwise annihilation by the last two terms as before due to the group structure of $\mathbb{Z}_2$. The set of all possible transitions can be visualized as follows:

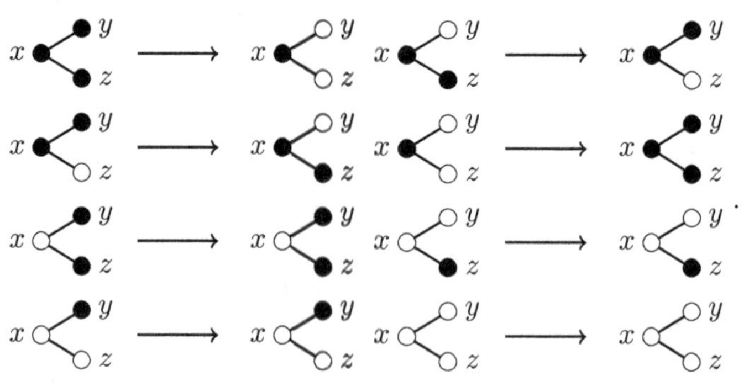

(3.14)

On the left-hand side of the first row of (3.14), all of the three sites x, y, z are occupied by 1-individuals, but after the update, the child of x invading site y and the resident 1-individual at y annihilate each other so that an empty site is created. The same holds at site z. On the right-hand side of the first row of (3.14), site y becomes occupied by a 1-individual after the update, but pairwise annihilation still occurs at site z.

Now the transposed linear transformation in (3.8) defines a random walk with annihilation:

$$\begin{aligned} \xi \to \xi + \{x, y\} \times \{x\}\xi &\equiv [\xi - \xi(x)\mathbf{1}_x - \xi(y)\mathbf{1}_y] + [\xi(x) + \xi(x)]\mathbf{1}_x \\ &\quad + [\xi(x) + \xi(y)]\mathbf{1}_y \\ &= [\xi - \xi(x)\mathbf{1}_x - \xi(y)\mathbf{1}_y] + [\xi(x) + \xi(y)]\mathbf{1}_y, \end{aligned} \tag{3.15}$$

so that the individual at site x moves to site y subject to pairwise annihilation:

$$\begin{array}{llll} x \bullet\!\!-\!\!\bullet\, y \longrightarrow & x \circ\!\!-\!\!\circ\, y & x \circ\!\!-\!\!\bullet\, y \longrightarrow & x \circ\!\!-\!\!\bullet\, y \\ x \bullet\!\!-\!\!\circ\, y \longrightarrow & x \circ\!\!-\!\!\bullet\, y & x \circ\!\!-\!\!\circ\, y \longrightarrow & x \circ\!\!-\!\!\circ\, y \end{array}. \tag{3.16}$$

On the left-hand side of the first row of (3.16), the 1-individual moves to site y, but this individual and the resident 1-individual at y annihilate each other. We get two empty sites in the end. On the right-hand side of the first row of (3.16), site x is empty and so has no effect on the type occupying y from the update event.

In contrast to the Neuhauser–Pacala model where a death event is immediately followed by a birth event at each updating step, the dual process is defined by reversing the orders of birth events and death events, and thus, is an invasion process. Also, it is possible to have multiple sites where a change of types occurs in both mechanisms defining the dual process.

To fully define the dual process in (3.12), we also need to specify the rates of the associated Poisson processes. The Poisson process corresponding to (3.13) jumps with the same rate as that for (3.6):

$$\widehat{r}(\{y, z\} \times \{x\}) \stackrel{\text{def}}{=} r(\{x\} \times \{y, z\}), \tag{3.17}$$

whereas the Poisson process corresponding to (3.15) jumps with rate

$$\widehat{r}(\{x, y\} \times \{x\}) \stackrel{\text{def}}{=} r(\{x\} \times \{x, y\}). \tag{3.18}$$

Taking expectations of both sides of (3.11), we have proved the following theorem when E is a finite set. It still holds on infinite sets.

Theorem 3.1: (Parity duality) *For any irreducible kernel (E,q) with a zero trace, finite subsets A,B of E, and $t\geq 0$, we have*

$$\mathbb{P}\big(\langle \mathbf{1}_B,\eta_t^A\rangle\equiv 1\big)=\mathbb{P}\big(\langle \widehat{\eta}_t^B,\mathbf{1}_A\rangle\equiv 1\big), \tag{3.19}$$

where $\equiv$ means an equality mod 2.

Remark 3.2: We can also view a population configuration η as the set of sites occupied by 1's in η. In this interpretation, $\langle \mathbf{1}_B,\eta\rangle$ is the same as $|B\cap\eta|$. Hence, (3.19) is also an equality for the probability that there is an odd number of sites with 1's in B under η_t^A and the probability that there is an odd number of sites with 1's in A under $\widehat{\eta}_t^B$. This interpretation is common in the literature due to its usefulness for geometric considerations of population configurations. □

The validity of Theorem 3.1 on infinite sets can be obtained by various methods. For example, one can check the following Feynman–Kac duality by generator calculations. The approach is to consider the forward equation satisfied by

$$\phi_A(B)=\phi_B(A)\stackrel{\rm def}{=}\mathbf{1}_{\{\langle \mathbf{1}_B,\mathbf{1}_A\rangle\equiv 1\}}$$

under the symmetric Neuhauser–Pacala model (with semigroup $P_t=\exp\{t\mathsf{L}^{\rm NP}\}$) and then to show that this forward equation is also a backward equation under the dual process (with semigroup ($Q_t=\exp\{t\mathsf{L}^{\rm dual}\}$):

$$\frac{d}{dt}P_t\phi_B(A)=P_t\mathsf{L}^{\rm NP}\phi_B(A)=Q_t\mathsf{L}^{\rm dual}\phi_A(B) \tag{3.20}$$

so that

$$P_t\phi_B(A)=Q_t\phi_A(B).$$

Here, $\mathsf{L}^{\rm NP}$ denotes the generator of the Neuhauser–Pacala model and $\mathsf{L}^{\rm dual}$ denotes the generator of its dual. In more detail, to do the algebra for the proof of the second equality in (3.20), one needs the possible transitions and rates defining $\mathsf{L}^{\rm NP}$ and the possible transitions and rates in (3.13), (3.15), (3.17) and (3.18) defining $\mathsf{L}^{\rm dual}$. We omit the details here, but in Section 4.2, we will illustrate this algebraic method by the interacting diffusions defined at the end of Section 2. See [22, Proposition 1] for this calculation and [18, Section III.4] or [15, 29] for more general discussions.

Harris's graphical representation [11] gives a different proof of Theorem 3.1 on infinite sets. To motivate the representation, first we consider the case that E is finite. We redefine (η_t^A) in (3.9) as follows. For each matrix J such that $\mathrm{Id}+J$ is a linear transformation in (3.6) or (3.7), we define a Poisson process $N(J) = \{N_t(J); t \geq 0\}$ with rate $r(J)$. We assume that these Poisson processes are independent. Order the arrival times of $\sum_J N(J)$ as

$$0 = T_0 < T_1 < T_2 < T_3 < \cdots, \tag{3.21}$$

and write J_n for the matrix J such that T_n is an arrival time of $N(J)$. Then by thinning and superposition of Poisson processes, the process (η_t^A) constructed before has the same distribution as the following process which we also denote by (η_t^A):

$$\eta_t^A = (\mathrm{Id} + J_n) \cdots (\mathrm{Id} + J_2)(\mathrm{Id} + J_1)\mathbf{1}_A \quad \text{if } T_n \leq t < T_{n+1}. \tag{3.22}$$

See also [23, Chapter 2] for the usual construction of continuous-time Markov chains.

We make some simple observations for matrix products to interpret the definition in (3.22). In general, for matrices $M_1, \ldots, M_n$ with entries in $\{0, 1\}$, write

$$\begin{aligned} &\mathbf{1}_B^\top M_n M_{n-1} \cdots M_1 \mathbf{1}_A \\ &= \sum_{x_n, x_{n-1}, \ldots, x_0 \in E} \mathbf{1}_B(x_n) M_n(x_n, x_{n-1}) \cdots M_1(x_1, x_0) \mathbf{1}_A(x_0). \end{aligned} \tag{3.23}$$

Since the matrices have entries only in $\{0, 1\}$, the summands are $\{0, 1\}$-valued with

$$\mathbf{1}_B(x_n) M_n(x_n, x_{n-1}) M_{n-1}(x_{n-1}, x_{n-2}) \cdots M_1(x_1, x_0) \mathbf{1}_A(x_0) = 1 \tag{3.24}$$

if and only if

$$1 = \mathbf{1}_A(x_0) = M_1(x_1, x_0) = \cdots = M_n(x_n, x_{n-1}) = \mathbf{1}_B(x_n).$$

If an $E \times E$ matrix M with entries in $\{0, 1\}$ is identified with a collection of oriented edges $x \to y$ in E such that $M(y, x) = 1$ and vice versa, then the product in (3.24) corresponds to the following oriented path:

$$x_0 \to x_1 \to \cdots \to x_n. \tag{3.25}$$

We stress that $M(y, x) = 1$ corresponds to the oriented path $x \to y$ rather than to the reversed one, since matrices are now multiplied from the right to the left as in (3.22). In summary, $\mathbf{1}_B^\top M_n M_{n-1} \cdots M_1 \mathbf{1}_A$ is the number of oriented paths from A to B.

For the purpose of (3.22), we set all of the matrices M_m in (3.23) to be of the following form

$$M \equiv \mathrm{Id} + J,$$

where the sum is read modulo 2 entry-wise as usual. Then on $\{T_n \leq t < T_{n+1}\}$, a nonzero term

$$\mathbf{1}_B(x_n)(\mathrm{Id}+J_n)(x_n,x_{n-1})(\mathrm{Id}+J_{n-1})(x_{n-1},x_{n-2})\cdots(\mathrm{Id}+J_1)(x_1,x_0)\mathbf{1}_A(x_0)$$

in the representation of η_t^A in (3.22) can be identified with an oriented path from $(x_0,0)$ to (x_n,t) in $E\times\mathbb{R}_+$. Over $[0,t)$, this space-time path is defined by the pair (x_j,s) when $s\in[T_j,T_{j+1})$. Hence, (3.22) can be written as

$$\eta_t^A(x) \equiv \#\{\text{oriented space-time paths from } A\times\{0\} \text{ to } \{x\}\times\{t\}\} \quad (3.26)$$

for all $x\in E$. We have obtained the graphical representation of (η_t^A). Moreover, for every fixed $t\in(0,\infty)$, the reversely oriented space-time paths between $A\times\{0\}$ and $B\times\{t\}$ define $\widehat{\eta}_t^{t,B}$ by the same counting procedure as in (3.26). In particular, $\langle\widehat{\eta}_t^{t,B},\mathbf{1}_A\rangle = \langle\mathbf{1}_B,\eta_t^A\rangle$.

Example 3.3: Let us visualize the graphical representation of (η_t^A) in the following simple setting. For convenience, we write ● = 1 and ● = 0. If $E=\{x,y,z\}$, the initial condition $\mathbf{1}_y$ after the update $\{x\}\times\{y,z\}$ can be visualized by

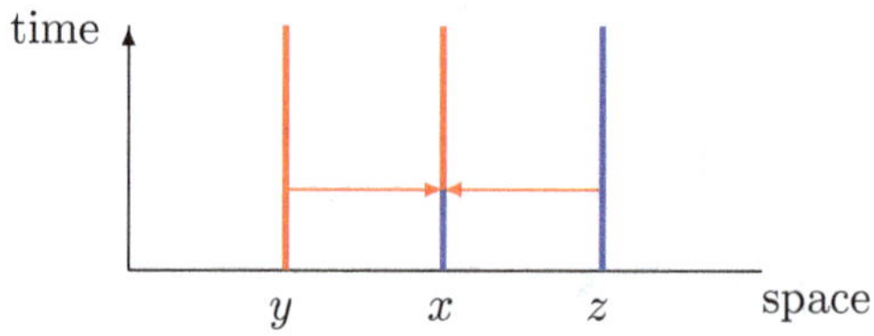

Similarly, for $E=\{x,y\}$, the initial condition $\eta_0=\mathbf{1}_y$ after the voting update of $\eta\mapsto\eta+\{x\}\times\{x,y\}\eta$ can be visualized by

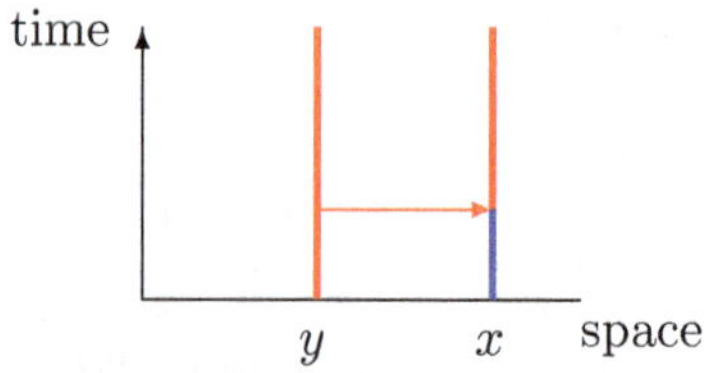

□

Now we consider the case that E is an infinite set. We define independent Poisson processes $N(J)$ as before. In this case, since $\sum_J r(J)$ is infinite, the arrival times of $\sum_J N(J)$ cannot be ordered as in (3.21). Moreover, (3.22) shows that we have to deal with an infinite product of matrices now. These issues can be easily circumvented since the following picture holds with probability one: *Locally*, there are only finitely many such arrival times which trigger nontrivial updates, and the infinite product of matrices can be reduced to a finite product of matrices.

Let us define these reductions precisely. For a general $E \times E$ matrix M with entries in $\{0,1\}$, we define the *range* of M by

$$\begin{aligned}\mathcal{R}(M) \stackrel{\text{def}}{=} &\{y : M(y,x) = 1 \text{ for some } x \in E \text{ and } x \neq y\}\\ &\cup \{x : M(y,x) = 1 \text{ for some } y \in E \text{ and } y \neq x\}.\end{aligned} \tag{3.27}$$

In terms of the above interpretation, $\mathcal{R}(M)$ is the set of endpoints of oriented edges in E defined by the nonzero entries of M, but we disregard oriented edges which are self-loops. Write $0 = T_0(J) < T_1(J) < \cdots$ for the arrival times of $N(J)$. Then the local picture follows from the observation that for all $x \in E$,

$$\sum_{J:x\in\mathcal{R}(\mathrm{Id}+J)} r(J) < \infty \Longrightarrow \sum_{J:x\in\mathcal{R}(\mathrm{Id}+J)} N_t(J) < \infty \text{ a.s. for all } t \geq 0. \tag{3.28}$$

We draw oriented edges induced by $\mathrm{Id} + J$ at all of the times $T_n(J)$ and require that these oriented edges are not self-loops. Then the property of the Poisson processes in (3.28) implies that up to a fixed time $t \in (0,\infty)$, there are only finitely many oriented edges in E which are not self-loops and use x as the endpoints. Therefore, a definition of η^A by (3.26) and an analogous definition for $\widehat{\eta}^{t,B}$ apply again. We still have $\langle \widehat{\eta}_t^{t,B}, \mathbf{1}_A\rangle = \langle \mathbf{1}_B, \eta_t^A\rangle$.

The usefulness of the parity duality in Theorem 3.1 follows from the fact that parity events $\{\eta; \langle \mathbf{1}_B, \eta\rangle \equiv 1\}$, $B \subseteq E$, uniquely determine a measure.

Proposition 3.4: *For finite measures ν_1 and ν_2 with the same total mass, $\nu_1\{\eta; \langle \mathbf{1}_B, \eta\rangle \equiv 1\} = \nu_2\{\eta; \langle \mathbf{1}_B, \eta\rangle \equiv 1\}$ for all finite subsets B of E implies that $\nu_1 = \nu_2$.*

Proof: First, for all finite subsets A of E, the test functions $\prod_{x\in A}[2\eta(x) - 1]$ can be written as

$$\prod_{x\in A}[2\eta(x)-1] = (-1)^{\langle \mathbf{1}_A, \eta^{\complement}\rangle} = 1 - 2\mathbf{1}_{\{\eta; \langle \mathbf{1}_A, \eta^{\complement}\rangle \equiv 1\}}.$$

($\prod_{x\in\varnothing}[2\eta(x) - 1]$ and other products over an empty set used below are equal to 1 by convention.) Hence, the assumption on ν_1 and ν_2 shows that for all finite subsets A of E,

$$\int \prod_{x\in A}[2\eta(x)-1]\nu_1(d\eta) = \int \prod_{x\in A}[2\eta(x)-1]\nu_2(d\eta).$$

Also, expanding $\prod_{x\in A}[2\eta(x)-1]$ yields

$$2^{|A|}\prod_{x\in A}\eta(x)=\prod_{x\in A}[2\eta(x)-1]-\sum_{B:B\subsetneq A}2^{|B|}(-1)^{|A-B|}\prod_{x\in B}\eta(x).$$

Hence, by the last two displays and an induction on $|A|$, we deduce that

$$\int\prod_{x\in A}\eta(x)\nu_1(d\eta)=\int\prod_{x\in A}\eta(x)\nu_2(d\eta)$$

for all finite subsets A of E. (We use the assumption that ν_1 and ν_2 have the same total mass for the initial induction step with $|A|=0$.) Since $\eta(x)$ is $\{0,1\}$-valued for all $x\in E$, the foregoing equality can be restated as

$$\nu_1\left(\bigcap_{x\in A}\{\eta;\eta(x)=1\}\right)=\nu_2\left(\bigcap_{x\in A}\{\eta;\eta(x)=1\}\right)$$

and we have $\{\eta;\eta(x)=1\}=\{\eta;\eta(x)=0\}^{\complement}$ for all $x\in E$. Hence, applying the inclusion-exclusion principle for the sets $\{\eta;\eta(x)=1\}$ shows that the two measures ν_1 and ν_2 are equal. □

3.2. *Invariance of equiparity coexistence*

If we start the symmetric Neuhauser–Pacala model with the Bernoulli product measure $\beta_{1/2}$ with density 1/2, then the corresponding parity duality from Theorem 3.1 can be written as

$$\mathbb{P}_{\beta_{1/2}}(\langle\mathbf{1}_B,\eta_t\rangle\equiv 1)=\mathbb{P}\left(\sum_{m=1}^{\langle\widehat{\eta}_t^B,\mathbf{1}\rangle}X_m\equiv 1\right).\tag{3.29}$$

Here, $X_1,X_2,\dots$ are i.i.d. $\{0,1\}$-valued Bernoulli variables with mean 1/2 and independent of $\widehat{\eta}_t^B$. They come from the random spins at all sites under $\beta_{1/2}$. By conditioning the event on the right-hand side of (3.29) on $\widehat{\eta}_t^B$, one sees that

$$\mathbb{P}_{\beta_{1/2}}(\langle\mathbf{1}_B,\eta_t\rangle\equiv 1)=\frac{1}{2}\mathbb{P}(\widehat{\eta}_t^B\neq\mathbf{0}).\tag{3.30}$$

Hence, the problem of proving existence of the limit of $\mathbb{P}_{\beta_{1/2}}(\langle\mathbf{1}_B,\eta_t\rangle\equiv 1)$ as $t\to\infty$ is equivalent to the problem of calculating the survival probability in $(\widehat{\eta}_t^B)$.

In the rest of Section 3, we give a brief discussion of calculating the limiting distribution of (η_t^A) as $t\to\infty$. The method is to refine what we just observed under the Bernoulli initial condition $\beta_{1/2}$. In this regard, the reader may wish to recall the discussion of (2.12). Since $\lambda=1$ and $\alpha_{01}=$

$\alpha_{10} = \alpha$ for the symmetric Neuhauser–Pacala model, it seems reasonable to expect that by suitable definitions, equilibria of the model can be related to the value p_0^* defined by (2.12), which is 1/2 under the present choice of $\lambda, \alpha_{01}, \alpha_{10}$. This number 1/2 is ubiquitous in the rest of this section.

The following is our first theorem about 1/2 in equilibria of the model.

Theorem 3.5: *If* $\alpha = 0$, *then* $\beta_{1/2}$ *is invariant for the symmetric Neuhauser–Pacala model.*

Proof: Recall that the parameter α from (3.1) enters the weights of the two mechanisms in (η_t^A) (see (3.6) and (3.7)) and the dual process $(\widehat{\eta}_t^B)$ (see (3.17) and (3.18)). If $\alpha = 0$, then there is no random walk with annihilation in $(\widehat{\eta}_t^B)$. In this case, observe that $(\widehat{\eta}_t^B)$ cannot die out unless $B = \varnothing$. Indeed, any empty site has no effect on any other sites. Algebraically, this is attributable to the fact that 0 is the additive identity in $\mathbb{Z}_2$. On the other hand, any 1-site x can only change its neighbors in three ways:

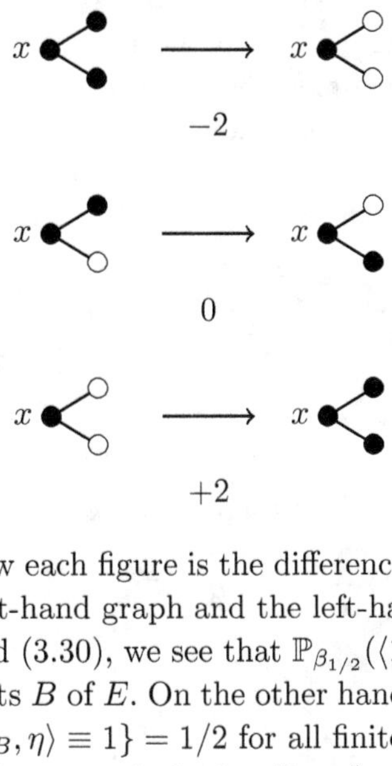

(3.31)

Here, the integer below each figure is the difference of the numbers of filled sites between the right-hand graph and the left-hand graph.

Now, by (3.29) and (3.30), we see that $\mathbb{P}_{\beta_{1/2}}(\langle \mathbf{1}_B, \eta_t \rangle \equiv 1) = 1/2$ for all finite nonempty subsets B of E. On the other hand, a symmetric argument shows that $\beta_{1/2}\{\eta; \langle \mathbf{1}_B, \eta \rangle \equiv 1\} = 1/2$ for all finite nonempty subsets B of E. By Proposition 3.4, we conclude that $\mathbb{P}_{\beta_{1/2}}(\eta_t \in \cdot) = \beta_{1/2}$ for all $t \geq 0$, as required. $\square$

There are two natural questions stemming from this simple theorem.

Question 3.6: Is $\beta_{1/2}$ also the unique limiting distribution whenever we start with initial conditions different from the all-1 configuration and the all-0 configuration?

Question 3.7: What can be said about the set of invariant distributions of the symmetric Neuhauser–Pacala model if $\alpha \in (0,1)$?

These questions are closely related to the following conjecture due to Neuhauser and Pacala [22, Conjecture 1].

Conjecture 3.8: There is coexistence in the symmetric Neuhauser–Pacala model on $\mathbb{Z}^d$ for any $d \geq 2$.

If $\alpha > 0$ and $q(x,y)$ defines the nearest-neighbor random walk on $\mathbb{Z}$, coexistence in the sense that one can see the populations of both types occupy a fixed region after large times fails with probability one, and hence, clustering holds. See [22, Theorem 2 (b)] for the precise statement. On the other hand, in view of the solution (2.12) under the classical Lotka–Volterra model and the convergence of the symmetric Neuhauser–Pacala model on large complete graphs (recall the discussion below (2.14)), it seems reasonable to expect that a mean-field phenomenon should be valid under these stochastic spatial generalizations in general. Coexistence in suitable notions should hold except in very few, if not pathological, cases. Some resolutions of Conjecture 3.8 are achieved in [5–7,28], since [22, Theorem 1] takes the first few steps to do so. Yet to the knowledge of the authors, the full resolution of Conjecture 3.8 remains open.

The role of 1/2 in equilibria can be illustrated more generally by the following theorem.

Theorem 3.9: *Suppose that (η_t) has initial condition given by β_u for $u \in (0,1)$ and that along some subsequence (t_k) tending to infinity,*

$$\langle \widehat{\eta}^B_{t_k}, \mathbf{1} \rangle \xrightarrow[t_k\to\infty]{\text{(d)}} Z_B \quad \text{in } \mathbb{Z}_+ \cup \{+\infty\} \tag{3.32}$$

for some random variable Z_B. Then

$$\mathbb{P}_{\beta_u}\big(\langle \mathbf{1}_B, \eta_{t_k}\rangle \equiv 1\big) \xrightarrow[t_k\to\infty]{} \frac{1}{2}\mathbb{E}\left[1-(1-2u)^{Z_B}\right]$$

with the convention $0^0 = 1$ when $u = 1/2$.

This theorem is a generalization of what we discuss above for the case $u = 1/2$ in Theorem 3.5. The following corollary is immediate from Theorem 3.9.

Corollary 3.10: *Suppose that (3.32) holds and* $u \neq 1/2$. *Then* $\mathbb{P}_{\beta_u}\big(\langle \mathbf{1}_B, \eta_{t_k}\rangle \equiv 1\big) \to \frac{1}{2}$ *if and only if the dual process survives:*

$$\lim_{t_k\to\infty} \mathbb{P}(\langle \widehat{\eta}^B_{t_k}, \mathbf{1}\rangle \geq 1) = 1$$

and there is ***extinction versus unbounded growth*** *in the dual process:*

$$\lim_{t_k\to\infty} \mathbb{P}(1 \leq \langle \widehat{\eta}^B_{t_k}, \mathbf{1}\rangle \leq L) = 0, \quad \forall\, L \in \mathbb{N}. \tag{3.33}$$

The notion of extinction versus unbounded growth is introduced by Sturm and Swart in [28]. We will discuss this notion in more detail later on.

The proof of Theorem 3.9 is an application of the parity duality and the following lemma.

Lemma 3.11: (Parity deviation) *Let* $X_1, X_2, \ldots, X_N$ *be independent* $\mathbb{Z}_+$*-valued random variables with* $\mathbb{P}(X_m \equiv 1) = u_m$. *Then it holds that*

$$\mathbb{P}\left(\sum_{m=1}^N X_m \equiv 0\right) - \mathbb{P}\left(\sum_{m=1}^N X_m \equiv 1\right) = \prod_{m=1}^N (1 - 2u_m).$$

Proof: The identity follows upon writing the left-hand side as $\mathbb{E}\big[(-1)^{\sum_{m=1}^N X_m}\big]$. □

Proof of Theorem 3.9: With a slight abuse of notation, we also write β_u as a random configuration independent of $(\widehat{\eta}^B_t)$. By the parity duality in Theorem 3.1, we have

$$\begin{aligned}\lim_{k\to\infty} \mathbb{P}_{\beta_u}\big(\langle \mathbf{1}_B, \eta_{t_k}\rangle \equiv 1\big) &= \lim_{k\to\infty} \mathbb{P}\big(\langle \widehat{\eta}^B_{t_k}, \beta_u\rangle \equiv 1\big)\\ &= \lim_{k\to\infty} \frac{1}{2}\mathbb{E}\left[1 - (1-2u)^{\langle \widehat{\eta}^B_{t_k}, \mathbf{1}\rangle}\right]\\ &= \frac{1}{2}\mathbb{E}\left[1 - (1-2u)^{Z_B}\right]\end{aligned}$$

by (3.32), where the second equality follows from Lemma 3.11 by conditioning on $\widehat{\eta}^B_{t_k}$ since

$$\begin{aligned}\mathbb{P}\left(\sum_{m=1}^N X_m \equiv 1\right) &= \frac{1}{2}\left[1 - \mathbb{P}\left(\sum_{m=1}^N X_m \equiv 0\right) + \mathbb{P}\left(\sum_{m=1}^N X_m \equiv 1\right)\right]\\ &= \frac{1}{2}\left(1 - \prod_{m=1}^N (1-2u_m)\right).\end{aligned}$$

□

For Theorem 3.9, further generalizations of the initial conditions of (η_t) are possible. For example, on a finite set, $(\beta_u)_{0<u<1}$ can explicitly generate uniform distributions over sets of configurations with fixed numbers of 1's [4, pages 655–656] and links to distributions defined by exchangeable random variables over sites by de Finetti's representation theorem [17, Theorem 11.10]. More generally, the proof of Theorem 3.9 can be modified in an obvious way to show a similar formula if we use Bernoulli product measures with site-dependent densities, where the densities $\neq 0, 1$ are bounded away from 0 and 1.

Theorem 3.9 and Lemma 3.11 should provide the basic idea behind the related methods in [3, 22, 28]. There the method to resolve the main technical issue for not starting with Bernoulli measures is to identify 'Bernoulli subsystems' which have sizes growing to infinity as $t \to \infty$. The frameworks along this direction in [3, 22, 28] can be roughly fit into the following principle, where we generalize the setting of Lemma 3.11 to conditional probabilities in a time-dependent setting. Let $(\mathscr{F}_t)$ be a filtration and $\mathcal{I}_t, \mathcal{B}_t$ be $\mathcal{F}_t$-measurable subsets of $\mathbb{N}$. For each $t \geq 0$, the random variables X_m in Lemma 3.11 are now replaced by a collection $\{X_m(t);\ m \in \mathcal{I}_t\}$ of $\{0, 1\}$-valued random variables. We assume that $X_m(t)$, $m \in \mathcal{B}_t$, are conditionally independent given $\mathscr{F}_t$ and $\sum_{m \in \mathcal{I}_t \setminus \mathcal{B}_t} X_m(t) \in \mathscr{F}_t$. Then Lemma 3.11 implies

$$\mathbb{P}\left(\sum_{m\in\mathcal{I}_t} X_m(t) \equiv 1 \,\middle|\, \mathscr{F}_t\right) \tag{3.34}$$

$$= \mathbb{P}\left(\sum_{m\in\mathcal{B}_t} X_m(t) \equiv 1 - \sum_{m\in\mathcal{I}_t\setminus\mathcal{B}_t} X_m(t) \,\middle|\, \mathscr{F}_t\right)$$

$$= \begin{cases} \dfrac{1}{2}\left(1 - \prod\limits_{m\in\mathcal{B}_t}(1-u_m(t))\right), & \text{on } \left\{\sum\limits_{m\in\mathcal{I}_t\setminus\mathcal{B}_t} X_m(t) \equiv 0\right\}, \\ \dfrac{1}{2}\left(1 + \prod\limits_{m\in\mathcal{B}_t}(1-u_m(t))\right), & \text{on } \left\{\sum\limits_{m\in\mathcal{I}_t\setminus\mathcal{B}_t} X_m(t) \equiv 1\right\}, \end{cases} \tag{3.35}$$

where

$$u_m(t) = \mathbb{P}(X_m(t) \equiv 1 | \mathscr{F}_t).$$

Hence, to make the conditional probability in (3.34) get close to 1/2 with high conditional probability, it is necessary to verify that with high probability, $|\mathcal{B}_t|$ is large and $u_m(t)$ are bounded away from 0 and 1.

We close the discussion of this section with a very brief account of two applications of (3.34)–(3.35) in [3, 22, 28]. There, the underlying spaces are integer lattices.

Extinction versus unbounded growth. The method in [28, Section 3] takes the following route. We consider the parity duality in the following form: for fixed $s > 0$,

$$\mathbb{P}(\langle \mathbf{1}_B, \eta^A_{s+t}\rangle \equiv 1) = \mathbb{P}(\langle \widehat{\eta}^B_t, \eta^A_s\rangle \equiv 1) = \mathbb{E}\big[\mathbb{P}(\langle \mathbf{1}_C, \eta^A_s\rangle \equiv 1)\big|_{\mathbf{1}_C=\widehat{\eta}^B_t}\big], \quad (3.36)$$

where (η^A_s) and $(\widehat{\eta}^B_t)$ are independent as above. The first step is to show that if there are sufficiently many types of matrices J such that the corresponding Poisson processes with rates $r(J)$ can trigger a change in $\langle \mathbf{1}_C, \eta^A_r\rangle$ for $r \in [0, s]$ by *one* occurrence, then we have

$$\mathbb{P}(\langle \mathbf{1}_C, \eta^A_s\rangle \equiv 1) \simeq \frac{1}{2}.$$

These matrices J's are chosen such that $\langle \mathbf{1}_C, (\mathrm{Id} + J)\mathbf{1}_A\rangle = 1$, and their ranges are sufficiently far apart so that one can write $\langle \mathbf{1}_C, \eta^A_s\rangle \equiv 1$ as a sum of these inner products and other terms. Then the principle in (3.34) comes into play because the Bernoulli variables $X_m(t)$, $m \in \mathcal{B}_t$, are defined from conditioning these Poisson processes with rates $r(J)$ to ring at most once by time s.

The second step then aims to find the good A's by using appropriate initial conditions and the good C's for (3.36) by using the extinction versus unbounded growth condition in Corollary 3.10. (The following initial conditions μ are enough for the required convergence: μ is translation invariant and is supported in the set of configurations $\mathbf{1}_A$ where η^A_t can be any configuration with positive probability when restricted to an arbitrary finite set for every $t > 0$.) See [28, Section 3.2] for these two steps.

The work in [28] also shows conditions for the extinction versus unbounded growth condition in terms of the survival of (η_t) and a certain recurrence property of $(\widehat{\eta}_t)$ when $\alpha \in (0, 1)$ [28, Sections 3.3 and 3.5]. □

Domination by oriented percolation. The method in [22], extended from [3], uses the following parity duality:

$$\mathbb{P}(\langle \mathbf{1}_B, \eta^A_{2t}\rangle \equiv 1) = \mathbb{P}(\langle \widehat{\eta}^B_t, \eta^A_t\rangle \equiv 1), \quad (3.37)$$

where $(\widehat{\eta}^B_t)$ is independent of (η^A_t). ((3.37) is a simple consequence of the Markov property of (η^A_t) and the parity duality in Theorem 3.1.) We identify configurations with sets of vertices occupied by 1-individuals. Then one writes $\langle \widehat{\eta}_t^B, \eta_t^A \rangle = |\widehat{\eta}_t^B \cap \eta_t^A|$ as a sum of the $\{0,1\}$-valued random variables

$$X_m(t) \equiv |\widehat{\eta}_t^B \cap \eta_t^A \cap E_m|, \quad m \in \mathcal{I}_t = \mathbb{Z},$$

where $\{E_m\}$ is a partition of the whole space. In this case, the method in [3] uses oriented percolations [8, 9] to show that the products in (3.35) are close to 1/2 for suitably chosen $\mathcal{B}_t$. □

4. Interacting diffusions on lattices

In this section, we turn to coexistence in the system of interacting Wright-Fisher diffusions defined in (2.17), where from now on we set $N = 1$.

In addition to the connection to the Lotka-Volterra model explained in Section 2, the model can be considered by itself as a model for the evolution of gene frequencies in a spatially structured two-type population subject to selection, where $p_t(x)$ describes the proportion of 0-individuals at site $x \in \mathbb{Z}^d$ and time $t \geq 0$. In this interpretation, at each site x, for $\mu < 1$ there is (directional) selection in favor of the 0-type if $s > 0$ and in favor of the 1-type if $s < 0$. If $\mu > 1$, then a type which is locally rare (corresponding to $p_t(x)$ being close to 0 or 1) has a selective advantage respectively disadvantage according to whether $s > 0$ or $s < 0$, thus for $s > 0$ we have selection in favor of heterozygosity and for $s < 0$ in favor of homozygosity. (Here, the term 'heterozygosity' refers to the degree of genetic variation in the population.) It is easy to see that $(1 - p_t)$ satisfies (2.17) with s and μ replaced by $(-s)(1 - \mu)$ and $\frac{\mu}{\mu-1}$, respectively. In particular, if $\mu = 2$, then $(1 - p_t)$ follows the same dynamics as (p_t), thus the two types evolve symmetrically. Also note that for the neutral case $s = 0$, the system (2.17) reduces to the well-known stepping stone model, see [24].

It is natural to conjecture that long-term coexistence of the two types is possible in the heterozygosity ('balancing') selection case, at least if s is large enough. Indeed, based on comparison arguments with oriented percolation, in [1, Theorem 1.4] a coexistence result in the following sense is proven: for $\mu > 1$ and fixed small $\varepsilon > 0$, there exists $s_0 \geq 0$ such that for all $s > s_0$ and all initial conditions p_0 with $p_0(x) \in (\varepsilon, 1 - \varepsilon)$ for all $x \in \mathbb{Z}^d$, we have

$$\liminf_{t \to \infty} \mathbb{P}\left(\varepsilon < p_t(x) < 1 - \varepsilon\right) > 0, \quad \forall\, x \in \mathbb{Z}^d.$$

Note that [1, Theorem 1.4] does not claim that coexistence fails for $s < s_0$, nor does it specify the value of s_0. But in fact, one would conjecture the following (see [1, Conjectures 2.2 and 2.3]):

Conjecture 4.1: Suppose $\mu > 1$. There exists a critical value $s_0 \geq 0$ such that we have coexistence for $s > s_0$ and non-coexistence for $s < s_0$.

As in Conjecture 3.8, the full resolution of this conjecture is still open to the knowledge of the authors.

Our discussion below will continue the point of view from Section 3 and relate coexistence to the question of survival for a suitable dual process. This time, the duality we can use is a *moment duality* extended from the well-known stepping stone model [24]. It requires a bit of algebra to see and, unfortunately, seems to be available for restricted values of the parameters s and μ only. In the 'symmetric' case $\mu = 2$, the moment dual is a *branching annihilating random walk*, which is a non-monotone system and difficult to analyze mathematically. Based on non-rigorous results in the physics literature, [1] actually conjectures the value of s_0 in Conjecture 4.1 to be equal to 0 in dimensions $d \geq 2$ and to be strictly positive in $d = 1$.

4.1. *Duality and coexistence*

We start with the particular case $s = 0$, the stepping stone model, for which there is a well-known moment duality due to Shiga [25] with a system of coalescing random walks $(\xi_t) = (\xi_t(x); x \in \mathbb{Z}^d)$ on $\mathbb{Z}_+^{\mathbb{Z}^d}$ defined with the following rates:

$$\begin{aligned} &\text{migration:} \begin{cases} \xi(x) \to \xi(x) - 1 \\ \xi(y) \to \xi(y) + 1 \end{cases} \text{with rate } \xi(x) m_{xy}, \\ &\text{coalescence: } \xi(x) \to \xi(x) - 1 \;\; \text{with rate } \frac{\xi(x)[\xi(x)-1]}{2}. \end{aligned} \tag{4.1}$$

The duality reads as follows: for each $p_0 \in [0,1]^{\mathbb{Z}^d}$ and initial condition $\xi_0 \in \mathbb{Z}_+^{\mathbb{Z}^d}$ such that $\sum_{x \in \mathbb{Z}^d} \xi_0(x) < \infty$, we have

$$\mathbb{E}\left[\prod_{x \in \mathbb{Z}^d} p_t(x)^{\xi_0(x)}\right] = \mathbb{E}\left[\prod_{x \in \mathbb{Z}^d} p_0(x)^{\xi_t(x)}\right]. \tag{4.2}$$

The migration mechanism in (4.1) shows that individuals at site x can migrate to site y with rate m_{xy}, whereas the coalescence mechanism shows that only two of the individuals at site x coalesce at a time.

The moment duality (4.2) can be used to quickly settle the question whether typically the underlying two types of individuals can coexist when the migration matrix m is such that starting from finitely many particles, the process (ξ_t) will almost surely end up with a single particle due to coalescence, that is

$$|\xi_t| \stackrel{\text{def}}{=} \sum_{x\in\mathbb{Z}^d} \xi_t(x) \xrightarrow[t\to\infty]{\text{a.s.}} 1.$$

In this case, if we consider the homogenous initial condition $p_0(x) = 1/2$ for all $x \in \mathbb{Z}^d$, then the duality (4.2) gives convergence of all mixed moments:

$$\mathbb{E}\left[\prod_{x\in\mathbb{Z}^d} p_t(x)^{\xi_0(x)}\right] \xrightarrow[t\to\infty]{} \frac{1}{2}, \quad \forall\, \xi_0 \in \mathbb{Z}_+^{\mathbb{Z}^d} \text{ with } \sum_{x\in\mathbb{Z}^d} \xi_0(x) < \infty,$$

which implies

$$p_t \xrightarrow[t\to\infty]{(\text{d})} \tfrac{1}{2}\delta_{\mathbf{1}} + \tfrac{1}{2}\delta_{\mathbf{0}}.$$

The two types cannot coexist.

For the case $s > 0$ and $\mu = 2$, the key idea from [1] is to consider the transformed process:

$$\sigma_t \stackrel{\text{def}}{=} 1 - 2p_t \in [-1, 1].$$

It satisfies the following system of SDEs:

$$\begin{aligned} d\sigma_t(x) = \sum_{y\in\mathbb{Z}^d} m_{xy}\big(\sigma_t(y) - \sigma_t(x)\big)\,dt + \frac{s}{2}\big(\sigma_t(x)^3 - \sigma_t(x)\big)\,dt \\ - \sqrt{1 - \sigma_t(x)^2}\,dW_t(x), \quad x \in \mathbb{Z}^d. \end{aligned} \tag{4.3}$$

It is stated in [1, Lemma 2.1] that the moment duality in (4.2) still applies for σ_t in place of p_t, but instead of a coalescing random walk, the dual process is now a *branching annihilating random walk.*

Definition 4.2: Fix $s \geq 0$. Define a Markov process (ξ_t) taking values $\xi_t \in \mathbb{Z}_+^{\mathbb{Z}^d}$ starting from finitely many particles at time zero (that is, $\sum_{x\in\mathbb{Z}^d} \xi_0(x) < \infty$) and with the following transition rates:

$$\begin{aligned} &\text{migration:} \begin{cases} \xi(x) \to \xi(x) - 1 \\ \xi(y) \to \xi(y) + 1 \end{cases} \text{with rate } \xi(x)m_{xy}, \\ &\text{double branching: } \xi(x) \to \xi(x) + 2 \quad \text{with rate } s\xi(x), \\ &\text{annihilation: } \xi(x) \to \xi(x) - 2 \quad \text{with rate } \frac{\xi(x)[\xi(x) - 1]}{2}. \end{aligned} \tag{4.4}$$

This process is called a *double branching annihilating random walk* (DBARW) with branching rate s.

Lemma 4.3: ([1]) *Fix* $\mu = 2$ *and* $s \geq 0$. *Then we have the following moment duality between the transformed process* (σ_t) *defined by (4.3) and the DBARW* (ξ_t) *from Definition 4.2 with branching rate* $s/2$*: for all* $\sigma_0 \in [-1,1]^{\mathbb{Z}^d}$ *and* $\xi_0 \in \mathbb{Z}_+^{\mathbb{Z}^d}$ *with* $\sum_{x\in\mathbb{Z}^d} \xi_0(x) < \infty$,

$$\mathbb{E}_{\sigma_0}\left[\prod_{x\in\mathbb{Z}^d} \sigma_t(x)^{\xi_0(x)}\right] = \mathbb{E}_{\xi_0}\left[\prod_{x\in\mathbb{Z}^d} \sigma_0(x)^{\xi_t(x)}\right]. \tag{4.5}$$

We will give the proof of Lemma 4.3 below in Section 4.2. As an application of the moment duality, we now show the equivalence of coexistence for the system (2.17) (for $\mu = 2$) and survival of DBARW. See also [28, Lemma 1] for a related result in the context of spin systems.

Proposition 4.4: *Fix* $s > 0$. *Let* (p_t) *denote the solution to the system of SDEs in* (2.17) *for* $\mu = 2$, *and let* (ξ_t) *denote the DBARW with branching rate* $s/2$ *defined above. Then the following four statements are equivalent:*

(a) For all initial conditions p_0 *such that* $p_0(x) \in (\varepsilon, 1-\varepsilon)$ *for all* x *for some small* $\varepsilon > 0$, *we have long-term coexistence of* (p_t) *with positive probability in the sense that there exists* $\kappa \in (0,1)$ *such that*

$$\liminf_{t\to\infty} \mathbb{P}\left(\kappa < p_t(0) < 1-\kappa\right) > 0.$$

(b) There exists some initial condition p_0 *for which we have long-term coexistence of* (p_t) *with positive probability.*

(c) The DBARW started with exactly two particles at the origin at time zero survives for all time with positive probability. That is, for the initial condition $\xi_0 = 2\mathbf{1}_{\{0\}}$, *we have*

$$\mathbb{P}\left(\xi_t \neq 0, \ \forall\, t \geq 0\right) > 0.$$

(d) The DBARW started with any even number of particles at time zero survives for all time with positive probability.

Proof: The equivalence of (c) and (d) is clear from the double branching mechanism in (4.4), and it is plain that (a) implies (b). Below, we show that (b) implies (c) and (c) implies (a). We work with the transformed process $\sigma_t = 1 - 2p_t$ taking values in $[-1,1]$, which satisfies (4.3). Note that coexistence for (p_t) as in (a) and (b) is equivalent to the following condition on the process (σ_t). For all initial conditions σ_0 such that $\sup_{x\in\mathbb{Z}^d} |\sigma_0(x)| < 1 - 2\varepsilon$ for some small $\varepsilon > 0$, there exists $\kappa > 0$ such that

$$\liminf_{t\to\infty} \mathbb{P}\left(|\sigma_t(0)| < 1 - 2\kappa\right) > 0.$$

The proof that (b) implies (c) is as follows. Suppose by way of contradiction that (c) is false. This means that if we start the DBARW with two particles at the origin at time zero, that is $\xi_0 = 2\mathbf{1}_{\{0\}}$, then the process will die out with probability one, that is there exists $t > 0$ with $\xi_t = \mathbf{0}$ almost surely. Consider any initial condition p_0 for which the process (p_t) coexists. Then by the moment duality and dominated convergence, we have

$$\mathbb{E}\left[\sigma_t(0)^2\right] = \mathbb{E}\left[\prod_{x\in\mathbb{Z}^d} \sigma_t(x)^{\xi_0(x)}\right] = \mathbb{E}\left[\prod_{x\in\mathbb{Z}^d} \sigma_0(x)^{\xi_t(x)}\right] \xrightarrow[t\to\infty]{} 1. \tag{4.6}$$

On the other hand, by the assumed long-term coexistence in (b), we know that there is some $\kappa \in (0,1)$ such that

$$\delta \stackrel{\text{def}}{=} \liminf_{t\to\infty} \mathbb{P}\left(\kappa < p_t(0) < 1-\kappa\right) = \liminf_{t\to\infty} \mathbb{P}\left(|\sigma_t(0)| < 1-2\kappa\right) > 0,$$

which implies

$$\begin{aligned}\mathbb{E}\left[\sigma_t(0)^2\right] &= \mathbb{E}\left[\sigma_t(0)^2\left(\mathbf{1}_{\{|\sigma_t(0)|<1-2\kappa)\}} + \mathbf{1}_{\{|\sigma_t(0)|\ge 1-2\kappa)\}}\right)\right]\\ &\le (1-2\kappa)^2\,\mathbb{P}\left(|\sigma_t(0)| < 1-2\kappa\right) + \mathbb{P}\left(|\sigma_t(0)| \ge 1-2\kappa\right)\end{aligned}$$

and thus

$$\liminf_{t\to\infty} \mathbb{E}\left[\sigma_t(0)^2\right] \le (1-2\kappa)^2\,\delta + (1-\delta) < 1,$$

which is a contradiction to (4.6). We have proved that (b) implies (c).

Next, we prove that (c) implies (a). Assume that (c) holds. Consider $\varepsilon > 0$ and any 'permissible' initial condition σ_0 in (a), that is $|\sigma_0(x)| \le 1-2\varepsilon$ for all $x \in \mathbb{Z}^d$. Suppose by way of contradiction that for all $\kappa > 0$ we have

$$\liminf_{t\to\infty} \mathbb{P}\left(|\sigma_t(0)| < 1-2\kappa\right) = 0.$$

Then for any $\kappa > 0$ we get

$$\begin{aligned}\mathbb{E}\left[\sigma_t(0)^2\right] &= \mathbb{E}\left[\sigma_t(0)^2\left(\mathbf{1}_{\{|\sigma_t(0)|<1-2\kappa)\}} + \mathbf{1}_{\{|\sigma_t(0)|\ge 1-2\kappa)\}}\right)\right]\\ &\ge 0 + (1-2\kappa)^2\,\mathbb{P}\left(|\sigma_t(0)| \ge 1-2\kappa\right)\\ &= (1-2\kappa)^2\left(1 - \mathbb{P}\left(|\sigma_t(0)| < 1-2\kappa\right)\right),\end{aligned}$$

from which we infer that

$$\limsup_{t\to\infty} \mathbb{E}\left[\sigma_t(0)^2\right] \ge (1-2\kappa)^2\left(1 - \liminf_{t\to\infty} \mathbb{P}\left(|\sigma_t(0)| < 1-2\kappa\right)\right) = (1-2\kappa)^2.$$

Since $\kappa > 0$ was arbitrary, this implies

$$\limsup_{t\to\infty} \mathbb{E}\left[\sigma_t(0)^2\right] = 1.$$

On the other hand, since the DBARW survives by (c), we have $\delta = \mathbb{P}(\xi_t \neq \mathbf{0},\ \forall\, t \geq 0) > 0$. Then by the duality we have for all $t > 0$

$$\begin{aligned} 0 \leq \mathbb{E}\left[\sigma_t(0)^2\right] &= \mathbb{E}\left[\prod_{x\in\mathbb{Z}^d} \sigma_t(x)^{\xi_0(x)}\right] = \mathbb{E}\left[\prod_{x\in\mathbb{Z}^d} \sigma_0(x)^{\xi_t(x)}\right] \\ &\leq \mathbb{E}\left[\prod_{x\in\mathbb{Z}^d} |\sigma_0(x)|^{\xi_t(x)}\right] \\ &= \mathbb{E}\left[\prod_{x\in\mathbb{Z}^d} |\sigma_0(x)|^{\xi_t(x)} \left(\mathbf{1}_{\{(\xi_t) \text{ survives}\}} + \mathbf{1}_{\{(\xi_t) \text{ dies out}\}}\right)\right] \\ &\leq (1-2\varepsilon)\,\delta + 1 - \delta < 1, \end{aligned}$$

again giving a contradiction. We have proved that (c) implies (a). The proof is complete. □

By Proposition 4.4, coexistence for just one initial condition p_0 implies coexistence for *all* initial conditions of the form considered in (a). Hence, to determine coexistence, it suffices to work with nice initial conditions such as $p_0(x) = 1/2$ for all $x \in \mathbb{Z}^d$, which we considered when discussing the case $s = 0$ above. Recall that using a comparison with oriented percolation, [1] proves coexistence for sufficiently large values of s and so, in view of Proposition 4.4, they obtain as a corollary the survival of DBARW if the branching rate is large enough.

There is another range of parameters for which solutions to (2.17) admit nice duals and so the question whether coexistence occurs can be settled again under appropriate conditions.

Lemma 4.5: *For $\mu \in [-1, 0]$ and $s \leq 0$, the solution (p_t) to (2.17) is dual to a* ***branching coalescing random walk*** *(BCRW) (ξ_t) taking values in $\mathbb{Z}_+^{\mathbb{Z}^d}$ with transition rates given by (4.1), and in addition, the following two:*

$$\begin{aligned} \xi(x) &\to \xi(x) + 1 \quad &&\textit{with rate } (-s)(\mu+1)\xi(x), \\ \xi(x) &\to \xi(x) + 2 \quad &&\textit{with rate } (-s)(-\mu)\xi(x). \end{aligned}$$

Clearly, BCRW survives for all time with probability one whenever the initial condition is not trivial. Note that Lemma 4.5 contains the coalescing random walk dual for the stepping stone model (that is the model with $s = 0$) as a special case. We will also sketch the proof of Lemma 4.5 in Section 4.2.

Proposition 4.6: *Let (p_t) denote the solution to (2.17) with parameters $s < 0$ and $\mu \in [-1, 0]$, and let (ξ_t) denote the BCRW with branching rates defined in Lemma 4.5. Assume that for each (non-trivial) initial condition ξ_0 with $0 < \sum_{x \in \mathbb{Z}^d} \xi_0(x) < \infty$, we have*

$$|\xi_t| = \sum_{x \in \mathbb{Z}^d} \xi_t(x) \xrightarrow[t \to \infty]{\text{a.s.}} +\infty. \tag{4.7}$$

Then for each initial condition p_0 such that $p_0(x) < 1 - \varepsilon$ for all $x \in \mathbb{Z}^d$ and some $\varepsilon > 0$, it holds that

$$p_t \xrightarrow[t \to \infty]{\text{a.s.}} \mathbf{0}.$$

Proof: By Lemma 4.5, we have

$$\mathbb{E}\left[\prod_{x \in \mathbb{Z}^d} p_t(x)^{\xi_0(x)}\right] = \mathbb{E}\left[\prod_{x \in \mathbb{Z}^d} p_0(x)^{\xi_t(x)}\right] \leq \mathbb{E}\left[(1 - \varepsilon)^{|\xi_t|}\right] \xrightarrow[t \to \infty]{} 0$$

by dominated convergence. This shows that all mixed moments of (p_t) converge to zero, which is enough for our assertion. □

Of course, this result is not surprising given that under the choice of the parameters μ and s in Proposition 4.6, type 0 has a selective disadvantage. We note that condition (4.7) is typically satisfied, see [27, Thm. 4.2]. Since $(1 - p_t)$ satisfies (2.17) with the same migration matrix m and s resp. μ replaced by $(-s)(1 - \mu)$ resp. $\frac{\mu}{\mu - 1}$, we may conclude that $p_t \xrightarrow[t \to \infty]{\text{a.s.}} \mathbf{1}$ for the parameter regime $s > 0$ and $\mu \in [0, \frac{1}{2}]$.

4.2. *Proof of the moment duality*

In this subsection, we give the proofs of the two duality results in Lemmas 4.3 and 4.5.

Proof of Lemma 4.3: Let L^ξ denote the generator of the DBARW (ξ_t) from Definition 4.2 and L^σ denote the generator of (σ_t) from (4.3). The bivariate duality function in use is now given by

$$H(\sigma, \xi) = \sigma^\xi \stackrel{\text{def}}{=} \prod_{x \in \mathbb{Z}^d} \sigma(x)^{\xi(x)}.$$

Then (4.3) implies that

$$
\begin{aligned}
&\mathsf{L}^{\sigma} H(\cdot,\xi)(\sigma) \\
&= \sum_{x\in\mathbb{Z}^d}\left(\sum_{y\in\mathbb{Z}^d} m_{xy}\big(\sigma(y)-\sigma(x)\big)\right)\frac{\partial}{\partial\sigma(x)}\sigma^{\xi} \\
&\quad+\frac{s}{2}\sum_{x\in\mathbb{Z}^d}\left(\sigma(x)^3-\sigma(x)\right)\frac{\partial}{\partial\sigma(x)}\sigma^{\xi} \\
&\quad+\frac{1}{2}\sum_{x\in\mathbb{Z}^d}\left(1-\sigma(x)^2\right)\frac{\partial^2}{\partial\sigma(x)^2}\sigma^{\xi} \\
&= \sum_{x\in\mathbb{Z}^d:\xi(x)\geq 1}\left(\sum_{y\in\mathbb{Z}^d} m_{xy}\big(\sigma(y)-\sigma(x)\big)\right)\xi(x)\sigma(x)^{\xi(x)-1}\prod_{y\neq x}\sigma(y)^{\xi(y)} \\
&\quad+\frac{s}{2}\sum_{x\in\mathbb{Z}^d:\xi(x)\geq 1}\left(\sigma(x)^3-\sigma(x)\right)\xi(x)\sigma(x)^{\xi(x)-1}\prod_{y\neq x}\sigma(y)^{\xi(y)} \\
&\quad+\frac{1}{2}\sum_{x\in\mathbb{Z}^d:\xi(x)\geq 2}\left(1-\sigma(x)^2\right)\xi(x)(\xi(x)-1)\sigma(x)^{\xi(x)-2}\prod_{y\neq x}\sigma(y)^{\xi(y)} \\
&= \sum_{x\in\mathbb{Z}^d:\xi(x)\geq 1}\xi(x)\left(\sum_{y\in\mathbb{Z}^d} m_{xy}\big(\sigma(y)\sigma(x)^{-1}-1\big)\right)\sigma^{\xi} \\
&\quad+\frac{s}{2}\sum_{x\in\mathbb{Z}^d:\xi(x)\geq 1}\xi(x)\left(\sigma(x)^2-1\right)\sigma^{\xi} \qquad (4.8)\\
&\quad+\frac{1}{2}\sum_{x\in\mathbb{Z}^d:\xi(x)\geq 2}\xi(x)(\xi(x)-1)\left(\sigma(x)^{-2}-1\right)\sigma^{\xi}.
\end{aligned}
$$

On the other hand, the definition of (ξ_t) shows that

$$
\begin{aligned}
&\mathsf{L}^{\xi} H(\sigma,\cdot)(\xi) \\
&= \sum_{x\in\mathbb{Z}^d}\sum_{y\in\mathbb{Z}^d}\xi(x)m_{xy}\big(H(\sigma,\xi-\delta_x+\delta_y)-H(\sigma,\xi)\big) \\
&\quad+\frac{s}{2}\sum_{x\in\mathbb{Z}^d}\xi(x)\big(H(\sigma,\xi+2\delta_x)-H(\sigma,\xi)\big) \\
&\quad+\frac{1}{2}\sum_{x\in\mathbb{Z}^d}\xi(x)(\xi(x)-1)\big(H(\sigma,\xi-2\delta_x)-H(\sigma,\xi)\big)
\end{aligned}
$$

$$= \sum_{x\in\mathbb{Z}^d:\xi(x)\geq 1}\sum_{y\in\mathbb{Z}^d}\xi(x)m_{xy}\big(\sigma(y)\sigma(x)^{-1}-1\big)\sigma^\xi$$
$$+\frac{s}{2}\sum_{x\in\mathbb{Z}^d:\xi(x)\geq 1}\xi(x)\big(\sigma(x)^2-1\big)\sigma^\xi$$
$$+\frac{1}{2}\sum_{x\in\mathbb{Z}^d:\xi(x)\geq 2}\xi(x)\big(\xi(x)-1\big)\big(\sigma(x)^{-2}-1\big)\sigma^\xi,$$

which is equal to the right-hand side of (4.8). Hence, by the arguments in [10, pages 188–193] (see also [15, Prop. 1.2]), the required moment duality (4.5) holds. □

Proof of Lemma 4.5: In this proof, L^ξ denotes the generator of the BCRW (ξ_t) with transition rates as defined in the statement of Lemma 4.5 and L^p denotes the generator of (p_t) from (2.17). The duality function is now changed to

$$H(p,\xi)=p^\xi\stackrel{\text{def}}{=}\prod_{x\in\mathbb{Z}^d}p(x)^{\xi(x)}.$$

Then by analogous calculations as in the proof of Lemma 4.3, we see that

$$\mathsf{L}^pH(\cdot,\xi)(p)$$
$$= \sum_{x\in\mathbb{Z}^d:\xi(x)\geq 1}\sum_{y\in\mathbb{Z}^d}\xi(x)m_{xy}\big(p(y)p(x)^{-1}-1\big)p^\xi$$
$$+s\sum_{x\in\mathbb{Z}^d:\xi(x)\geq 1}\xi(x)\big(1-(\mu+1)p(x)+\mu p(x)^2\big)p^\xi$$
$$+\frac{1}{2}\sum_{x\in\mathbb{Z}^d:\xi(x)\geq 2}\xi(x)\big(\xi(x)-1\big)\big(p(x)^{-1}-1\big)p^\xi$$
$$= \sum_{x\in\mathbb{Z}^d:\xi(x)\geq 1}\sum_{y\in\mathbb{Z}^d}\xi(x)m_{xy}\big(p(y)p(x)^{-1}-1\big)p^\xi$$
$$+(-s)(-\mu)\sum_{x\in\mathbb{Z}^d}\xi(x)\big(p(x)^2-1\big)p^\xi+(-s)(\mu+1)\sum_{x\in\mathbb{Z}^d}\xi(x)\big(p(x)-1)\big)p^\xi$$
$$+\frac{1}{2}\sum_{x\in\mathbb{Z}^d:\xi(x)\geq 2}\xi(x)\big(\xi(x)-1\big)\big(p(x)^{-1}-1\big)p^\xi$$
$$=\mathsf{L}^\xi H(p,\cdot)(\xi).$$

□

References

1. J. Blath, A. Etheridge and M. Meredith, "Coexistence in locally regulated competing populations and survival of branching annihilating random walk", *Annals of Applied Probability,* **17** (2007), 1474–1507.
2. B. M. Bolker and S. W. Pacala, "Spatial moment equations for plant competition: understanding spatial strategies and the advantages of short dispersal", *American Naturalist,* **153** (1999), 575–602.
3. M. Bramson, W. Ding and R. Durrett, "Annihilating branching processes", *Stochastic Processes and their Applications,* **37** (1991), 1–17.
4. Y.-T. Chen, "Sharp benefit-to-cost rules for the evolution of cooperation on regular graphs", *Annals of Applied Probability,* **23** (2013), 637–664.
5. J. T. Cox and E. A. Perkins, "Rescaled Lotka-Volterra models converge to super-Brownian motion", *Annals of Probability,* **33** (2005), 904–947.
6. J. T. Cox and E. A. Perkins, "Survival and coexistence in stochastic spatial Lotka-Volterra models", *Probability Theory and Related Fields,* **139** (2007), 89–142.
7. J. T. Cox, M. Merle and E. A. Perkins, "Coexistence in a two-dimensional Lotka-Volterra model", *Electronic Journal of Probability,* **15** (2010), paper no. 38, 1190–1266.
8. R. Durrett, "Oriented percolation in two dimensions", *Annals of Probability,* **12** (1984), 999–1040.
9. R. Durrett, "Ten lectures on particle systems", in: P. Bernard (ed.), *Lectures on Probability Theory,* Lecture Notes in Mathematics, vol. **1608**, Springer, Berlin, Heidelberg, 1995.
10. S. N. Ethier and T. G. Kurtz, *Markov Processes: Characterization and Convergence,* 2nd edition, John Wiley & Sons, New York, 2005.
11. T. E. Harris, "Additive set-valued Markov processes and graphical methods", *Annals of Probability,* **6** (1978), 355–378.
12. J. Hofbauer and K. Sigmund, *Evolutionary Games and Population Dynamics,* Cambridge University Press, Cambridge, 1998.
13. E. E. Holmes, M. A. Lewis, J. E. Banks and R. R. Veit, "Partial differential equations in ecology: spatial interactions and population dynamics", *Ecology,* **75** (1994), 17–29.
14. S. Jansen and N. Kurt, "Graphical representation of certain moment dualities and application to population models with balancing selection", *Electronic Communications in Probability,* **18** (2013), paper no. 14, 1–15.
15. S. Jansen and N. Kurt, "On the notion(s) of duality for Markov processes", *Probability Surveys,* **11** (2014), 59–120.
16. J. Jacod and A. N. Shiryaev, *Limit Theorems for Stochastic Processes,* 2nd edition, Grundlehren der mathematischen Wissenschaften, vol. **288**, Springer, Berlin, 2003.
17. O. Kallenberg, *Foundations of Modern Probability,* 2nd edition, Probability and Its Applications, Springer, Berlin, 2002.
18. T. M. Liggett, *Interacting Particle Systems,* Classics in Mathematics, Springer, Berlin, 2005.

19. A. J. Lotka, "The growth of mixed populations: two species competing for a common food supply", *Journal of Washington Academy of Sciences*, **23** (1932), 461–469.
20. C. Mueller and R. Tribe, "A phase diagram for a stochastic reaction diffusion system", *Probability Theory and Related Fields*, **149** (2011), 561–637.
21. D. J. Murrell and R. Law, "Heteromyopia and the spatial coexistence of similar competitors", *Ecology Letters*, **6** (2003), 48–59.
22. C. Neuhauser and S. W. Pacala "An explicitly spatial version of the Lotka-Volterra model with interspecific competition", *Annals of Applied Probability*, **9** (1999), 1226–1259.
23. J. R. Norris, *Markov Chains*, Cambridge Series in Statistical and Probabilistic Mathematics, Cambridge University Press, Cambridge, 1998.
24. T. Shiga, "An interacting system in population genetics", *Journal of Mathematics of Kyoto University*, **20** (1980), 213–242.
25. T. Shiga, "Stepping stone models in population genetics and population dynamics", in: S. Albeverio, P. Blanchard, M. Hazewinkel, L. Streit (eds), *Stochastic Processes in Physics and Engineering*, Mathematics and Its Applications, vol. **42**, Springer, Dordrecht, 1988.
26. T. Shiga and A. Shimizu, "Infinite dimensional stochastic differential equations and their applications", *Journal of Mathematics of Kyoto University*, **20** (1980), 395–416.
27. T. Shiga and K. Uchiyama, "Stationary states and their stability of the stepping stone model involving mutation and selection", *Probability Theory and Related Fields*, **73** (1986), 87–117.
28. A. Sturm and J. Swart, "Voter models with heterozygosity selection", *Annals of Applied Probability*, **18** (2008), 59–99.
29. A. Sturm, J. Swart and F. Völlering, "The algebraic approach to duality: an introduction", Preprint (2018), arXiv:1802.07150.
30. V. Volterra, "Variations and fluctuations of the number of individuals in animal species living together", *ICES Journal of Marine Science*, **3** (1928), 3–51.

Stochastic Evolution of Genealogies of Spatial Populations: State Description, Characterization of Dynamics and Properties

Andrej Depperschmidt* and Andreas Greven†

Department Mathematik, Universität Erlangen-Nürnberg
Cauerstr. 11, 91058 Erlangen, Germany
* *depperschmidt@math.fau.de*
† *greven@math.fau.de*

We survey results on the description of stochastically evolving genealogies of populations and marked genealogies of multitype populations or spatial populations via tree-valued Markov processes on (marked) ultrametric measure spaces. In particular we explain the choice of state spaces and their topologies, describe the dynamics of genealogical Fleming-Viot and branching models by well-posed martingale problems, and formulate the typical results on the longtime behavior. Furthermore we discuss the basic techniques of proofs and sketch as two key tools of analysis the different forms of duality and the Girsanov transformation.

Contents

1. Background

The aim of this contribution is to present the work of the authors and coworkers Peter Pfaffelhuber, Rongfeng Sun and Anita Winter, on *evolving genealogies* in *spatial* population models in a systematic way and to explain its role in the context of the field of tree-valued Markov processes which is a field with many facets which is developing in many directions and with many applications from population genetics to computer science. For reasons of consistency in this survey we deviate from the original notation at various points.

In the 1970s the study of spatial population models such as voter model, branching random walk, contact process began and the focus was on the understanding of the longtime behavior of these systems. In particular the occupation measures were studied, that is, the configurations of numbers of particles/types per site was viewed as a measure on geographic and/or types space. This process of *occupation measures* is a Markov process and can be studied as *measure-valued processes* on infinite geographic spaces; see e.g. [5, 54, 26, 13]. Nevertheless inspired by the graphical construction often used, the results were interpreted in terms of *genealogical* information on individuals.

In this context tracing of *ancestral paths* of an individual from the current population backwards in time to its "ancestor" played an increasing role and was formalized by Dawson and Perkins in [18] via the so-called *historical process*, i.e., a measure-valued process with measures on paths which were the ancestral paths of individuals currently alive. The approach worked well for branching random walks and the Dawson-Watanabe super processes. In the latter case these paths encoded indeed the genealogical distance between individuals using the information on location or type. However, the genealogies were *not* explicitly part of the state.

The techniques of treatment of historical processes used infinite divisibility. Conceptually the historical processes of the ancestral paths are also of importance in other population models as in the voter model, contact process or Moran and Fleming-Viot models, where results could be obtained using the technique of the *hierarchical mean-field limit*, see [16] or latter using ideas of the *lookdown construction* of Donnelly and Kurtz [25] and *particle system techniques* of Liggett and Spitzer [55] in [37]. In the Fleming-Viot model on $\mathbb{Z}^1$ the construction of the Brownian web [4, 59, 32] can be used to obtain the processes of ancestral path, [43]. The historical process can be combined with the coding of genealogical information as genealogical distances in quite some generality, even in case of branching processes and super random walks getting again (generalized) branching processes, provided that the genealogical information is coded suitably [41]. We will discuss this coding later on.

The genealogical point of view gained more attention in the analysis of infinite particle systems which exhibit a *regime of clustering* in low dimensions. For example in the case of voter model there are growing clumps of a single opinion as time diverges, an opinion which originates from a single individual with that opinion. The formation of such clusters appears in Fleming-Viot models and in branching models as clumps of survivors in the sea of unpopulated sites. In many papers the growth in time and space of such "mono-ancestor" clumps was studied in detail starting from Cox and Griffeath [12] with [14, 31, 46, 47, 61]. In this series of works it became more and more apparent that the *genealogy* of the population and their evolution is behind all the effects.

At this point Aldous introduced in [1, 2, 3] the *continuum random tree* which made the genealogical distances in a branching population explicit in the state description. The approach focused on all individuals ever alive and was working with *metric trees*. In this context the description of the genealogy based on an embedding of the genealogical trees in *Brownian excursions*, such as in [58, 51] for example, gave an impulse to the ideas in the theory on continuum random trees. More systematically the concept of metric spaces, more precisely ultrametric spaces, was applied by Evans in [29] to study the genealogy associated with the entrance law from an infinite population of the Kingman coalescent describing the individuals currently alive with their genealogy in a Fleming-Viot population. This was taken up by Le Gall in [52, 53] in the work on the *Brownian snake*. Here the location marks on the path of descent are constructed as well. However, here also the population alive at some time was modeled as marked labeled tree.

The ideas of tracing ancestries and genealogical relations was a key point in the work of Donnelly and Kurtz [24, 25]. In their *lookdown construction* the evolution of the population in, for example a Fleming-Viot model, is represented by the evolution of a countably infinite ordered population. The state of the original population is obtained through the state of the ordered population and the latter is constructed by breaking the symmetry of the dynamics.

Also taking up the work of Donnelly and Kurtz [25, 24], another method to code genealogical information of an evolving population was invented by Bertoin and LeGall in [9, 10, 11], generalized by Dawson and Li in [17] and studied further in a number of concrete cases in [33, 50, 45]. This method leads to so called *"flows of bridges"* which might be seen as a specific technique to construct an *explicit representation* of a *labeled* tree which models the evolution of the genealogy of an evolving population.

In the program "Genealogies of interacting particle systems" in Singapore 2017 a learning session "Tree-valued Markov processes" (held by the authors of this contribution) was devoted to flows of bridges. Some other learning sessions where devoted to Aldous' work on continuum random trees and variants of the lookdown construction of Donnelly and Kurtz. The approach via flows of bridges works extremely nicely in the case of non-spatial models whereas introducing space is a very difficult matter. We don't have the space to spell out the details in this survey.

In 2009 Greven, Pfaffelhuber and Winter started to describe the *evolution of genealogies* of the individuals alive at the current time t of populations in the large populations limit [39, 40]. This was continued in a series of papers [20, 21, 43, 35, 19]. This was motivated first of all by trying to understand some effects in the longtime behavior of spatially interacting systems and the growth in space of monotype clusters in such systems as time diverges, a phenomenon discussed above. Further motivation came from the interplay between the mechanisms of migration and selection in Fleming-Viot models. As these models are infinite population models, where also the representation by lookdown constructions gets a bit intransparent, we saw the necessity to work systematically in a framework with *martingale problems* on state spaces encoding the genealogies to an extent needed to obtain the observed effects. To understand the genealogy of voter models exhibiting the monotype clusters in the scaling limit on spatial scale $\sqrt{t}$ starting with [4], later [32, 60], went through a long struggle. We think that this model now can be treated best in the framework of processes with values in ultrametric measure spaces using the tools of the theory on the Brownian web [4, 32, 59, 60]; see [43].

The description of the genealogy of a population and its dynamics is by means of *well-posed martingale problems* with values in the *equivalence classes of ultrametric measure spaces* respectively *marked*-ultrametric measures spaces in the case of multitype and/or spatial populations. Using methods of stochastic calculus *path properties*, longtime behavior and properties of *equilibria* could be proved for population models driven by *Fleming-Viot resampling* or *Feller branching*, i.e. in the case of two large classes of models which arise as limits of individual based models.

In the study of the evolution of genealogies of *branching* processes the idea of infinite divisibility plays also on the level of genealogies an important role, as it does for branching processes on $\mathbb{R}$. On $\mathbb{R}$ infinite divisibility was based on the concept of convolution and hence it makes use of the *semigroup* structure of $\mathbb{R}$, which is known to be sufficient to derive Lévy-Khintchine formulas. Thus, on the state space for the random genealogies one needs to have similar concepts and therefore we need here a binary operation with suitable algebraic properties, to be able to obtain the *Lévy-Khintchine* and *Cox point process representations* of the random genealogies in branching populations. Algebraic and topological methods play an important role since for the genealogies one can define *semigroup* operations, see here Evans and Molchanov [30] respectively collections of such different operations introduced by Glöde, Greven and Rippl in [35], [41], with the *concatenation* as binary operation of *truncated* trees, truncation generating a collection indexed in the truncation level, leading to a consistent structure of connected semigroups.

We will explain in this survey how states are modeled in the approaches mentioned in the last three paragraphs, introduce the martingale problems for the most basic processes and describe their properties. We consider *tree-valued Fleming-Viot* and *tree-valued Feller diffusions* and later the corresponding *spatial* versions. More precisely, we consider the genealogical processes associated with the following classical diffusion models for populations which are the many individuals/small mass limits of individual based models:

- the *Fleming-Viot diffusion*, cf. [40],
- the *branching Feller diffusion*, cf. [19],
- the *selfcatalytic branching diffusion*, cf. [34],
- *Fleming-Viot diffusion with selection and mutation*, cf. [21],
- *interacting Fleming-Viot diffusions* on countable geographic spaces, cf. [43],
- *interacting branching Feller diffusions*, cf. [19], and *interacting logistic Feller-branching* diffusions, cf. [38],
- *interacting Fleming-Viot diffusions* on the continuum, i.e. $\mathbb{R}$, cf. [43].

The work in this directions continues; see [56, 48, 57, 45, 6]. Furthermore, the approach can be extended to genealogies of individuals ever alive up to the current time horizon. This leads to the study of the evolution of genealogies of *fossils*, cf. [42, 41].

We choose here to outline two directions of our current research, namely first *logistic branching* processes to model populations in competition for limited resources and the effects on genealogies and second Fleming-Viot models with *recombination*. We will explain therefore at the end in Section 7 how *infinitely divisible* random genealogies can be studied and how the mechanism of *recombination* can be incorporated to give some perspectives.

We would like to mention that the field develops also in various other directions with further types of evolutions requiring new points of view. Recently Kliem and Winter [49] and Athreya, Löhr, Winter [6, 7] have studied different types of evolution than in the list above which however we cannot discuss here in more detail.

2. State spaces

In the discussion of state spaces we start with the simplest case, pass then to refinements and finally come to some topological issues which will be needed later on.

2.1. *Ultrametric probability measure spaces and random genealogies*

The key problem in passing from individual based models and their genealogies to the "continuum" limit is the need to find a suitable Polish space which can serve as the state space for the evolution of *random genealogies*. We next describe the solution to that problem.

The state space $\mathbb{U}_1$ The first objective is to find a description of the essential features of the *genealogy* of the population of currently alive individuals of a population evolving in time and then to find a *Polish space* in which these states can be embedded. This will allow to apply the standard methods of the theory of stochastic processes, such as Markov process theory and martingale problems.

The key concept of the genealogical information is that of an *ancestor*. In particular the most recent common ancestor (*MRCA*) of two individuals allows to define the *genealogical tree distance* as twice the time back to the MRCA. Hence, our approach is to describe the genealogy by a pair $(\widetilde{U}, r)$ where the set $\widetilde{U}$ labels the individuals of the population currently alive and r is the pseudo-ultrametric encoding the genealogical distance. Note that an individual having offspring generates individuals whose distance is zero at the time of their birth. Therefore it is convenient to declare them as equivalent points in $(\widetilde{U}, r)$, obtaining an ultrametric measure space (U, r)

and encode the information, having here *several individuals*, by introducing a Borel probability measure on (U, r) which is called the *sampling measure*. This measure allows to *sample* typical individuals from the population and is also useful in infinite populations to obtain *observable data* from the genealogy by drawing finite samples. Hence we obtain the *ultrametric probability measure space*

$$(U, r, \mu) \tag{2.1}$$

associated with the genealogy of the population of individuals currently alive. We do not want to keep the individual names as part of the description and therefore pass to *equivalence classes* of ultrametric probability measure spaces, which we denote by

$$\mathcal{U} = [(U, r, \mu)], \tag{2.2}$$

where two spaces (U, r, μ) and (U', r', μ') are said to be *equivalent* if there exists a bijective map

$$\varphi : \text{supp}(\mu) \to \text{supp}(\mu'), \tag{2.3}$$

with

$$r'\big(\varphi(x_1), \varphi(x_2)\big) = r(x_1, x_2), \quad \mu\text{-a.s.,} \quad \text{and} \quad \varphi_*\mu = \mu'. \tag{2.4}$$

Here $\varphi_*\mu$ denotes the image measure of μ under the map φ. Note that for finite genealogical trees with sampling measure this equivalence relation means that any measure preserving renaming of the vertices gives a population with the same genealogy. We denote the set of all equivalence classes (to avoid set theoretic paradoxes, we require $U \subseteq \mathbb{R}$ here, for example) by

$$\mathbb{U}_1. \tag{2.5}$$

This is the *state space* for the process of genealogies of the population alive at a given time t.

In order to obtain a decent state space for stochastic processes, we have to equip $\mathbb{U}_1$ with a *topology* $\mathcal{G}$ so that the topological space $(\mathbb{U}_1, \mathcal{G})$ is *Polish*, and is therefore suitable to accommodate limits of sequences of finite trees. All populations we consider here arise as *limits* of *finite individual based* models.

We choose here the so called *Gromov-weak topology*. The idea to define convergence, and thus $\mathcal{G}$, is as follows. Suppose we consider for all $\mathcal{U} \in \mathbb{U}_1$ a sequence of *finite sampled trees*. Then we define the sequence $(\mathcal{U}_n)_{n\in\mathbb{N}}$ as convergent to $\mathcal{U}_\infty$, if $\mu^{\otimes\mathbb{N}}$-a.s. all sampled finite trees converge in the sense of finite graphs with a metric to the corresponding object in $\mathcal{U}_\infty$ in law. In other words the $(n \times n)$-matrix of distances induces a family of distributions $(\nu_m^{(n)})_{m\in\mathbb{N}}$ on $(\mathbb{R}_+)^{\binom{n}{2}}$ which converges weakly to $\nu_\infty^{(n)}$:

$$\nu_m^{(n)} \Longrightarrow \nu_\infty^{(n)}, \text{ for all } n \in \{2,3,\dots\} \text{ as } m \to \infty. \tag{2.6}$$

This convergence defines a topology, the *Gromov-weak topology*, which has the following property.

Theorem 2.1: *(Polish state space* [39]*)*
The space $\mathbb{U}_1$ *equipped with the Gromov-weak topology is Polish.*

Remark 2.2: (Metric on $\mathbb{U}_1$) The metric can be introduced similarly to the Gromov-Hausdorff metric using the Prohorov metric of probability measures (instead of Hausdorff metric). More precisely, to compare $[(U,r,\mu)]$ and $[(U',r',\mu')]$ we consider *isometric embeddings* (of representatives) into a third metric space, say (Z, r_Z)

$$U \hookrightarrow Z,\ U' \hookrightarrow Z \tag{2.7}$$

giving image probability measures $\mu_U^*, \mu_{U'}^* \in \mathcal{P}((Z,r_Z))$ and then consider the Prohorov metric $d_{\mathrm{P}}(\mu_U^*, \mu_{U'}^*)$. Then take the infimum over all such embeddings to define the distance. Since the quantity is independent of the choice of representatives of $[(U,r,\mu)]$ respectively $[(U',r',\mu')]$ this defines a *metric* on $\mathbb{U}_1$. It can be shown that this metric generates the Gromov-weak topology; see [39].

$\mathbb{U}_1$-valued random variables Next we turn to random $\mathbb{U}_1$-valued variables, which we denote by

$$\mathfrak{U}. \tag{2.8}$$

The genealogy of a stochastically evolving population generates a $\mathbb{U}_1$-valued stochastic process, which we denote by

$$\mathfrak{U} = (\mathfrak{U}_t)_{t\geq 0} \tag{2.9}$$

slightly abusing the notation. The corresponding path spaces

$$D([0,\infty),\mathbb{U}_1) \quad \text{and} \quad C([0,\infty),\mathbb{U}_1) \tag{2.10}$$

are again *Polish spaces* and the theory of stochastic processes, Markov processes and martingale problems can be applied.

In order to study the $\mathbb{U}_1$-valued stochastic process by means of martingale problems or studying the longtime behavior we need a description via *test functions* on $\mathbb{U}_1$ whose expectations are law determining. The key idea is to draw finite samples of trees and then study these finite samples by test functions.

We consider therefore test functions of the following form: For $\mathcal{U} = [(U, r, \mu)] \in \mathbb{U}_1$ we set

$$\Phi^{n,\varphi}(\mathcal{U}) = \int_{\mathcal{U}} \varphi\left((r(x_i, x_j))_{1 \le i < j \le n}\right) \mu^{\otimes n}(\mathrm{d}x_1, \dots, \mathrm{d}x_n), \tag{2.11}$$

where $\varphi \in C_b(\mathbb{R}^{\binom{n}{2}}, \mathbb{R})$ for $n \ge 2$, and φ constant function for $n = 1$. Alternatively, for $n \ge 2$ we could write

$$\Phi^{n,\varphi}(\mathcal{U}) = \langle \nu_{\mathcal{U}}^{(n)}, \varphi \rangle, \tag{2.12}$$

where $\nu^{(n)}(\mathcal{U})$ is the *distance matrix distribution* of order n of $\mathcal{U}$. We refer to such test functions as *polynomials*. The set of all polynomials (together with the constant functions) will be denoted by

$$\Pi. \tag{2.13}$$

This set forms an *algebra* of bounded continuous functions.

Theorem 2.3: *(Polynomials are law and convergence determining)*

(i) The set Π *is separating on* $\mathbb{U}_1$ *and* $\{\mathbb{E}[F(\mathfrak{U})], F \in \Pi\}$ *is law determining.*
(ii) The set Π *is convergence determining on the set of probability measures on* $\mathbb{U}_1$.

For proofs of the above results see Theorem 5 in [20] or Corollary 2.8 in [56].

Now that we have introduced the polynomials let us note that the topology on $\mathbb{U}_1$, as defined in (2.6), could also be defined equivalently in terms of polynomials (see [39]):

$$\mathcal{U}_n \xrightarrow{n\to\infty} \mathcal{U} \text{ in } \mathbb{U}_1 \quad \text{iff} \quad \Phi(\mathcal{U}_n) \xrightarrow{n\to\infty} \Phi(\mathcal{U}) \text{ for all } \Phi \in \Pi. \tag{2.14}$$

This will be used next for extending and enriching the state spaces.

2.2. *Extensions: Marks, finite and infinite populations, fossils*

The concepts introduced in the previous section have to be extended in two ways. First to handle *spatial* and *multitype* populations we have to pass to *marked* genealogies. Second to incorporate *varying population sizes* we have to allow sampling measures which are not necessarily probability measures and which could even have infinite total mass, at least globally.

Genealogies of multitype and spatial populations If we have a multitype population which is spatially distributed we have to associate with the individuals types from some *type set* $\mathbb{I}$ and the types may influence the reproduction mechanism. Similarly we may have a *geographic space* $\mathbb{G}$ in which the individuals are located, i.e., we assign them geographic locations. We assume that the metric spaces $(\mathbb{I}, r_{\mathbb{I}})$ and $(\mathbb{G}, r_{\mathbb{G}})$ are complete and separable.

We incorporate this in our formalism by considering a measurable map

$$\widetilde{\kappa} : U \to \mathbb{V}, \quad \mathbb{V} = \mathbb{I} \times \mathbb{G}, \tag{2.15}$$

where $\mathbb{V}$ is the mark space and $\widetilde{\kappa}(u)$ is the mark of individual $x \in U$. Here $\mathbb{V}$ equipped with the product metric $r_{\mathbb{V}}$ is also complete and separable. This results in the structure

$$(U, r, \mu, \widetilde{\kappa}) \tag{2.16}$$

describing a $\mathbb{V}$*-marked genealogy*. We now introduce the Borel-measure ν on $(U \times \mathbb{V}, r \otimes r_{\mathbb{V}})$:

$$\nu = \mu \otimes \kappa, \quad \kappa(x, \mathrm{d}v) = \delta_{\widetilde{\kappa}(x)}(\mathrm{d}v), \ x \in U. \tag{2.17}$$

Then abstracting again from the individuals names (of course not their types or locations) we consider the *equivalence classes* of $\mathbb{V}$-marked ultrametric probability measures

$$[(U \times \mathbb{V}, r \otimes r_{\mathbb{V}}, \nu)], \tag{2.18}$$

where we say that $(U \times \mathbb{V}, r \otimes r_{\mathbb{V}}, \nu)$ and $(U' \times \mathbb{V}, r' \otimes r_{\mathbb{V}}, \nu')$ are equivalent if there is a map $\varphi : U \times \mathbb{V} \to U' \times \mathbb{V}$ such that (here π_U and $\pi_{\mathbb{V}}$ are the projections on U respectively $\mathbb{V}$)

$$\begin{aligned} &\pi_U \circ \varphi \text{ is an isometry between } \operatorname{supp}((\pi_U)_* \nu) \text{ and } \operatorname{supp}((\pi_{U'})_* \nu') \\ &\quad \varphi((x, v)) = ((\pi_U \varphi)(x, v), v) \quad \nu\text{-a.s.,} \quad \text{and} \\ &\quad\quad \varphi_* \nu = \nu'. \end{aligned} \tag{2.19}$$

To obtain later a *complete* space it is necessary to allow here instead of mark functions more generally *mark kernels* κ. Then we denote the set of all equivalence classes of $\mathbb{V}$-marked ultrametric probability measure spaces by

$$\mathbb{U}_1^{\mathbb{V}}, \tag{2.20}$$

which will be the state space for the marked genealogies.

On this space we introduce the $\mathbb{V}$-*marked Gromov-weak topology* as follows. Again we use the idea that finite samples converge as finite marked trees. Let

$$\Pi^{\mathbb{V}} \tag{2.21}$$

be the set of polynomials of the form

$$\Phi^{n,\varphi,\chi}([(U \times \mathbb{V}, r \otimes r_{\mathbb{V}}, \nu)]) = \int_{(U\times\mathbb{V})^n} \varphi(\underline{\underline{r}})\xi(\underline{v})\nu^{\otimes n}(\mathrm{d}(x_1,v_1),\dots,\mathrm{d}(x_n,v_n)), \tag{2.22}$$

where $\varphi \in C_b([0,\infty)^{\binom{n}{2}},\mathbb{R})$, $\xi \in C_b(\mathbb{V}^n,\mathbb{R})$ and $n \in \mathbb{N}_0$. Using these polynomials we can again define convergence on $\mathbb{U}_1^{\mathbb{V}}$ by the requirement:

$$\mathcal{U}_n \xrightarrow{n\to\infty} \mathcal{U} \text{ in } \mathbb{U}_1^{\mathbb{V}} \quad \text{iff} \quad \Phi(\mathcal{U}_n) \xrightarrow{n\to\infty} \Phi(\mathcal{U}) \text{ for all } \Phi \in \Pi^{\mathbb{V}}. \tag{2.23}$$

The corresponding topology is called the $\mathbb{V}$-*marked* Gromov-weak topology and we obtain a *Polish space* as a state space for the random marked genealogies. It is metrizable by a metric which generalizes the one described in Remark 2.2; see [20].

Theorem 2.4: *(Polish state space)*
The space $\mathbb{U}_1^{\mathbb{V}}$ equipped with the $\mathbb{V}$-marked Gromov-weak topology is a Polish space.

Now we can define evolving random genealogies $(\mathfrak{U}_t)_{t\geq 0}$ of spatial multitype populations as $\mathbb{U}_1^{\mathbb{V}}$-valued stochastic processes. Again for $\mathbb{U}_1^{\mathbb{V}}$ and $\Pi^{\mathbb{V}}$ the analogue of Theorem 2.3 holds; see [20]. As a prototype we will discuss later the evolving genealogies of the spatial Fleming-Viot process with selection.

Genealogies of populations with varying population sizes If the *population size* of a population is fluctuating we have to incorporate this size explicitly in the state description, think of a multitype Feller diffusion for example. This means we consider now an element

$$\mathcal{U} = (\bar{\mathcal{U}}, \widehat{\mathcal{U}}), \quad \bar{\mathcal{U}} \in \mathbb{R}, \quad \widehat{\mathcal{U}} \in \mathbb{U}_1^{\mathbb{I}}, \tag{2.24}$$

where $\bar{\mathcal{U}}$ describes the population size and $\widehat{\mathcal{U}}$ the $\mathbb{I}$-marked genealogy from above (without the spatial component). We note that now we consider the map

$$\mathcal{U} = (\bar{\mathcal{U}}, \widehat{\mathcal{U}}) \mapsto [(U \times \mathbb{I}, r \otimes r_{\mathbb{I}}, \bar{\mathcal{U}}\nu)] \tag{2.25}$$

and we obtain the equivalence class of an ultrametric measure space where the measure on $U \times \mathbb{I}$ is now in $\mathcal{M}_{\text{fin}}((U \times \mathbb{I}), \mathcal{B}(U \times \mathbb{I}))$. We denote this set of states by

$$\mathbb{U}_{\text{fin}}^{\mathbb{I}}. \tag{2.26}$$

Here we see that if we introduce a topology we have to settle, how we want to handle the elements $\{(0, \widehat{\mathcal{U}}) : \widehat{\mathcal{U}} \in \mathbb{U}_1^{\mathbb{I}}\}$. Without marks these are extinct populations where genealogy could at most make sense as a limiting object, genealogy at extinction. (In fact below we shall discuss also the role of extinct types.)

Consider first the case without marks. Then we could use on $\mathbb{U}_{\text{fin}}$ simply the topology based on the convergence of polynomials (we extend the definition of polynomials from (2.11) to $\mathbb{U}_{\text{fin}}$). In this case a sequence of elements where the population sizes converge to zero is automatically converging to the 0-element and in particular all elements $\{(0, \widehat{\mathcal{U}}) : \widehat{\mathcal{U}} \in \mathbb{U}_1\}$ are *identified* with the 0-element, i.e., give one object.

Alternatively we could define convergence for a sequence with total masses converging to zero as follows: *the sequence* $(\bar{\mathcal{U}}_n, \widehat{\mathcal{U}}_n)$ *converges if* $\bar{\mathcal{U}}_n$ *converges in* $\mathbb{R}$ *and* $\widehat{\mathcal{U}}_n$ *converges in* $\mathbb{U}_1$. In the latter case we have the product topology of $[0, \infty) \times \mathbb{U}_1$ and we distinguish elements $\{[(0, \widehat{\mathcal{U}})] : \widehat{\mathcal{U}} \in \mathbb{U}_1\}$ rather than identifying them all with the 0-element. *On the set where* $\bar{\mathcal{U}} > 0$ *both topologies agree.*

We denote the two topological spaces by (cf. [19, 41])

$$\mathbb{U}_{\text{fin}} \text{ respectively } \mathbb{U}_{\text{fin}}^*. \tag{2.27}$$

In $\mathbb{U}^*$ we can study the behavior of a *population at extinction* more precisely.

In the case with *marks* a suitable description is more subtle. We assume in the following that the geographic space $\mathbb{G}$ is countable. The types could be extinct temporarily during the evolution, or geographic locations could be vacant during some time periods and may be recolonized again during the others. If we choose to define convergence by requiring that polynomials of the states converge we get (we extend the definition of polynomials from (2.22)):

$$\mathbb{U}_{\text{fin}}^{\mathbb{V}}. \tag{2.28}$$

However, in spatial models it is often of interest to study the genealogy on the way to extinction. What about the analogue of above (2.24) representation? We proceed as follows.

We consider the situation where for example the state at each site is given as a pair (population size, genealogy), i.e., an element in $\mathbb{R}_+ \times \mathbb{U}_1$ and the complete system as a measure on the geographic space $\mathbb{V} = \mathbb{G}$ and the genealogy as an ultrametric measure space which is an element of $[(\bigcup_{g\in\mathbb{G}} U_g, r, \sum_{g\in\mathbb{G}} \mu_g)]$ so that we have pairs in $\mathcal{M}_{\text{fin}}(\mathbb{V}) \otimes \mathbb{U}_{|\mathbb{V}|}$. Then the product topology is the generalization leading to the state space

$$\mathbb{U}^{\mathbb{V},*}, \tag{2.29}$$

by choosing the product topology which is Polish.

In this situation of varying population sizes we obtain the random variables and stochastic processes

$$\mathfrak{U} = (\bar{\mathfrak{U}}, \widehat{\mathfrak{U}}), \tag{2.30}$$

where we can distinguish *population sizes* and where $\bar{\mathfrak{U}}$ is $[0,\infty)$-valued or $\mathcal{M}(\mathbb{V})$-valued where $\mathcal{M}$ denotes the set of Borel measures and *genealogies* $\widehat{\mathfrak{U}}$ of the evolving population.

Theorem 2.5: *(Polish state space)*
In both constructions the spaces $\mathbb{U}^{\mathbb{V}}_{\text{fin}}$ *and* $\mathbb{U}^{\mathbb{V},*}_{\text{fin}}$ *are Polish.*

Again polynomials play an important role via moments even though now their expectations for stochastic processes do not exist in general or are not law determining, however in case we discuss later they will, see [34, 23]. We have no space to explain the issues in further details here.

Globally infinite locally finite population sizes If we consider for example a population in an infinite geographic space as for example $\mathbb{Z}^d$ or $\mathbb{R}^d$ then it is often necessary to consider populations with an *overall infinite population size* which is however *locally finite*, that is, at single sites in $\mathbb{Z}^d$ or in a compact subsets of $\mathbb{R}^d$. We now modify our description of the genealogy to account for this possibility and distinguish a *measure-valued part* $(\bar{\mathcal{U}}_t)_{t\geq 0}$ describing occupation measures of the locations and $(\widehat{\mathcal{U}}_t)_{t\geq 0}$ describing genealogies so that for example for $\mathbb{G}$ countable we may have a collection of ultrametric probability measure spaces and corresponding population sizes, which we view again as a measure on $\mathbb{G}$.

One important case is where this measure is always the counting measure (locally) fixed population as in Fleming-Viot models, while in super random walk these sizes fluctuate and we have to work with the full space of locally finite measures. In other words we consider sampling measures such that the projection on $\mathbb{G}$ satisfies:

$$\nu_{\mathbb{G}}(A) < \infty \text{ for all bonded Borel sets } A \subset \mathbb{V}. \tag{2.31}$$

We denote this space by

$$\mathbb{U}^{\mathbb{V},\#}. \tag{2.32}$$

The topology is defined by considering the spaces corresponding to bounded and closed subsets of the geographic space $\mathbb{G}_n \subseteq \mathbb{G}$ and $\mathbb{G}_n \uparrow \mathbb{G}$. We say that a sequence of elements in $\mathbb{U}^{\mathbb{V},\#}$ converges if all restrictions to $\mathbb{G}_n$ converge in the sense described before. This defines the *Gromov-weak-#-topology* which is independent of the choice of the sequence $(\mathbb{G}_n)_{n\in\mathbb{N}}$; see [43].

Theorem 2.6: *(Polish state space)*
The space $\mathbb{U}^{\mathbb{V},\#}$ equipped with the Gromov-weak-#-topology is Polish.

This again means that we have a suitable state space also for *spatial* multitype populations with *infinite overall mass.*

Genealogies including fossils So far we have focused on the description of the genealogy of the population *currently alive*, leading to *ultrametric* measure spaces. Sometimes however one is interested in the evolution in time t of the genealogy of the population alive up to time t which includes the individuals alive at *some time* $s < t$ but which do not necessarily have descendants at time t. In other words we include the *fossils* generated up to the present time t and see how this object now evolves as t increases.

This will then lead to equivalence classes of metric measure spaces denoted generically by $\mathbb{M}$ whose projections on subspace of *leaves* gives the object $\mathbb{U}$ we looked at above. Here $\mathbb{U}$ denotes either $\mathbb{U}_1$ or $\mathbb{U}_{\text{fin}}$. The set $\mathbb{M}$ equipped with the Gromov-weak topology is a Polish space which contains $\mathbb{U}$ as a subspace (this holds also for the $\mathbb{V}$-marked case), see [39, 20]. However in order to model the states of the genealogy in an evolving population of all the individual which have been alive up to the present time T (and for equilibrium considerations then also forever) it is useful to work with a subspace of $\mathbb{M}^{\text{root}}$ of rooted versions of $\mathbb{M}$. By rooted we mean that to the elements of the spaces a root is added as distinguished element which has to be preserved under forming of the equivalence classes. The root allows to pin down time 0 in the genealogy.

If $T \geq 0$ is the present time of the population, then the basic set of individuals is of the form

$$M = \{(s,u) \mid u \in U_s,\ s \in [0,T]\} \cup \{\varrho\}, \tag{2.33}$$

where $\varrho = (0, \varrho^*)$ is the *root* used as reference point for time and U_s describes the population *alive at time* s. Between the individuals alive at or before time t we have a *partial order* induced by the *ancestor-descendant* relation between any two elements of the set M, which arise as children of children of children etc. of earlier living individuals. Such a structure arises in branching, Fleming-Viot models, in the contact process, the voter model and many more. We obtain in such a setup a metric measure space

$$(M, \widetilde{r}, \widetilde{\mu}), \tag{2.34}$$

with $\widetilde{r}(i,i')$ being the genealogical distance of the individuals i and i' in M, which is the sum of the times back to their MRCA and points in U_s have distance $2s$ to the root. The root allows to define the position on the time axes uniquely. Further we think of the population alive at time s, for each s as equipped with a sampling measure $\widetilde{\mu}_s$ on (U_s, r_s), where r_s is the genealogical distance restricted to $U_s \times U_s$. this we lift to a measure $\widetilde{\mu}_s$ on $[0,T] \times U_s$ by extension. We note that with T denoting the present time the individuals (i,T) with $(i \in U_T)$ play a special role as population currently alive as the one driving the further evolution. Therefore the sampling measure of the whole structure M must be able to sample here a finite sample, whereas from the fossils, (s,i) with $s < T$ we should simply sample according to an aggregated (over time) sampling measure from $\widetilde{\mu}_s$, $s \in [0,T]$ which are the measures allowing to sample from U_s. Hence $\widetilde{\mu}$ is chosen of the form:

$$\widetilde{\mu} = \int_0^T \mu_s \,\mathrm{d}s + \mu^{\mathrm{top}}, \quad \mu_s \text{ is a measure on } [0,T] \times U_s \tag{2.35}$$

and $s \mapsto \mu_s$ is assumed measurable as a map $[0,T] \to \bigcup_{s \in [0,T]} [0,T] \times U_s$ supported on $\{s\} \times U_s$, $s \in [0,T]$ and μ^{top} on $\{T\} \times U_T$.

Then we consider again the *equivalence classes* of such rooted metric measure spaces w.r.t. measure preserving and root preserving isometries. We obtain for each T a Polish space and the union over all $T \geq 0$ gives

$$\mathbb{M}_{(T,\varrho)} \text{ and with marks } \mathbb{M}^{\mathbb{V}}_{(T,\varrho)}. \tag{2.36}$$

If we also introduce marks, we introduce the topology again via polynomials. This is being currently worked out in [42], see also [41].

This leads to stochastic processes describing the genealogy of a population up to time t, where this structure evolves then with t:

$$(\bar{\mathfrak{M}}_t, \widehat{\mathfrak{M}}_t)_{t\in[0,T]}, \tag{2.37}$$

which is a $\mathbb{M}^{\mathbb{V}}_{(T,\varrho)}$-valued stochastic process.

This structure allows now to describe the genealogy of a population in time, which is an enrichment of the process $(\mathfrak{U}_t)_{t\geq 0}$ with states in $\mathbb{U}$, which now appear in the slices $[U_s, r_s, \mu_s]$ but include the additional information on genealogical relations between the individuals alive at different times *before* time T.

Including now an infinite time horizon is possible, but needs some thought to deal with possibly infinite distances, if we try to take the closure of $\bigcup_{T\geq 0.} \mathbb{M}_{(T,\varrho)}$.

3. Martingale problems for evolving genealogies

3.1. *Concepts and examples*

The processes of evolving genealogies are defined by *well-posed martingale problems.* First we recall the notion of martingale problems that we use here; see [27].

Definition 3.1: (Martingale problem)
Let E be a Polish space and let $\mathcal{B}(E)$ denote the space of bounded measurable functions on E. Furthermore let $\mathrm{P}_0 \in \mathcal{M}_1(E)$, $\mathcal{F} \subset \mathcal{B}(E)$ and let L be a linear operator on $\mathcal{B}(E)$ with domain $\mathcal{F}$. The law P of an E-valued stochastic process $\mathcal{X} = (X_t)_{t\geq 0}$ is called a solution of the $(\mathrm{P}_0, L, \mathcal{F})$-martingale problem if X_0 has distribution P_0, $\mathcal{X}$ has paths in the space $\mathcal{D}_E([0,\infty))$, P-almost surely, and for all $F \in \mathcal{F}$,

$$\left(F(X_t) - \int_0^t LF(X_s)\,\mathrm{d}s\right)_{t\geq 0} \tag{3.1}$$

is a P-martingale with respect to the canonical filtration. Moreover, the $(\mathrm{P}_0, L, \mathcal{F})$-martingale problem is said to be *well-posed* if there is a unique solution P.

To define a process by means of a martingale problem we need to define an operator L acting on an appropriate subset of the set of test functions $\mathcal{F}$. We do this in the following paragraphs for several examples of tree-valued processes by adding more and more evolutionary mechanisms to the processes.

The neutral tree-valued process without types The most basic evolutionary mechanisms involving genealogies are

- *aging*, i.e., growth of genealogical distances with time;
- the *splitting* of a genealogical line in two due to resampling (i.e. birth) events.

Now the idea is to describe the evolution by describing the evolution of all finite samples drawn by the resampling measure. This means we consider the following test functions for the martingale problem.

For $\mathcal{U} = [(U, r, \mu)] \in \mathbb{U}_1$ consider the polynomials

$$\Phi^{n,\varphi}(\mathcal{U}) = \int_{U^n} \varphi(\underline{\underline{r}}) \mu^{\otimes n}(\mathrm{d}\underline{x}) \\ \text{with} \quad \varphi : \mathbb{R}^{\binom{n}{2}} \to \mathbb{R} \quad \text{continuously differentiable,} \tag{3.2}$$

where $\underline{x} = (x_1, \ldots, x_n) \in U^n$ and $\underline{\underline{r}} := (r_{k,\ell})_{1 \le k < \ell \le n} = (r(x_k, x_\ell))_{1 \le k < \ell \le n}$.

Then the above mechanisms are described by the generator

$$\Omega = \Omega^{\text{grow}} + \Omega^{\text{res}}. \tag{3.3}$$

The *growth operator* acting on polynomials, see (3.2) is given by

$$\Omega^{\text{grow}} \Phi^{n,\varphi}(\mathcal{U}) = 2 \int_{U^n} \mu^{\otimes n}(\mathrm{d}\underline{x}) \sum_{1 \le k < \ell \le n} \frac{\partial}{\partial r_{k,\ell}} \varphi(\underline{\underline{r}}) \tag{3.4}$$

and the *resampling generator* by

$$\Omega^{\text{res}} \Phi^{n,\varphi}(\mathcal{U}) = d \int_{U^n} \mu^{\otimes n}(\mathrm{d}\underline{x}) \sum_{1 \le k < \ell \le n}^{n} (\varphi \circ \theta_{k,\ell} - \varphi)(\underline{\underline{r}}). \tag{3.5}$$

Here, $d > 0$ is the *resampling rate* and the *resampling operator* $\theta_{k,\ell}$ replaces the distances of the ℓ-th component by those of k-th. More precisely it is defined by $\theta_{k,\ell}(\underline{\underline{r}}) = \underline{\underline{\tilde{r}}}$ and

$$\tilde{r}_{i,j} := \begin{cases} r_{i,j}, & \text{if } i, j \neq \ell, \\ r_{i \wedge k, i \vee k}, & \text{if } j = \ell, \\ r_{j \wedge k, j \vee k}, & \text{if } i = \ell. \end{cases} \tag{3.6}$$

The $\mathbb{U}_1$-valued process with generator consisting of the growth and resampling parts, the $\mathbb{U}_1$*-valued Fleming-Viot processes* $\mathfrak{U}^{\text{FV}}$, was constructed and studied in [40]. The approach can be extended to *Cannings processes* or so called Λ-Fleming-Viot models, see [36]. *Mutation* and *selection* have been included in [21] leading to $\mathbb{U}_1^{\mathbb{I}}$-valued processes. *Spatial* models giving

$\mathbb{U}^{\mathbb{G}}$-valued Fleming-Viot processes were considered in [43]. Both the latter are successively explained next in two paragraphs.

Similarly one can introduce on $\mathbb{U}_{\text{fin}}$ with these above two operators the process $\mathfrak{U}^{\text{Fel}}$ describing the evolution of the genealogies of the continuum state Feller branching diffusion and the self-catalytic branching diffusion; see [34, 35, 41, 19].

Tree-valued process with mutation and migration Let the type of an individual be given by its genotype and its spatial location so that we may assume that the type space is $\mathbb{V} = \mathbb{G} \times \mathbb{I}$ with $\mathbb{G}$ being the geographical space and $\mathbb{I}$ the type space. Think for instance of $\mathbb{G}$ as a (finite) subset of $\mathbb{Z}^d$ and $\mathbb{I}$ as a compact space.

For $\mathbb{U}_1^{\mathbb{V}}$-valued processes we can take test functions of the following form: For $\mathcal{U} = [(U \times \mathbb{V}, r \otimes r_{\mathbb{V}}, \nu)]$ set

$$\Phi^{n,\varphi,\psi,\eta}(\mathcal{U}) = \int_{(U\times\mathbb{G}\times\mathbb{I})^n} \varphi(\underline{\underline{r}})\psi(\underline{g})\eta(\underline{u})\nu^{\otimes n}(\mathrm{d}(\underline{x},\underline{g},\underline{u})), \tag{3.7}$$

with φ as in (3.2), $\psi : \mathbb{G}^n \to \mathbb{R}$ and $\eta : \mathbb{I}^n \to \mathbb{R}$, $\underline{g} = (g_1, \dots, g_n) \in \mathbb{G}^n$ and $\underline{u} = (u_1, \dots, u_n) \in \mathbb{I}^n$.

As in the basic case, due to aging the genealogical distances grow with time and do not affect the types or locations of individuals. Thus, the *growth operator* action on $\Phi^{n,\varphi,\psi,\eta}$ is given by

$$\begin{aligned}&\Omega^{\text{grow}}\Phi^{n,\varphi,\psi,\eta}(\mathcal{U})\\ &\quad = 2\int_{(U\times\mathbb{G}\times\mathbb{I})^n} \nu^{\otimes n}(\mathrm{d}(\underline{x},\underline{g},\underline{u}))\psi(\underline{g})\eta(\underline{u}) \sum_{1\le k<\ell\le n} \frac{\partial}{\partial r_{k,\ell}}\varphi(\underline{\underline{r}}).\end{aligned} \tag{3.8}$$

The *resampling generator* does replace distances as before but now it also replaces the types of individuals but not their location. It is given by

$$\begin{aligned}\Omega^{\text{res}}\Phi^{n,\varphi,\psi,\eta}(\mathcal{U}) = d\int_{(U\times\mathbb{G}\times\mathbb{I})^n} &\nu^{\otimes n}(\mathrm{d}(\underline{x},\underline{g},\underline{u}))\psi(\underline{g})\\ &\cdot\sum_{1\le k<\ell\le n} \mathbb{1}_{\{g_k=g_\ell\}}((\varphi\otimes\eta)\circ\theta_{k,\ell} - \varphi\otimes\eta)(\underline{\underline{r}},\underline{u}).\end{aligned} \tag{3.9}$$

Here, generalizing the resampling operator from (3.9), $\theta_{k,\ell}$ is defined by $\theta_{k,\ell}(\underline{\underline{r}},\underline{u}) = (\underline{\underline{\tilde{r}}},\underline{\tilde{u}})$ with $\underline{\underline{\tilde{r}}}$ as in (3.6) and

$$\tilde{u}_i := \begin{cases} u_i & \text{if } i \neq \ell, \\ u_k & \text{if } i = \ell. \end{cases} \tag{3.10}$$

For *mutation*, let $\vartheta \geq 0$ be the *mutation rate* and let $\beta(\cdot,\cdot)$ be a Markov transition kernel on $\mathbb{I} \times \mathbb{I}$ and set

$$\begin{aligned}\Omega^{\text{mut}}\Phi^{n,\varphi,\psi,\eta}(\mathcal{U})\\ = \vartheta \int_{(U\times\mathbb{G}\times\mathbb{I})^n} \nu^{\otimes n}(\mathrm{d}(\underline{\underline{x}},\underline{g},\underline{u}))\varphi(\underline{\underline{r}})\psi(\underline{g}) \sum_{k=1}^{n} (\beta_k\eta - \eta)(\underline{u}) \quad (3.11)\end{aligned}$$

where

$$(\beta_k\eta)(\underline{u}) = \int \eta(\underline{u}_k^v)\beta(u_k, \mathrm{d}v), \tag{3.12}$$

with $\underline{u}_k^v = (u_1, \ldots, u_{k-1}, v, u_{k+1}, \ldots, u_n)$.

For *migration* let A be a transition kernel on $\mathbb{G} \times \mathbb{G}$ and for $\psi : \mathbb{G}^n \to \mathbb{R}$ we set

$$(M_{g_j,g'}\psi)(g_1, \ldots, g_n) = \psi(g_1, \ldots, g_{j-1}, g', g_{j+1}, \ldots, g_n). \tag{3.13}$$

Then,

$$\begin{aligned}\Omega^{\text{migr}}\Phi^{n,\varphi,\psi,\eta}(\mathcal{U}) = \int_{(U\times\mathbb{G}\times\mathbb{I})^n} &\nu^{\otimes n}(\mathrm{d}(\underline{\underline{x}},\underline{g},\underline{u}))\varphi(\underline{\underline{r}})\eta(\underline{u}) \\ &\cdot \sum_{k=1}^{n} \sum_{g'\in G} A(g_j, g')(M_{g_j,g'}\psi - \psi)(\underline{g}).\end{aligned} \tag{3.14}$$

Adding selection For *selection* let $\alpha \geq 0$ be the *selection coefficient* and let $\chi : \mathbb{V} \to [0,1]$ be a function that we refer to as *fitness function*. Then the selection operator is given by

$$\begin{aligned}\Omega^{\text{sel}}\Phi^{n,\varphi,\psi,\eta}(\mathcal{U}) = \alpha \int_{(U\times\mathbb{G}\times\mathbb{I})^{n+1}} &\nu^{\otimes(n+1)}(\mathrm{d}(\underline{\underline{x}},\underline{g},\underline{u}))\varphi(\underline{\underline{r}})\psi(\underline{g}) \\ &\cdot \sum_{k=1}^{n} \mathbb{1}_{\{g_k=g_\ell\}}(\eta\cdot\chi_k - \eta\cdot\chi_{n+1})(\underline{u}).\end{aligned} \tag{3.15}$$

Here $\chi_k(\underline{u}) = \chi(u_k)$. And the functions on ψ and g are extended to domains $\mathbb{R}^{\binom{n+1}{2}}$ and $\mathbb{G}^{n+1}$. Note that it is possible to consider more general fitness functions to model *diploid selection* and *kin selection*, i.e. the fitness may depend on the distance of individuals involved in the selection event. We refer to Section 2.4. in [21] for background on selection that we have in mind here.

The branching world Similarly to the above we can introduce $\mathbb{U}_{\text{fin}}$-valued branching diffusions, for example continuous state branching processes as $\mathbb{U}_{\text{fin}}$-valued Feller diffusion, or the spatial analog the $\mathbb{U}^{\mathrm{G}}$-valued super random walk. Here one has as generator $\Omega^{\text{res}} + \Omega^{\text{grow}}$ and additionally Ω^{migr} in the spatial case acting now on polynomials arising by allowing for finite measures instead of probability measures in (3.7) and adding the functions $\mathfrak{u} \mapsto \bar{\mathfrak{u}}$ and $\mathfrak{u} \mapsto c$, where c is a constant. As in the Fleming-Viot model we may consider drift terms, here of interest are sub/supercritical branching instead of just the critical case (which is the second order part) or we may consider the analog of selection which gives logistic branching with a *competition term* acting only on the total masses.

3.2. *Results*

In this section we state some general results concerning solutions of martingale problems and their basic properties in the case that the linear operator is given by the "full" generator

$$\Omega = \Omega^{\text{grow}} + \Omega^{\text{res}} + \Omega^{\text{mut}} + \Omega^{\text{migr}} + \Omega^{\text{sel}} \tag{3.16}$$

acting on functions from $\Pi^{\mathbb{V}}$ which are differentiable with respect to the distance part so that the action of Ω^{grow} in (3.8) is defined. Depending on the evolutionary forces to be included in the considered process one can of course omit some parts of the generator. In this case the set of polynomials $\Pi^{\mathbb{V}}$ should also be modified accordingly. The corresponding proofs of Theorem 3.2 and Proposition 3.3 can be found in [40], [21] and [43] and the proof in the case of "full" generator can be deduced by combination of the relevant results in the cited papers.

Theorem 3.2: *(Martingale problem is well-posed)*
For all $\mathrm{P}_0 \in \mathcal{M}_1(\mathbb{U}_1^{\mathbb{V}})$, *the* $(\mathrm{P}_0, \Omega, \Pi^{\mathbb{V}})$*-martingale problem is well-posed.*

A solution P of the $(\mathrm{P}_0, \Omega, \Pi^{\mathbb{V}})$ martingale problem from the above theorem is referred to as *tree-valued Fleming-Viot dynamics with mutation, migration and selection* or $\mathbb{U}_1^{\mathbb{V}}$*-valued Fleming-Viot dynamics* with initial distribution P_0 and has the following basic properties.

Proposition 3.3: *(Properties of* $\mathbb{U}_1^{\mathbb{V}}$*-valued Fleming-Viot dynamics)*
Let $\mathfrak{U} = (\mathfrak{U}_t)_{t \geq 0}$ *be a* $\mathbb{U}_1^{\mathbb{V}}$*-valued Fleming-Viot dynamics with distribution* P. *Then the following properties hold:*

(i) The sample paths of $\mathfrak{U}$ *are in* $C_{\mathbb{U}_1^{\mathbb{V}}}([0,\infty))$, P*-a.s.*

(ii) *For all $t > 0$ we have $\mathfrak{U}_t \in \mathbb{U}^{\mathbb{V}}_{1,\text{comp}}$, P-a.s., where $\mathbb{U}^{\mathbb{V}}_{1,\text{comp}}$ refers to marked metric measure spaces in which the metric space components (U, r) are compact.*
(iii) *The process $\mathfrak{U}$ is a Feller and a strong Markov process.*

The meaning of assertions (i) and (iii) in the above proposition is self explaining. Assertion (ii) states that all genealogies except maybe the initial one can be encoded by a compact ultrametric space. On the level of genealogies that means that for each $\varepsilon > 0$ the population at time $t > 0$ can be decomposed in finitely many families so that the time to MRCA of individuals within each of the families is bounded by ε. This should remind the reader of the "coming down from infinity"-property of some coalescent processes.

The same statements as above are or will be proven on $\mathbb{U}_{\text{fin}}$ respectively $\mathbb{U}^{\mathbb{V}}_{\text{fin}}$ with $\mathbb{V} = \mathbb{G}$ and $\mathbb{G}$ a countable, i.e., mostly infinite, Abelian group for $\mathbb{U}_{\text{fin}}$-valued Feller diffusion and $\mathbb{U}^{\mathbb{G}}_{\text{fin}}$-valued super random walk in [34], [19], [41] and in [38] for logistic spatial branching.

Remark 3.4: (Existence of solutions) A standard method to construct a solution to the martingale problem, as for example the class in (3.16) is by approximations with individual based models. In a number of the models appearing in population genetics as in (3.16) the solution to such a martingale problem as in (3.16) can be obtained by graphical construction of the ancestral relations in time, see for example the *Brownian web* in [43] or by *lookdown constructions* see [37, 36, 45], or by *flows of bridges* constructions [9, 33]. However for complicated selection or recombination models this becomes intransparent.

4. Longtime behavior

In this section we discuss first under which conditions $\mathbb{U}_1$ and more generally $\mathbb{U}^{\mathbb{V}}$-valued processes have *unique invariant distributions* and are *ergodic*. In presence of migration we assume that the geographical space is *finite*.

Let us consider the case without marks first. To this end we briefly recall the *Kingman coalescent measure tree*. Consider Kingman's N-coalescent tree defined as follows: Starting with N lines coalesce each pair of lines which are present at rate d, which is the resampling rate of the forwards in time process, and continue until all lines have merged into one. Alternatively we can say that when $k \in \{2, \dots, N\}$ lines are present the total coalescence rate is $d\binom{k}{2}$ and at a coalescence event a random pair is chosen to coalesce. Let $U_N = \{1, \ldots, N\}$ be the set of leaves and let the (random) metric r_N given by the tree distance of a Kingman's N-coalescent tree, that is $r_N(i,j)$ is twice the distance to the MRCA of the individuals i and j, $1 \leq i, j \leq N$. Let μ_N be the uniform measure on U_N and consider the metric measure space $\mathcal{U}_N := [(U_N, r_N, \mu_N)]$. By Theorem 4 in [39] there is an $\mathbb{U}_1$-valued random variable $\mathcal{U}_\infty$ so that

$$\mathcal{U}_N \xrightarrow{N\to\infty} \mathcal{U}_\infty. \tag{4.1}$$

The limiting random variable was introduced in [29] and is called the *Kingman coalescent measure tree.*

The following result is Theorem 3 in [40] and shows in particular that the distribution of the Kingman coalescent measure tree is the unique invariant distribution of the neutral Fleming-Viot dynamics without marks.

Theorem 4.1: *(Longtime behavior of the basic process)* *Let $\mathfrak{U} = (\mathfrak{U}_t)_{t\geq 0}$ be an $\mathbb{U}_1$-valued process with $\mathfrak{U}_0 = \mathcal{U} \in \mathbb{U}_1$ and let $\mathcal{U}_\infty \in \mathbb{U}_1$ be the Kingman coalescent measure tree. Then we have*

$$\mathfrak{U}_t \xrightarrow{t\to\infty} \mathfrak{U}_\infty. \tag{4.2}$$

Now let us turn to the general case. Let $\pi_{\mathbb{V}}$ be the projection from $X \times \mathbb{V}$ on $\mathbb{V}$. Given $\mathfrak{U}_t = [(U_t, r_t, \mu_t)]$, $t \geq 0$, we define the process $\zeta := (\zeta_t)_{t\geq 0}$ by projecting the sampling measures to the type space, i.e., we set

$$\zeta_t := (\pi_{\mathbb{V}})_* \mu_t, \quad t \geq 0. \tag{4.3}$$

Then ζ is the corresponding *measure-valued Fleming-Viot process* and conditions for its ergodicity are available in the literature. For example, under neutral evolution, i.e. $\alpha = 0$, ergodicity of ζ has been shown if the Markov pure jump mutation process on $\mathbb{I}$ with has a unique equilibrium distribution; see [13]. In the case $\alpha > 0$ and $\chi \neq 0$, the process ζ is ergodic if mutation has a parent independent component. In the case with no parent-independent component in the mutation operator ergodicity of ζ has been shown in [28] using coupling techniques. In [15] a set-valued dual allows to prove quite general ergodic theorems, even in the spatial context under some conditions on the mutation process.

Obviously, an $\mathbb{U}^{\mathbb{V}}$-valued process cannot be ergodic if its projection on the type space, i.e., the corresponding measure-valued process, is not ergodic. Also it is easy to see that ergodicity of ζ is implied by the ergodicity of $\mathfrak{U}$. The following result shows that also the converse implication holds.

Theorem 4.2: *(Longtime behavior of the process with marks)*
Let $\mathfrak{U} = (\mathfrak{U}_t)_{t\geq 0}$ be an $\mathbb{U}_1^{\mathbb{V}}$-valued process with $\mathfrak{U}_0 = \mathcal{U}$ and let ζ be as above. Then, there exists a unique invariant distribution on $\mathbb{U}_{1,\text{comp}}^{\mathbb{V}}$ and for a random variable $\mathfrak{U}_\infty$ with this distribution for all $\mathcal{U} \in \mathbb{U}_1^{\mathbb{V}}$ we have

$$\mathfrak{U}_t \overset{t\to\infty}{\Longrightarrow} \mathfrak{U}_\infty \tag{4.4}$$

if and only if ζ has a unique equilibrium distribution.

Note that Theorem 4.2 tells us that ergodic distribution of types is given by the ergodic distribution of the corresponding measure-valued Fleming-Viot process, but does say little about the "geometry" of the limiting genealogies. In the neutral case with mutation a combination of Theorem 4.1 and Theorem 4.2 provides a more complete picture.

We next discuss processes on *infinitive* geographic space where new features occur. Indeed spatial populations on infinite geographic spaces, for example countable Abelian groups like $\mathbb{Z}^d$ or continuum space models like $\mathbb{R}^d$ show in their longtime behavior an interesting *dichotomy* between low and high dimensions, or to be more precise, between *transient* and *recurrent* symmetrized migration arising from the difference of the positions of two tagged "individuals". For example the Fleming-Viot model in its neutral form and compact type space has for $t \to \infty$ limiting states which are monotype if the symmetrized migration is recurrent. This is the case for nearest neighbor migration for $d = 1, 2$, whereas for the transient case, i.e., $d \geq 3$, we have equilibria with coexistence or more precisely for every intensity measure θ on $\mathbb{I}$ we have a translation invariant shift ergodic equilibrium with this intensity. What happens on the level of genealogies?

In the transient case a common ancestor may not exist locally as we obtain in the limit $t \to \infty$ infinite distances. We introduce therefore the transformation of the ultrametric (into another ultrametric) given by:

$$r \mapsto r', \quad r'(x, x') = 1 - \exp(-r(x, x')), \tag{4.5}$$

where the distance 1 corresponds exactly to infinite distance in the original metric r. We denote the transformed process by $(\widetilde{\mathfrak{U}}_t^{FV})_{t\geq 0}$.

Then we obtain the following. Consider initial distributions of the population on a countable (infinite) Abelian group $\mathbb{G}$ such that the projection on $\mathbb{V} = \mathbb{I} \times \mathbb{G}$ gives a translation invariant shift ergodic law with $E\big[((\pi_{\mathbb{V}})_*\nu)(\{g\} \times \cdot)\big] = \theta(\cdot) \in \mathcal{M}_1(\mathbb{I})$. Note that this means that if $\mathbb{G}$ is infinite then the "sampling" measure is infinite as well.

Theorem 4.3: *(Longtime behavior of spatial $\mathbb{U}^{\mathbb{V}}$-valued Fleming-Viot process $\mathfrak{U}^{FV}$)*

(a) If the kernel $\widehat{a}(g,g') = \frac{1}{2}(a(g,g') + a(g',g))$ is recurrent then

$$\mathcal{L}[\mathfrak{U}_t^{FV}] \underset{t\to\infty}{\Longrightarrow} \Gamma \in \mathcal{M}_1(\mathbb{U}^{\mathbb{V}}), \tag{4.6}$$

where Γ is concentrated on mono-ancestor and monotype configurations, i.e., the states have a.s. a finite essential diameter and $\nu(U\times\cdot) = \int_{\mathbb{I}}(\lambda \otimes \delta_u)\theta(du)$.

(b) For every $\theta \in \mathcal{M}_1(\mathbb{I})$ there exists an equilibrium measure $\widetilde{\Gamma}_\theta$ which is translation-invariant and shift-ergodic, satisfies $\mathrm{E}[\nu(U \times \cdot)] = \lambda \otimes \theta$, λ being the Haar measure on $\mathbb{G}$ normed to $\lambda(\{0\}) = 1$, such that

$$\mathcal{L}[\widetilde{\mathfrak{U}}_t^{FV}] \underset{t\to\infty}{\Longrightarrow} \widetilde{\Gamma}_\theta. \tag{4.7}$$

The essential diameter of the states attains 1 *a.s.*

This means that in the recurrent case we have *mono-ancestor* and *monotype* populations developing in any finite spatial window whereas in the transient case, locally we have *coexistence* of descendants of different ancestors and, if θ is not the point measure we also have coexistence of different types. This is proved in [36] for the more general case of Λ-Fleming-Viot models, see [43] for information on spatial tree-valued Fleming-Viot.

Similar questions can be addressed for *spatial branching models* where convergence to equilibria can be established. Here a transformation of the metric as in (4.5) may be needed to obtain equilibria in spatial models where with positive probability no common ancestor of the population exists. See [19], [38] for more information.

5. Two techniques of analysis for our processes

There are two major tools to investigate our processes for example to prove uniqueness or existence of martingale problems, namely *duality* and *Girsanov transformation*. The former is also crucial for analyzing the longtime behavior and the latter is useful for obtaining path properties of more complicated processes from corresponding path properties of simpler processes. We explain the techniques in the following in more detail.

5.1. *Duality*

We explain here the general concept of *duality* and give the results for genealogy-valued processes. Then we continue by discussing *strong duality*, a stronger concept allowing representations of the state. Finally we discuss the so-called *conditional duality*.

5.1.1. *The concept of duality*

Let $(X_t)_{t\geq 0}$ be a Markov process with state space $(E, \mathcal{B})$, where E is a Polish space and $\mathcal{B}$ the Borel σ-algebra. The process X is the solution of a *martingale problem* with operator L acting on the test functions $\mathcal{F} \subseteq C_b(E, \mathbb{R})$, the $(L, \mathcal{F}, \delta_x)$-martingale problem, with $x \in E$.

Then we have a *dual process* $(Y_t)_{t\geq 0}$ with state space $(E', \mathcal{B}')$ contained in a Polish space, $\mathcal{B}'$ being the Borel σ-algebra. This process is the solution of the $(L', \mathcal{F}', \delta_y)$-martingale problem, $\mathcal{F}' \subseteq C_b(E', \mathcal{B}')$, $y \in E'$. Often L' and $\mathcal{F}'$ are chosen with $\mathcal{F}' \subseteq C_b(E', \mathbb{R})$ such that for $f' \in \mathcal{F}'$, $Lf' \in C_b(E', \mathbb{R}')$ and for the solution Y the random variables $f'(Y_t)$ are integrable for all t.

In order to define a *Feynman-Kac duality* we define a functional

$$\beta_t = \int_0^t B(Y_s)\,\mathrm{d}s \tag{5.1}$$

where B is a bounded continuous function on E', the so called *potential.*

Definition 5.1: (Duality relation, Feynman-Kac duality)

(a) We say that X and Y are in *duality* w.r.t. the duality function $H : E \times E' \to \mathbb{R}$ if the following holds:

(i) $H(\cdot,\cdot) \in C_b(E \times E', \mathbb{R})$,
(ii) $\{H(\cdot, y) : y \in E'\}$ is separating on E,
(iii) the following equality holds for all $t \geq 0$:

$$\mathrm{E}_x[H(X_t, y)] = \mathrm{E}_y[H(x, Y_t)], \quad \forall (x, y) \in E \times E'. \tag{5.2}$$

(b) The process X and Y are in *Feynman-Kac duality* if for all $t \geq 0$, (i) and (ii) above hold and (iii) is replaced by:

$$\mathrm{E}_x[H(X_t, y)] = \mathrm{E}_y[H(x, Y_t) \exp(\beta_t)] \tag{5.3}$$

and if additionally the r.h.s. is integrable for all $t \geq 0$.

The duality relation can be established by showing the following relations:

$$(LH(\cdot,y))(x) = (L'H(x,\cdot))(y) + B(y)H(x,y), \quad (x,y) \in E \times E', \tag{5.4}$$

$$LH(\cdot,y) \in \mathcal{F},\ L'H(x,\cdot) \in \mathcal{F}', \quad \forall (x,y) \in E \times E' \tag{5.5}$$

and establishing the integrability condition at the end of the definition above.

Remark 5.2: (General applications) The duality relations have some consequences for processes X and Y from the above definition:

- If a solution for the $(L', \mathcal{F}', \delta_y)$-martingale problem *exists*, then a solution of the $(L, \mathcal{F}, \delta_x)$-martingale problem is *unique*. Note that it is sometimes possible to drop the boundedness and/or continuity assumptions on the duality functions and potentials and still be able to use duality to establish uniqueness.
- If $\mathcal{L}[(X_t)_{t\geq 0}]$ is tight and if $\mathcal{L}[(Y_t)_{t\geq 0}] \Longrightarrow \mathcal{L}[Y_\infty]$ as $t \to \infty$ then $\mathcal{L}[X_t] \underset{t\to\infty}{\Longrightarrow} \pi$ and π is an *equilibrium distribution* of the Markov process X.

Furthermore in many cases:

- The existence of the process Y as solution to its martingale problem and fulfillment of (5.4) and (5.5) allows to obtain the *existence* of a process X solving the $(L, \mathcal{F}, \delta_x)$-martingale problem.
- The duality might be helpful to deduce the Feller property of processes when their duals fulfil certain properties. Roughly speaking this is case when the dual process is a "finite" particle system. For one of the many examples see for instance Theorem 1(c) in [21] and its proof on p. 2602.

Of particular importance for us will be the case of *moment duals*, this means for population models that the function $H(\cdot, y)$ can be interpreted as *testing a sample of size* n from the population. In the case of the Fisher-Wright diffusion or the Feller diffusion this gives with $E = \mathbb{R}$ and $E' = \mathbb{N}$ indeed duality functions $H(x,n) = x^n$ but the appropriate version will hold for the case of the tree-valued Fleming-Viot process or tree-valued Feller process.

5.1.2. *Dualities and Feynman-Kac dualities for $\mathbb{U}_{\text{fin}}$-valued processes*

We will explain now how we obtain for the evolving genealogies of the Fleming-Viot model respectively the Feller diffusion a *moment duality* respectively *Feynman-Kac moment duality* based on an enriched version of the *Kingman coalescent*, namely the version *enriched* by a component of *distance matrices* for the individuals. Later we explain how this carries over to the spatial case or the multitype case with selection and mutation.

We introduce now the *dual process* and the *duality function.*

(1) Consider the Kingman coalescent starting with n individuals. The state of this process is an element of $\mathbb{S}_n$ the set of all partitions of the basic set $\{1,\dots,n\}$, i.e., a tuple $(p_1,\dots,p_{|p|})$ with $p_i \subseteq \{1,\dots,n\}$, $i \in 1,\dots,|p|$, $\bigcup_{i=1}^{|p|} p_i = \{1,\dots,n\}$ and $p_i \cap p_j = \varnothing$ for $i,j \in \{1,\dots,|p|\}$ with $i \neq j$.

We denote by $\mathbb{D}_n$ the set of all $n \times n$ distance matrices, i.e., $\mathbb{D}_n \subset \mathbb{R}_+^{\binom{n}{2}}$ so that for $(d_{i,j})_{1 \le i \le j \le n} \in \mathbb{D}_n$ we have $d_{i,j} \le d_{i,k} + d_{j,k}$ and $d_{i,i} = 0$ for all $i,j,k = 1,\dots,n$. We abbreviate $E'_n = \mathbb{S}_n \times \mathbb{D}_n$. Then the sate space of this process is

$$E' = \bigcup_{n=1}^{\infty} E'_n. \tag{5.6}$$

The *dynamics* is that every pair of partition elements independently coalesces at rate d, i.e., $(p_1,\dots,p_{|p|}) \to (p_1,\dots,\widehat{p}_i,\dots,p_i \cup p_j,\dots,p_{|p|})$ where $\widehat{p}_i$ denotes deletion at the ith position. The distances between elements of the basic set grow at speed 2 as long as they belong to different partition elements. The resulting process is denoted by $\mathfrak{C}$. For a given $n \in \mathbb{N}$, $\varphi \in C_b(\mathbb{R}_+^{\binom{n}{2}},\mathbb{R})$ we later want to view (φ,n) as part of the state of the dual. In the neutral case the element (φ,n) will *not* change. To this end, we extend the state space E' by setting $E'_n = \mathbb{S}_n \times \mathbb{D}_n \times C_b(\mathbb{R}_+^{\binom{n}{2}},\mathbb{R}) \times \{n\}$.

(2) To define the duality function for a given $n \in \mathbb{N}$, $\varphi \in C_b(\mathbb{R}_+^{\binom{n}{2}},\mathbb{R})$ we consider the function

$$H\left([U,r,\mu],(p,(\varphi,n))\right) = \int_U \cdots \int_U \varphi(\underline{\underline{r}} + \underline{\underline{r}}')\, \mathrm{d}\mu^{\otimes n}, \quad [U,r,\mu] \in \mathbb{U}_1, (p,\underline{\underline{r}}') \in E'_n. \tag{5.7}$$

We define the potential B

$$B\left((p,\underline{\underline{r}}'),(\varphi,n)\right)=d\cdot\binom{|p|}{2}. \tag{5.8}$$

Consider now the $\mathbb{U}_1$-valued Fleming-Viot diffusion $\mathfrak{U}^{FV}$ and the $\mathbb{U}_{\text{fin}}$-valued Feller diffusion $\mathfrak{U}^{\text{Fel}}$ with branching rate b, i.e., the resampling operator has coefficient b, then we have the following result.

Theorem 5.3: *(Duality-relation)*

(a) The processes $\mathfrak{U}^{FV}$ and $\mathfrak{C}$ are in duality w.r.t. H from (5.7).
(b) The process $\mathfrak{U}^{\text{Fel}}$ and $\mathfrak{C}$ are in Feynman-Kac duality w.r.t. to H and potential B from (5.7) *and* (5.8)*, where $d=b$.*

In the case of *spatial* models we consider the spatial coalescent where the state is enriched by a map

$$\Omega\to\Omega^{|p|}, \tag{5.9}$$

which associates with every partition element a *location*, i.e. an element in $\mathbb{G}$. Accordingly the test function φ will be replaced by $\varphi\cdot\psi$, where ψ is a bounded function on $\mathbb{G}^n$, i.e. $\psi\in C_b(\mathbb{G}^n,\mathbb{R})$. In the case of infinite geographic spaces, i.e. on $\mathbb{U}^{\mathbb{V},\#}$, we restrict to ψ with bounded support in $\mathbb{G}$. The dynamics is modified by adding a new mechanism. The partition elements perform independent random walks with rates $a(\cdot,\cdot)$ until they coalesce and then the individuals in the new partition element now follow the same random walk. Coalescence occurs only for pairs of partition elements sharing the same geographical location.

Corollary 5.4: *(Duality: spatial model)*
For the $\mathbb{U}_1^{\mathbb{G}}$-valued interacting Fleming-Viot diffusion and the $\mathbb{U}^{\mathbb{G},\#}$-valued super random walk the duality theorem above holds with the modified ingredients.

Similarly we can consider *multitype* spatial populations with $\mathbb{V}=\mathbb{I}\times\mathbb{G}$ type frequency change by resampling respectively branching, but with the principle that types are inherited. If we have also types we extend ψ, i.e., $\psi\in C_b\left((\mathbb{I}\times\mathbb{G})^n,\mathbb{R}\right)$. In this case we have to assign the $\mathbb{I}$-types to the partition element by choosing a type for a partition element for the duality relation at time t by sampling the type from the sampling measure of the initial state of the forward process restricted to the position of the partition elements position at time t. Then partition elements will be assigned a type when we evaluate the duality function where the type is chosen according to the projection of μ on $\mathbb{I}$ at the location $g \in \mathbb{G}$ at the partition element at time T, i.e., $\mu_0(U \times \{\cdot\} \times \{g\})$.

Somewhat more complicated are the multitype processes in particular if we include *selection, mutation* and *recombination* in the mechanism where already the measure-valued model requires a *function-valued dual* which is more subtle to explain already on that level, but in fact this constructions can be carried out on $\mathbb{U}_1^{\mathbb{V}}$ and even in the spatial setup.

The function-valued part is needed to have a duality once we include *selection* and *mutation* or even *recombination* in the model. The dual process is built upon a spatial coalescent enriched with a distance matrix in addition to coalescence, migration and distance growth we have a *birth* mechanism which drives a *function-valued process*, i.e. at the "birth" transition a selection operation acts on the function. Furthermore there is action on the function via a mutation operation. Function-valued means here that we consider φ and ψ as part of the dual state where ψ now follows a dynamic itself. See [15] for an extensive treatment of the measure-valued case. We cannot give an account of this theory here. However we show in [21] that one can extend this to the $\mathbb{U}^{\mathbb{I}}$-valued case. The key point is to take the same driving particle system as in the measure-valued case but to incorporate as well the function φ of the distances into this picture.

5.1.3. *Strong dual representation and conditional duality*

Strong dual representation It turns out that we can extend the dual process in many of our models to an $\mathbb{U}_{\text{fin}}$-valued state by passing to the *entrance law* of the coalescent starting from countably many individuals. Then we introduce a metric $r'(\cdot,\cdot)$ on $\mathbb{N}$ by defining $T_{i,j}$ as the first time when i, j are in the same partition element if the time is less than the finite time horizon T. Otherwise we set $T_{i,j}$ equal to T. Then we set

$$r'(i,j) = 2T_{i,j}; \quad i,j \in \mathbb{N}. \tag{5.10}$$

Furthermore we define on $\mathbb{N}|_{[1,n]}$ the uniform distribution and extend this to all subsets which have a frequency. That way we obtain a sequence of ultrametric probability measure spaces $[(U_n, r'_n, \mu_n)]$. These families of equivalence classes form a tight sequence under the law of the Kingman coalescent and we get a family of consistent laws. Therefore we have a state $[(U', r', \mu')] \in \mathbb{U}_1$ on the completion of $(\mathbb{N}, r')$; see Section 4 in [39]. We denote this random state as $\mathfrak{C}_t^{\infty,T}$.

We need next the operation of connecting two trees in a certain way which allows to use the fact that in the evolution of $\mathbb{U}_{\text{fin}}$-valued processes of the type $\mathfrak{U}^{\mathrm{Fel}}$, $\mathfrak{U}^{\mathrm{FV}}$ is special in the sense that the *initial state* only enters by the result of the evolution being *glued* onto it. This procedure connecting an ultrametric measure $\widetilde{\mathfrak{U}}$ space onto a given one $\mathfrak{U}$, denoted $\mathfrak{U} \perp \widetilde{\mathfrak{U}}$, which corresponds to grafting $\widetilde{\mathfrak{U}}$ onto $\mathfrak{U}$, we make precise next. Assume that $\widetilde{\mathfrak{U}}$ is the state of a process evolving for time t of diameter $2t$ from $[(U, \underline{\underline{0}}, \mu)]$.

To build the glued object $[(U, r \perp \widetilde{r}, \mu)] = [(U, r, \mu)] \perp [(\widetilde{U}, \widetilde{r}, \widetilde{\mu})]$ we proceed as follows. We represent the ultrametric measure space $\widetilde{\mathfrak{U}}$ by embedding it into a *marked weighted* $\mathbb{R}$*-tree*, such that the leaves represent $\mathrm{supp}(\mu)$, compare [40] with it the $\widetilde{\mathfrak{U}}$-state. Then we can define the ancestors of a leave ι at time 0 denotes by $\mathrm{anc}_t(\iota)$ as the point in the tree having distance t to the leave. These ancestors we then associate with an element $\beta(\iota)$ in U to obtain the new distance function. Then we want to extend the ancestral path of the $\mathbb{R}$-tree associated with $\widetilde{\mathfrak{U}}$ using the one of $\mathfrak{U}$, to read off the distance $r \perp \widetilde{r}$ on $\widetilde{U}$. Let $(r \perp \widetilde{r})(\iota, \iota') = \widetilde{r}(\iota, \iota')$ if this value is less than $2t$ and define the distances $\geq 2t$ by $(r \perp \widetilde{r})(\iota, \iota') = r(\beta(\mathrm{anc}_t(\iota)), \beta(\mathrm{anc}_t(\iota')))$. How to define $\beta(\cdot)$?

We sample from U according to μ points $x_1, x_2, \ldots$. Then we want to associate these elements with the $\widetilde{U}$ founding ancestors. How to match this with the at most countable set of all $\widetilde{U}$-ancestors? We can for any such ancestor execute for every element here a μ-draw to match.

This works if the population has a dynamic where we can embed the $\mathbb{R}$-trees associated with the states in $\mathbb{U}_{\mathrm{fin}}$ all into each other. This is satisfied for the Fleming-Viot dynamics; see [44].

Theorem 5.5: *(Strong duality)*
Let $(\mathfrak{U}_t^{FV})_{t \geq 0}$ be the $\mathbb{U}_1$-valued Fleming-Viot process. Then for every $T > 0$ we have

$$\mathcal{L}[\mathfrak{U}_T^{FV}] = \mathcal{L}[\mathfrak{U}_0 \perp \mathfrak{C}_T^{\infty,T}]. \tag{5.11}$$

This representation allows to generate the complete state from the dual dynamic in form of the entrance law from a countably infinite population and the initial state. This is the proper generalization of the *duality via graphical constructions* into our context, which has played an important role for individual based models.

A similar result holds for the $\mathbb{U}_{\mathrm{fin}}$-valued Feller diffusion. However here is an additional subtlety which we discuss next since this is only a *conditional* representation of the genealogy $\widehat{\mathfrak{U}}$ given the path of total masses.

Conditioned dualities For many models the evolution of the total masses is again a Markov process evolving autonomously, since the evolution does not depend explicitly on the genealogy. In models with varying population sizes it is therefore often possible to condition on the *complete* path $(\bar{\mathfrak{U}}_t)_{t\geq 0}$ and to then obtain for the conditioned law of the genealogy part $(\widehat{\mathfrak{U}}_t)_{t\geq 0}$ a dual process, for every realization of the total mass process.

This is the case for the $\mathbb{U}_{\text{fin}}$-valued Feller diffusion (which solves uniquely the $(\Omega^{\text{grow}} + \Omega^{\text{res}}, \Pi)$-martingale problem), where one obtains as a dual process the coalescent introduced above but where the coalescence rate b is replaced by a time-dependent one, namely

$$b \cdot (\bar{\mathfrak{U}}_t)^{-1}, \text{ resp. } 0 \text{ if } \bar{\mathfrak{U}}_t = 0, \tag{5.12}$$

which also identifies the conditioned genealogy process as a time-inhomogeneous Fleming-Viot diffusion; see [34, 19]. In other words besides the Feynman-Kac duality we also have a *conditioned duality.* This means we can now get the following analogue of (5.11).

Corollary 5.6: *(Conditioned strong duality: Feller)*
Consider the $\mathbb{U}_{\text{fin}}$*-valued Feller diffusion* $\mathfrak{U}^{\text{Fel}}$ *and write* $\mathfrak{U}^{\text{Fel}} = (\bar{\mathfrak{U}}^{\text{Fel}}, \widehat{\mathfrak{U}}^{\text{Fel}})$*, then*

$$\mathcal{L}\Big[\widehat{\mathfrak{U}}_T^{\text{Fel}} \mid (\bar{\mathfrak{U}}_t^{\text{Fel}})_{t\geq 0} = \bar{\mathfrak{u}}\Big] = \mathcal{L}\Big[\widehat{\mathfrak{U}}_0^{\text{Fel}} \perp \mathfrak{C}_T^{\infty,T}(\bar{\mathfrak{u}})\Big]. \tag{5.13}$$

This says in particular that the process $\widehat{\mathfrak{U}}$-conditioned on the process $\bar{\mathfrak{U}}$ is a time-inhomogeneous $\mathbb{U}^1$-valued Fleming-Viot process, so that we can obtain that the path properties of this process carry over to the Feller cases as for example that the states are non-atomic, or that we have a mark function if we consider multitype models, see Section 5.2 for these properties.

5.2. *Girsanov transformation*

The Girsanov transformation allows us to obtain certain evolutions of marked genealogies from some basic processes, Fleming-Viot models with *selection,* mutation and resampling from those with mutation and resampling alone being the most important example. In that case one can obtain the model with selection via Girsanov transform from the one with only resampling and mutation as we shall explain below.

If the geographic space in our population models is finite or if we even have a non-spatial model, certain of our evolutions have the property that an additional drift term in the generator as migration, selection, mutation or sub- and supercriticality terms in branching models have the property that they have a *density* w.r.t. the process without this terms on laws in path space over some finite time interval. In that case one can also often obtain this density *explicitly* with an appropriate *Girsanov formula.* This we explain next.

We note that on *infinite* geographic space this typically fails and one can obtain the path law only as limit of a sequence of approximate processes on suitable *finite* geographic spaces where the above reasoning then can be applied. Next we explain this for the $\mathbb{U}_1^{\mathbb{I}}$-valued Fleming-Viot process with selection.

We consider as basic process the $\mathbb{U}_1^{\mathbb{I}}$-valued Fleming-Viot diffusion with mutation. Then we consider the case with selection with fitness function $\alpha\Psi$ where $\alpha \in \mathbb{R}^+$. We denote the respective processes by $\mathfrak{U}^{\mathrm{FV,mut},0}$ and $\mathfrak{U}^{\mathrm{F,mut},\alpha\Psi}$.

For the Girsanov transform we need some ingredients. Based on the fitness function χ' consider

$$\Psi(\mathcal{U}) = \frac{\alpha}{\gamma}\langle \nu^{\mathcal{U}}, \chi'_{1,2}\rangle \text{ where } \chi'_{k,\ell}(\underline{\underline{r}}, \underline{u}) = \chi'(u_k, u_\ell, r_{k\wedge\ell}, r_{\ell,k}). \tag{5.14}$$

Define $\mathcal{M} = (M_t)_{t\geq 0}$ where

$$M_t = \Psi(\mathfrak{U}_t) - \Psi(\mathfrak{U}_0) - \int_0^t \Omega_\alpha \Psi(\mathfrak{U}_s)\,\mathrm{d}s, \quad t \geq 0. \tag{5.15}$$

Denoting by $[\,\cdot\,]$ the quadratic variation of a semimartingale we have the following result. Its proof is a combination of Theorem 2 and equation (3.38) in [21].

Theorem 5.7: *(Absolute continuity and Girsanov transform)*

(a) The law $\mathcal{L}\big[(\mathfrak{U}_t^{\mathrm{FV,mut},0})_{t\in[0,T]}\big]$ and $\mathcal{L}\big[(\mathfrak{U}_t^{\mathrm{FV,mut},\alpha\Psi})_{t\in[0,T]}\big]$ are equivalent for $\alpha \geq 0$, for $T > 0$.

(b) The Radon-Nikodym derivative is given by

$$\frac{\mathrm{d}Q_T^\alpha}{\mathrm{d}P_T}(\mathfrak{U}) = e^{M_T - \frac{1}{2}[\mathcal{M}]_T}, \tag{5.16}$$

where the quadratic variation of $\mathcal{M}$ is given by the formula:

$$[\mathcal{M}]_t = \frac{\alpha^2}{\gamma}\int_0^t \mathrm{Var}_{\nu_s}[\chi']\,\mathrm{d}s, \tag{5.17}$$

where we write $\nu_s = \nu^{\mathfrak{U}_s}$ and $\mathrm{Var}_{\nu_s}[\,\cdot\,]$ denotes variance w.r.t. measure ν_s.

Remark 5.8: (First and second order operators) The key point here is that the resampling operator is a second order operator whereas all other operators are first order operators; cf. Section 4 in [21]. This allows to use here some abstract theory on such operators due to Bakry and Émery; see [8].

The Girsanov representation has two mayor applications:

- reducing the *well-posedness* of the processes with selection to the one of the basic process,
- transferring *path properties* of the basic process to the one with selection.

For example the following path properties can be proven for the Fleming-Viot processes with selection by using the equivalence of laws with the neutral case, which is simpler to analyze (see [21, 22] for the results below):

- Continuity of paths.
- The asymptotics of the number of ancestors time ε back from the present time t is $2/\varepsilon$ as $\varepsilon \downarrow 0$.
- The non-atomicity of the states of the process for positive times, i.e., the fact that with probability one for $t > 0$ the sampling measure projected on U has no atoms.
- The Fleming-Viot process with selection has a mark function for $t > 0$ if this holds for $t = 0$, i.e., the sampling measure on $U \times \mathbb{I}$ can be written as $\mu_t(\mathrm{d}x, \mathrm{d}v) = \mu_t^U(\mathrm{d}x)\delta_{\kappa_t(x)}(\mathrm{d}v)$ for a measurable $\kappa : U \to \mathbb{I}$, which is known for the neutral process; see [48].

This technique is also of use trying to handle models with recombination and selections, since it is possible to focus first on building recombination into the neutral model and to then add selection, see [23].

6. Compactness and tightness of marked genealogies

We discuss here two points namely what are *compact* sets in the spaces $\mathbb{U}_1$, $\mathbb{U}_{\text{fin}}$ and their marked versions and how to formulate *tightness* criteria for probability measures on this spaces, respectively processes with values in these spaces.

(i) Most basic is the criterion for compactness of a set K in $\mathbb{U}_1$ where one needs for $\mathcal{U} = [(U, r, \mu)] \in K$:

- bounded distances,
- a no dust condition.

The first requires that pairwise distances in a set are bounded, i.e.

$$R(\mathcal{U}) := \operatorname*{ess\,sup}_{\mu\otimes\mu}[r(i,i')] \le M, \quad \forall i,i' \in U, \text{ for all } \mathcal{U} \in K, \tag{6.1}$$

while the second guarantees that the number $N_\varepsilon(\mathcal{U})$ of ε-balls needed such that the sampling measure of the subset of U covered by the union of the balls exceeds $(1-\varepsilon)$ is bounded for all $\mathcal{U} \in K$ and all $\varepsilon > 0$. Note that $N_\varepsilon(\mathcal{U})$ can be viewed as the number of *ancestors* of an $(1-\varepsilon)$*-fraction* of the population.

In $\mathbb{U}_{\text{fin}}$ we need in addition to the above properties that the total masses are bounded, i.e. for some $M < \infty$:

$$M(\mathcal{U}) = \bar{\mu} \le M, \quad \forall\, \mathcal{U} = [U,r,\mu] \in K. \tag{6.2}$$

If we want to extend this to $\mathbb{U}_{\text{fin}}^{\mathbb{V}}$ the key point is that we need for a set $K \subseteq \mathbb{U}_{\text{fin}}^{\mathbb{V}}$ to be compact that the projections on $\mathbb{U}_{\text{fin}}$ and on $\mathcal{M}_{\text{fin}}(\mathbb{V})$ are compact, the latter in the weak topology. (On $\mathbb{U}^{\mathbb{V},\#}$ one applies this to the localizations again to get the same picture.)

(ii) Next we pass to random variables with values in $\mathbb{U}_1$, $\mathbb{U}_{\text{fin}}$ and their marked versions, as well as the laws of the random variables. For *tightness* of a set of $\mathfrak{K} \subseteq \mathcal{M}_1(\mathbb{U})$ of laws we need that

$$\{\mathcal{L}_\Gamma\left[(R(\mathcal{U}), N_\varepsilon(\mathcal{U}), M(\mathcal{U}))\right], \Gamma \in \mathfrak{K}\} \tag{6.3}$$

is for all $\varepsilon > 0$ tight in $\mathcal{M}_1([0,\infty)^3)$.

(iii) The key task is now to pass to *stochastic* processes and to obtain the tightness of laws on *sets of path* of stochastic processes from the above facts. This requires to get (6.3) "uniformly in time" by representing the process by graphical constructions, lookdown constructions or using special analytic information, on the operator of the martingale problem. Here one uses typically more about the ancestral structure and its evolution in time. Compare for example [40, 43, 44, 36].

7. Perspectives

In this section we describe some current development in this area, in particular we describe some questions where the concepts introduced so far need to be further refined and generalized.

(1) The branching world One issue is to study the branching world in more detail, from classical branching to logistic branching.

(i) The first point is to establish some important properties of the $\mathbb{U}_{\text{fin}}$-valued processes of the branching type, as the $\mathbb{U}_{\text{fin}}$-valued version of the *branching property*, a generalized version of the concept of *infinite divisibility* and a corresponding *Lévy-Khintchine formula* and to then use this to obtain more specifics of the longtime behavior and structure of populations. This point is pursued in [35] and now in [41, 19].

We introduce the subsets of $\mathbb{U}_{\text{fin}}$ denoted

$$\mathbb{U}_{\text{fin}}(h) \text{ and } \mathbb{U}_{\text{fin}}(h)^{\sqcup} \tag{7.1}$$

as the elements with essential diameter $< 2h$ respectively $\leq 2h$ for $h > 0$.

The basic concept which is needed here is that of the *h-tree tops* of $\mathcal{U} = [(U, r, \mu)]$, denoted $\mathcal{U}(h)$ for $h > 0$. Namely consider for (U, r) now $(U, r \wedge 2h)$ and restrict μ to the corresponding σ-algebra to obtain the h-tree top

$$\mathcal{U}(h) \in \mathbb{U}_{\text{fin}}(h)^{\sqcup}. \tag{7.2}$$

We can then consider the *h-concatenation as* binary operation in $\mathbb{U}_{\text{fin}}(h)^{\sqcup}$:

$$[(U_1, r_1, \mu_1)] \sqcup [(U_2, r_2, \mu_2)] = [(U_1 \uplus U_2, r_1 \oplus r_2, \widetilde{\mu}_1 + \widetilde{\mu}_2)], \tag{7.3}$$

where

$$r_1 \oplus r_2(i_1, i_2) = \begin{cases} r_1(i_1, i_2) & \text{if } i_1, i_2 \in U_1, \\ r_2(i_1, i_2) & \text{if } i_1, i_2 \in U_2, \\ 2h & \text{if } i_1 \in U_1, i_2 \in U_2. \end{cases} \tag{7.4}$$

This operation defines for each $h > 0$ a

$$\textit{topological semigroup } (\mathbb{U}_{\text{fin}}(h)^{\sqcup}, \sqcup^h). \tag{7.5}$$

In particular we can associate with $\mathbb{U}_{\text{fin}}$ the collection of *semigroups* $\{(\mathbb{U}_{\text{fin}}(h)^{\sqcup}, \sqcup^h), h > 0\}$ and the collection $\{T_h, h > 0\}$ of *truncation maps*

$$T_h : \mathbb{U} \to \mathbb{U}_{\text{fin}}(h)^{\sqcup},\ \mathcal{U} \mapsto \mathcal{U}(h). \tag{7.6}$$

One has acting on the semigroup in (7.5) the property that

$$T_{h'}(\mathfrak{u} \sqcup^h \mathfrak{v}) = T_{h'}(\mathfrak{u}) \sqcup^{h'} T_{h'}(\mathfrak{v}) \ \textit{(consistency)}, \tag{7.7}$$

for all $h' \in [0, h), h > 0, \mathfrak{u}, \mathfrak{v} \in \mathbb{U}_{\text{fin}}(h)^{\sqcup}$.

Therefore we associate with $\mathbb{U}_{\text{fin}}$ the pair of collections of *topological semigroups and truncation maps*:

$$\left(\{(\mathbb{U}_{\text{fin}}(h)^{\sqcup}, \sqcup^h), h > 0\}, \{T_h; h > 0\}\right). \tag{7.8}$$

With these concepts one can generalize the *branching property* of an $\mathbb{U}_{\text{fin}}$-valued process. We can define *infinite divisibility* for $\mathbb{U}_{\text{fin}}$-valued random variables and prove a *Lévy-Khintchine representation* in terms of *concatenations of Cox point processes* on $\mathbb{U}_{\text{fin}}(h)^{\sqcup}$ for all $h > 0$.

An $\mathbb{U}_{\text{fin}}$-valued random variable $\mathfrak{U}$ is *t-infinitely divisible* if for every $n \in \mathbb{N}$ and $h \in [0, t]$:

$$\mathfrak{U}(h) \stackrel{(d)}{=} \mathfrak{U}^{(h,1)} \sqcup^h \cdots \sqcup^h \mathfrak{U}^{(h,n)} \tag{7.9}$$

for i.i.d. $(U^{(h,i)})_{i=1,\dots,n}$ with values in $\mathbb{U}_{\text{fin}}(h)^{\sqcup}$.

Then the *Lévy-Khintchine formula* on $\mathbb{U}_{\text{fin}}$ reads as follows. The formula represents the expectation of certain exponentials via the *Lévy measure*. We need the following ingredients.

We define the Laplace-transform of the random variable $\mathfrak{U}$:

$$L_{\mathfrak{U}}(\Phi^{n,\phi}) = \mathrm{E}\left[\exp(-\Phi^{n,\phi}(\mathfrak{U}))\right], \tag{7.10}$$

for a positive polynomial $\Phi^{n,\varphi}$. We also need the h-truncated polynomial

$$\Phi_h^{n,\phi} = \Phi^{n,\phi_h}, \quad \phi_h(\underline{\underline{r}}) = \phi(\underline{\underline{r}} \wedge 2h). \tag{7.11}$$

Then the *Lévy-Khintchine formula* for an *infinitely divisible random variable* $\mathfrak{U}$ says that there exists a unique *Lévy measure* $\lambda_\infty \in \mathcal{M}^{\#}(\mathbb{U} \setminus \{0\})$ with $\int(\bar{\mathfrak{u}} \wedge 1)\lambda_\infty(d\mathfrak{u}) < \infty$ such that for all $h > 0$ ($h \in (0, t]$ for t-inf.div.):

$$-\log L_{\mathfrak{U}}(\Phi_h) = \int_{\mathbb{U}_{\text{fin}}(h)^{\sqcup} \setminus \{0\}} (1 - e^{-\Phi_h(\mathfrak{u})})\lambda_h(d\mathfrak{u}), \tag{7.12}$$

where

$$\lambda_h(d\mathfrak{u}) = \int_{\mathbb{U} \setminus \{0\}} \lambda_\infty(d\mathfrak{v}) 1(\mathfrak{v}(h) \in d\mathfrak{u}) \in \mathcal{M}^{\#}(\mathbb{U}_{\text{fin}}(h)^{\sqcup} \setminus \{0\}). \tag{7.13}$$

This allows to write the h-tops of $\mathfrak{U}$ as follows:

$$\mathfrak{U}(h) = \bigsqcup_{\mathfrak{u} \in N^{\lambda_h}} \mathfrak{u}, \tag{7.14}$$

where N^{λ_h} is a Poisson point process on $\mathcal{M}^{\#} \setminus \{0\}$ with intensity measure λ_h.

The marginals of the $\mathbb{U}_{\text{fin}}$-valued Feller diffusion are infinitely divisible if we start in an infinitely divisible initial law, see so that we can ask for the Lévy measure for $\mathfrak{U}$ and its properties. This is carried out in [19].

Furthermore can we ask for the laws conditioned to survive forever or the size-biased law. These questions are studied in detail in [35, 41] and [19].

(ii) Another direction here is to consider the *analog of selection* for branching type processes where a total mass dependent sub-/supercriticality in the branching rate occurs which reflects the *carrying capacity* K of the environment where a population lives, introducing supercriticality for states where the total population size is below K and subcriticality if it is above K. This is studied in [38], where the $\mathbb{U}_{\text{fin}}^{\mathbb{G}}$-valued spatial logistic Feller branching diffusion is constructed and where the longtime behavior is studied via the *conditioned duality.* The goal is to understand the changes in the genealogy induced by the competition for resources.

(2) Tree-valued processes with recombination From the point of view of classical population genetics a natural step in generalizing the tree-valued processes with mutation migration and selection is to also include *recombination.* The resulting process then would involve all the important evolutionary forces. To this end we need to extend the notion of (marked) ultrametric measure spaces appropriately to incorporate *multi-locus genealogies.* This is work in progress, [23].

To better understand some of the difficulties that arise in the case with recombination, let us consider a processes without mutation, selection and migration and assume that the set of loci is given by a finite set $\mathbb{L}$. (All the forces that we exclude here for the sake of simplicity can be included and it is also possible to consider infinitely countable sets of loci.)

In presence of recombination with a finite set of loci $\mathbb{L}$ the genealogical distance of individuals along different loci might be different. The genealogies along all loci together can be encoded by a *multi-ultrametric space* $(U, (r^\ell)_{\ell \in \mathbb{L}})$. In the single locus case the sampling measure μ was a measure on the Borel-σ-algebra on U induced by the corresponding metric. In the multi-locus case the set of individuals is still given by U but the sampling measure should be somehow connected to the whole set of metrics $(r^\ell)_{\ell \in \mathbb{L}}$. As the domain for the sampling measures we choose the Borel-σ-algebra on $U^{\mathbb{L}}$ generated by the product measure induced by $(r^\ell)_{\ell \in \mathbb{L}}$ and we require that the sampling measures are concentrated on the diagonal of $U^{\mathbb{L}}$.

One can then introduce a concept of equivalence of such spaces and pass to the set of equivalence classes. Furthermore one can introduce a generalization of polynomials, distance matrix distributions and a generalized notion of Gromov-weak convergence. The corresponding topology can again be generated by a suitable Gromov-Prohorov metric generating a Polish state space for multi-loci-genealogies.

Finally, the tree-valued processes with recombination can be obtained as solutions to martingale problems.

Acknowledgments

We would like to thank Matthias Birkner, Rongfeng Sun and Jan Swart for organizing the very stimulating program "Genealogies of interacting particle systems" in Singapore 2017. We thank the Institute for Mathematical Sciences at the National University of Singapore for hospitality. Finally, we thank the referee for careful reading of the first version of the manuscript and helpful comments which improved the readability.

References

1. D. Aldous. The continuum random tree. I. *Ann. Probab.*, 19(1):1–28, 1991.
2. D. Aldous. The continuum random tree. II. An overview. In *Stochastic analysis (Durham, 1990)*, volume 167 of *London Math. Soc. Lecture Note Ser.*, pages 23–70. Cambridge Univ. Press, Cambridge, 1991.
3. D. Aldous. The continuum random tree. III. *Ann. Probab.*, 21(1):248–289, 1993.
4. R. A. Arratia. Coalescing Brownian motions and the voter model on $\mathbb{Z}$. Unpublished partial manuscript, 1981.
5. K. B. Athreya and P. E. Ney. *Branching processes*, volume 196 of *Grundlehren der mathematischen Wissenschaften*. Springer-Verlag, New York-Heidelberg, 1972.
6. S. Athreya, W. Löhr, and A. Winter. The gap between Gromov-vague and Gromov-Hausdorff-vague topology. *Stochastic Process. Appl.*, 126(9):2527–2553, 2016.
7. S. Athreya, W. Löhr, and A. Winter. State spaces of (continuum) trees: R-trees versus algebraic trees, 2017. In preparation. Talk by W. Löhr at Workshop: Genealogies of Interacting Particle Systems, 07 Aug 2017, IMS Singapore.
8. D. Bakry and M. Émery. Diffusions hypercontractives. In *Séminaire de probabilités, XIX, 1983/84*, volume 1123 of *Lecture Notes in Math.*, pages 177–206. Springer, Berlin, 1985.
9. J. Bertoin and J.-F. Le Gall. The Bolthausen-Sznitman coalescent and the genealogy of continuous-state branching processes. *Probab. Theory Related Fields*, 117(2):249–266, 2000.
10. J. Bertoin and J.-F. Le Gall. Stochastic flows associated to coalescent processes. *Probab. Theory Related Fields*, 126(2):261–288, 2003.
11. J. Bertoin and J.-F. Le Gall. Stochastic flows associated to coalescent processes. II. Stochastic differential equations. *Ann. Inst. H. Poincaré Probab. Statist.*, 41(3):307–333, 2005.

12. J. T. Cox and D. Griffeath. Diffusive clustering in the two-dimensional voter model. *Ann. Probab.*, 14(2):347–370, 1986.
13. D. A. Dawson. Measure-valued Markov processes. In *École d'Été de Probabilités de Saint-Flour XXI—1991*, volume 1541 of *Lecture Notes in Math.*, pages 1–260. Springer, Berlin, 1993.
14. D. A. Dawson and A. Greven. Hierarchical models of interacting diffusions: multiple time scale phenomena, phase transition and pattern of cluster-formation. *Probab. Theory Related Fields*, 96(4):435–473, 1993.
15. D. A. Dawson and A. Greven. *Spatial Fleming-Viot models with selection and mutation*, volume 2092 of *Lecture Notes in Mathematics*. Springer, Cham, 2014.
16. D. A. Dawson, A. Greven, and J. Vaillancourt. Equilibria and quasiequilibria for infinite collections of interacting Fleming-Viot processes. *Trans. Amer. Math. Soc.*, 347(7):2277–2360, 1995.
17. D. A. Dawson and Z. Li. Stochastic equations, flows and measure-valued processes. *Ann. Probab.*, 40(2):813–857, 2012.
18. D. A. Dawson and E. A. Perkins. Historical processes. *Mem. Amer. Math. Soc.*, 93(454):iv+179, 1991.
19. A. Depperschmidt and A. Greven. Tree-valued Feller diffusion. in preparation, 2018.
20. A. Depperschmidt, A. Greven, and P. Pfaffelhuber. Marked metric measure spaces. *Electron. Commun. Probab.*, 16:174–188, 2011.
21. A. Depperschmidt, A. Greven, and P. Pfaffelhuber. Tree-valued Fleming-Viot dynamics with mutation and selection. *Ann. Appl. Probab.*, 22(6):2560–2615, 2012.
22. A. Depperschmidt, A. Greven, and P. Pfaffelhuber. Path-properties of the tree-valued Fleming-Viot process. *Electron. J. Probab.*, 18(84):1–47, 2013.
23. A. Depperschmidt, A. Greven, and P. Pfaffelhuber. Evolving genealogies of Fleming-Viot processes with recombination. in preparation, 2018.
24. P. Donnelly and T. G. Kurtz. Genealogical processes for Fleming-Viot models with selection and recombination. *Ann. Appl. Probab.*, 9(4):1091–1148, 1999.
25. P. Donnelly and T. G. Kurtz. Particle representations for measure-valued population models. *Ann. Probab.*, 27(1):166–205, 1999.
26. R. Durrett. *Lecture notes on particle systems and percolation.* Wadsworth & Brooks/Cole Advanced Books & Software, Pacific Grove, CA, 1988.
27. S. N. Ethier and T. G. Kurtz. *Markov processes: Characterization and convergence.* John Wiley & Sons Inc., New York, 1986.
28. S. N. Ethier and T. G. Kurtz. Coupling and ergodic theorems for Fleming-Viot processes. *Ann. Probab.*, 26(2):533–561, 1998.
29. S. N. Evans. Kingman's coalescent as a random metric space. In *Stochastic models (Ottawa, ON, 1998)*, volume 26 of *CMS Conf. Proc.*, pages 105–114. Amer. Math. Soc., Providence, RI, 2000.
30. S. N. Evans and I. Molchanov. The semigroup of metric measure spaces and its infinitely divisible probability measures. *Trans. Amer. Math. Soc.*, 369(3):1797–1834, 2017.
31. K. Fleischmann and A. Greven. Diffusive clustering in an infinite system of hi-

erarchically interacting diffusions. *Probab. Theory Related Fields*, 98(4):517–566, 1994.
32. L. R. G. Fontes, M. Isopi, C. M. Newman, and K. Ravishankar. The Brownian web: characterization and convergence. *Ann. Probab.*, 32(4):2857–2883, 2004.
33. C. Foucart. Generalized Fleming-Viot processes with immigration via stochastic flows of partitions. *ALEA Lat. Am. J. Probab. Math. Stat.*, 9(2):451–472, 2012.
34. P. K. Glö de. *Dynamics of genealogical trees for autocatalytic branching processes.* PhD thesis, Department Mathematik, Erlangen, Germany, 2012. `http://nbn-resolving.de/urn:nbn:de:bvb:29-opus-45453`.
35. A. Greven, P. K. Glöde, and T. Rippl. Branching trees I: Concatenation and infinite divisibility. *ArXive 1612.01265*, submitted June 2017. `http://arxiv.org/abs/1612.01265`.
36. A. Greven, A. Klimovsky, and A. Winter. Evolving genealogies of spatial Λ-Cannings processes with mutation. in preparation, 2018.
37. A. Greven, V. Limic, and A. Winter. Representation theorems for interacting Moran models, interacting Fisher-Wright diffusions and applications. *Electron. J. Probab.*, 10:no. 39, 1286–1356, 2005.
38. A. Greven and C. Mukherjee. Genealogies of spatial logistic branching models. in preparation, 2018.
39. A. Greven, P. Pfaffelhuber, and A. Winter. Convergence in distribution of random metric measure spaces (Λ-coalescent measure trees). *Probab. Theory Related Fields*, 145(1-2):285–322, 2009.
40. A. Greven, P. Pfaffelhuber, and A. Winter. Tree-valued resampling dynamics Martingale problems and applications. *Probab. Theory Related Fields*, 155(3-4):789–838, 2013.
41. A. Greven, T. Rippl, and P. K. Glöde. Branching processes — a general concept. in preparation, 2018.
42. A. Greven, R. Sun, and A. Winter. The evolving genealogy of fossils: Unique characterization by martingale problems and applications. In preparation 2018.
43. A. Greven, R. Sun, and A. Winter. Continuum space limit of the genealogies of interacting Fleming-Viot processes on $\mathbb{Z}$. *Electron. J. Probab.*, 21:Paper No. 58, 64, 2016.
44. M. Grieshammer. *Measure Representations of Genealogical Processes and Applications to Fleming-Viot Models.* PhD thesis, Department Mathematik, Erlangen, Germany, 2017. `http://nbn-resolving.de/urn:nbn:de:bvb:29-opus4-85653`
45. S. Gufler. Pathwise construction of tree-valued Fleming-Viot processes. *Electron. J. Probab.*, 23:Paper No. 42, 58, 2018.
46. A. Klenke. Different clustering regimes in systems of hierarchically interacting diffusions. *Ann. Probab.*, 24(2):660–697, 1996.
47. A. Klenke. Multiple scale analysis of clusters in spatial branching models. *Ann. Probab.*, 25(4):1670–1711, 1997.
48. S. Kliem and W. Löhr. Existence of mark functions in marked metric measure spaces. *Electron. J. Probab.*, 20:no. 73, 24, 2015.

49. S. Kliem and A. Winter. Evolving phylogenies of trait-dependent branching with mutation and competition, part I: Existence. *Stochastic Process. Appl.*, 2018.
50. C. Labbé. From flows of Λ-Fleming-Viot processes to lookdown processes via flows of partitions. *Electron. J. Probab.*, 19:no. 55, 49, 2014.
51. J.-F. Le Gall. Marches aléatoires, mouvement brownien et processus de branchement. In *Séminaire de Probabilités, XXIII*, volume 1372 of *Lecture Notes in Math.*, pages 258–274. Springer, Berlin, 1989.
52. J.-F. Le Gall. The uniform random tree in a Brownian excursion. *Probab. Theory Related Fields*, 96(3):369–383, 1993.
53. J.-F. Le Gall. *Spatial branching processes, random snakes and partial differential equations.* Lectures in Mathematics ETH Zürich. Birkhäuser Verlag, Basel, 1999.
54. T. M. Liggett. *Interacting particle systems*, volume 276 of *Grundlehren der Mathematischen Wissenschaften.* Springer-Verlag, New York, 1985.
55. T. M. Liggett and F. Spitzer. Ergodic theorems for coupled random walks and other systems with locally interacting components. *Z. Wahrsch. Verw. Gebiete*, 56(4):443–468, 1981.
56. W. Löhr. Equivalence of Gromov-Prohorov- and Gromov's $\underline{\square}_\lambda$-metric on the space of metric measure spaces. *Electron. Commun. Probab.*, 18:no. 17, 10, 2013.
57. W. Löhr, G. Voisin, and A. Winter. Convergence of bi-measure $\mathbb{R}$-trees and the pruning process. *Ann. Inst. Henri Poincaré Probab. Stat.*, 51(4):1342–1368, 2015.
58. J. Neveu and J. Pitman. The branching process in a Brownian excursion. In *Séminaire de Probabilités, XXIII*, volume 1372 of *Lecture Notes in Math.*, pages 248–257. Springer, Berlin, 1989.
59. C. M. Newman, K. Ravishankar, and R. Sun. Convergence of coalescing nonsimple random walks to the Brownian web. *Electron. J. Probab.*, 10:no. 2, 21–60, 2005.
60. E. Schertzer, R. Sun, and J. M. Swart. The Brownian web, the Brownian net, and their universality. In *Advances in disordered systems, random processes and some applications*, pages 270–368. Cambridge Univ. Press, Cambridge, 2017.
61. A. Winter. Multiple scale analysis of spatial branching processes under the Palm distribution. *Electron. J. Probab.*, 7:No. 13, 72, 2002.

The Algebraic Approach to Duality: An Introduction

Anja Sturm
Institute for Mathematical Stochastics
Georg-August-Universität Göttingen
Goldschmidtstr. 7
37077 Göttingen, Germany
asturm@math.uni-goettingen.de

Jan M. Swart
The Czech Academy of Sciences
Institute of Information Theory and Automation
Pod vodárenskou věží 4, 18200 Prague 8, Czech Republic
swart@utia.cas.cz

Florian Völlering
Universität Leipzig
Mathematisches Institut
PF 10 09 20, D-04009 Leipzig, Germany
florian.voellering@math.uni-leipzig.de

This survey article gives an elementary introduction to the algebraic approach to Markov process duality, as opposed to the pathwise approach. In the algebraic approach, a Markov generator is written as the sum of products of simpler operators, which each have a dual with respect to some duality function. We discuss at length the recent suggestion by Giardinà, Redig, and others, that it may be a good idea to choose these simpler operators in such a way that they form an irreducible representation of some known Lie algebra. In particular, we collect the necessary background on representations of Lie algebras that is crucial for this approach. We also discuss older work by Lloyd and Sudbury on duality functions of product form and the relation between intertwining and duality.

MSC 2010: Primary: 82C22, Secondary: 60K35, 17B10, 22E46.

Keywords: Interacting particle system, duality, intertwining, representations of Lie algebras.

Acknowledgement. This work was sponsored by grant 16-15238S of the Czech Science Foundation (GA CR) and the German Research Foundation (DFG) within the DFG Priority Programme 1590 'Probabilistic Structures in Evolution', grant OR 310/1-1 and grant STU 527/1-2.

Contents

1. Introduction

1.1. *Outline*

The aim of the present text is to give an introduction to the algebraic approach to the theory of duality of Markov processes. In particular, we present some of the pioneering work done by Lloyd and Sudbury [15, 16, 17] and spend a lot of time explaining the more recent work of Giardinà, Redig, and others [4, 7]. The algebraic approach differs fundamentally from the pathwise approach propagated in e.g., [11, 14]. In principle, the algebraic approach is able to find a wider class of dualities, but the price we pay for this is that it may suggest dual operators that turn out not to be Markov generators.

In the remainder of this section, we quickly introduce the basic ideas behind the algebraic approach. In Subsection 1.2, we explain how Markov process duality can algebraically be viewed as an intertwining relation between the generator of one Markov process and the adjoint of the generator of another Markov process. As explained in Subsection 1.3, it is then natural to view a Markov generator as being built up out of sums and products of other, simpler operators. If all these building blocks have duals with respect to a duality function, then so has the original Markov generator.

A central idea of Giardinà, Redig, et al. [4, 7] is to choose these building blocks so that they form a representation of a Lie algebra. To understand why that may be a good idea, one needs quite a bit of background on Lie algebras. Since probabilists may not be familiar with this, after a small detour to pathwise duality in Subsection 1.4, we devote all of Section 2 to providing this background.

The study of Lie algebras and their representations is a huge subject with a venerable history. Although there exist good introductory texts, we will need some theory that is considered too advanced for the usual textbooks. In particular, this refers to the representation theory of noncompact Lie groups like SU(1,1) or the Heisenberg group. In order to squeeze the essential facts that we need for our purposes into little over 10 pages, we had to cut some corners and in some cases resign on full mathematical rigour. We also leave out a lot of background material (e.g., Lie groups, as opposed to Lie algebras, stay almost completely out of the picture). To partly compensate for this, we have added Appendix A which gives a somewhat more complete, but still sketchy picture.

After our little excursion into Lie algebras in Section 2, we come to the core of our text in Section 3. In Subsections 3.1, 3.3, and 3.4 we demonstrate the approach via Lie algebras on three examples, which are based on representation theory for the Heisenberg algebra, SU(2), and SU(1,1), respectively. In Subsection 3.1, we formulate a general principle and apply it to discover a self-duality of the Wright-Fisher diffusion from (1.8). After Subsection 3.2, which is needed to deal with infinite state spaces, in Subsection 3.3, we use the well-known representation theory of SU(2) to derive a duality for the symmetric exclusion process. This duality is not very interesting on its own, but serves as a preparation for the symmetric inclusion process in Section 3.4 which turns out to be very similar to the former, except that SU(2) is replaced by SU(1,1).

In Sections 3.5–3.7 we present results of Lloyd and Sudbury [15, 16, 17] that do not require knowledge of Lie algebras, but do use some facts about tensor products from Section 2.6. In particular, in Section 3.5 we discuss duality functions of product form, including q-duality, while in Section 3.6 we discuss intertwining of Markov processes, and in particular thinning relations which are closely connected to q-duality.

In Sections 3.8 and 3.9, finally, we discuss another observation from [7], who show that nontrivial dualities can sometimes be found by starting from a "trivial" duality which is based on reversibility, and then using a symmetry of the model to transform such a duality into a nontrivial one. Although Lie algebras are not strictly needed in this approach, writing generators in terms of the basis elements of a representation of a Lie algebra can help finding suitable symmetries.

1.2. *Markov duality and intertwining*

In Section 1, for technical simplicity, we mostly restrict ourselves to Markov processes with finite state spaces. As we will see in Section 3, many of the basic ideas discussed here can with some care be made to work also in infinite dimensional settings. How to do this is in part discussed in Section 3.2, but for brevity, we will not always go into the technical details and sometimes use the calculations of the present section merely as an inspiration.

The generator of a continuous-time Markov process with finite state space Ω is a matrix L such that

$$L(x,y) \geq 0 \quad (x \neq y) \quad \text{and} \quad \sum_y L(x,y) = 0. \tag{1.1}$$

Equivalently, we can identify L with the linear operator $L : \mathbb{R}^\Omega \to \mathbb{R}^\Omega$ defined by

$$Lf(x) := \sum_{y\in\Omega} L(x,y)f(y) \qquad (x \in \Omega), \tag{1.2}$$

where $\mathbb{R}^\Omega$ denotes the space of all functions $f : \Omega \to \mathbb{R}$. A linear operator $L : \mathbb{R}^\Omega \to \mathbb{R}^\Omega$ is a Markov generator (i.e., satisfies (1.1)) if and only if the semigroup[a] of operators $(P_t)_{t\geq 0}$ defined by

$$P_t := e^{tL} = \sum_{n=0}^{\infty} \frac{1}{n!} t^n L^n$$

is a Markov semigroup, i.e., P_t is a probability kernel for each $t \geq 0$. If L is a Markov generator, then $(P_t)_{t\geq 0}$ are the transition kernels of some Ω-valued Markov process $(X_t)_{t\geq 0}$.

Let Ω and $\hat{\Omega}$ be finite sets. We can view a function $D : \Omega \times \hat{\Omega} \to \mathbb{R}$ as a matrix

$$(D(x,y))_{x\in\Omega,\ y\in\hat{\Omega}}$$

that as in (1.2) corresponds to a linear operator $D : \mathbb{R}^{\hat{\Omega}} \to \mathbb{R}^\Omega$.

Let L and $\hat{L}$ be generators of Markov processes $(X_t)_{t\geq 0}$ and $(Y_t)_{t\geq 0}$ with state spaces Ω and $\hat{\Omega}$ and semigroups $(P_t)_{t\geq 0}$ and $(\hat{P}_t)_{t\geq 0}$, and let $D : \Omega \times \hat{\Omega} \to \mathbb{R}$ be a function. We make the following simple observation. Below, we let $A^\dagger(x,y) := A(y,x)$ (or $A^\dagger(x,y) := \overline{A(y,x)}$ for matrices over the complex numbers) denote the *adjoint* of a matrix A.[b]

Lemma 1.1: (Duality) *The following conditions are equivalent.*

(i) $LD = D\hat{L}^\dagger$,
(ii) $P_t D = D\hat{P}_t^\dagger$ *for all* $t \geq 0$,
(iii) $\mathbb{E}^x[D(X_t,y)] = \mathbb{E}^y[D(x,Y_t)]$ *for all* $x \in \Omega$, $y \in \hat{\Omega}$, *and* $t \geq 0$.

[a]The semigroup property says that $P_0 = I$ and $P_s P_t = P_{s+t}$.

[b]In other words, $\langle A^\dagger f|g\rangle := \langle f|Ag\rangle$ where $\langle f|g\rangle := \sum_{x\in\Omega} \overline{f(x)}g(x)$ denotes the usual inner product on $\mathbb{C}^\Omega$. For adjoints with respect to a general inner product on finite or infinite dimensional spaces we write A^*.

Proof: If (i) holds for L, then it also holds for any linear combination of powers of L. In particular, filling in the definition of P_t, we see that (i) implies (ii). Conversely, differentiating with respect to t, we see that (ii) implies (i). Condition (iii) is just a rewrite of (ii). □

If the conditions of Lemma 1.1 are satisfied, then we say that $(X_t)_{t\geq 0}$ and $(Y_t)_{t\geq 0}$ are *dual* with *duality function* D. If $L = \hat{L}$, then we speak of *self-duality*. We would like to point out that the duality relationship stated in Condition (iii) has many useful probabilistic consequences. In particular, if the family of functions $\{D(\cdot, y), y \in \hat{\Omega}\}$ is distribution or even convergence determining for probability measures on Ω then we can study the distribution of X_t, respectively its convergence, via Y_t. This is particularly useful if it turns out that the process Y is easier to analyze than X.

Condition (i) can also be written as

$$LD(\,\cdot\,, y)(x) = \hat{L}D(x, \,\cdot\,)(y) \qquad (x \in \Omega,\ y \in \hat{\Omega}). \tag{1.3}$$

Under suitable assumptions, the equivalence of (iii) and (1.3) can often also be established for Markov processes with infinite state space.

An algebraic relation of the form $AB = BC$ is called an *intertwining relation* between operators A and C. The operator B is called the *intertwiner*. Thus, Lemma 1.1 says that two Markov processes are dual if and only if there exists an intertwiner between the generator of one Markov process, and the adjoint of the generator of another Markov process. Note that if L is dual to $\hat{L}$ with duality function D, then $\hat{L}$ is dual to L with duality function $D^\dagger$. Thus, duality is a symmetric concept.

Closely related to Markov process duality is the concept of *intertwining of Markov processes*, which has a more narrow meaning than the algebraic concept of intertwining. Let, again, L and $\hat{L}$ be generators of Markov processes $(X_t)_{t\geq 0}$ and $(Y_t)_{t\geq 0}$ with state spaces Ω and $\hat{\Omega}$ and semigroups $(P_t)_{t\geq 0}$ and $(\hat{P}_t)_{t\geq 0}$. Let $K : \Omega \times \hat{\Omega} \to \mathbb{R}$ be a function. In what follows, we assume that K is a probability kernel, i.e., $K(x, y) \geq 0\ \forall x, y$ and $\sum_y K(x, y) = 1$ for each x.

Lemma 1.2: (Intertwining of Markov processes) *The following conditions are equivalent.*

(i) $LK = K\hat{L}$.
(ii) $P_tK = K\hat{P}_t \qquad (t \geq 0)$.
(iii) $\mu_0 K = \nu_0$ *implies* $\mu_0 P_t K = \nu_0 \hat{P}_t \qquad (t \geq 0)$.

Proof: The equivalence of (i) and (ii) follows by the same argument as in Lemma 1.1. Condition (ii) implies $\mu_0 P_t K = (\mu_0 K)\hat{P}_t$ $(t \geq 0)$. Setting $\mu_0 = \delta_x$ we see that (iii) implies (ii). □

In condition (iii), note that $\mu_0 P_t$ and $\nu_0 \hat{P}_t$ describe the laws at time t of the Markov processes $(X_t)_{t\geq 0}$ and $(Y_t)_{t\geq 0}$ started in initial laws μ_0 and ν_0, respectively. If the conditions of Lemma 1.2 are satisfied, then we say that the Markov processes $(X_t)_{t\geq 0}$ and $(Y_t)_{t\geq 0}$ are *intertwined.*

If K is invertible as a matrix, then $LK = K\hat{L}$ implies $\hat{L}K^{-1} = K^{-1}L$; however, K^{-1} will in general not be a probability kernel. In view of this, in an intertwining relation between Markov processes, the two processes do not play symmetric roles. To stress the different roles of X and Y, following [19], it is convenient to say that Y is an intertwined Markov process *on top* of X.

If the conditions of Lemma 1.2 are satisfied, then the Markov processes X and Y can actually be coupled such that $(X_t, Y_t)_{t\geq 0}$ is a Markov process and

$$\mathbb{P}[Y_t \in \cdot\, |(X_s)_{0\leq s\leq t}] = K(X_s, \cdot) \quad \text{a.s.} \quad (t \geq 0),$$

see [5, 19]. Note that this strengthens condition (iii) of Lemma 1.2.

1.3. *The algebraic approach*

We make the following simple observation.

Lemma 1.3: (Duality of building blocks) *Let $\Omega, \hat{\Omega}$ be finite spaces and let $A_i : \mathbb{R}^{\Omega} \to \mathbb{R}^{\Omega}$, $B_i : \mathbb{R}^{\hat{\Omega}} \to \mathbb{R}^{\hat{\Omega}}$ $(i = 1, 2)$, and $D : \mathbb{R}^{\hat{\Omega}} \to \mathbb{R}^{\Omega}$ be linear operators such that*

$$A_i D = D B_i^{\dagger} \qquad (i = 1, 2). \tag{1.4}$$

Then, for any $r_1, r_2 \in \mathbb{R}$,

$$(r_1 A_1 + r_2 A_2)D = D(r_1 B_1 + r_2 B_2)^{\dagger} \quad \textit{and} \quad (A_1 A_2)D = D(B_2 B_1)^{\dagger}. \tag{1.5}$$

Lemma 1.3 implies that if we can write a Markov generator L as a linear combination of products of "simpler" operators A_i, for example, (denoting the identity operator by I),

$$L = r_{\emptyset} I + r_1 A_1 + r_{23} A_2 A_3 + r_{113} A_1^2 A_3, \tag{1.6}$$

and these "building blocks" satisfy $A_i D = D B_i^{\dagger}$ for some duality function D, the L will be dual to the operator

$$\hat{L} = r_{\emptyset} I + r_1 B_1 + r_{23} B_3 B_2 + r_{113} B_3 B_1^2. \tag{1.7}$$

Note that in each term, we have not only replaced A_i by B_i but also reversed the order of the factors. If we are lucky, $\hat{L}$ is a Markov generator and we have discovered a Markov duality.

We demonstrate this approach on the Wright-Fisher diffusion with selection parameter $s \in \mathbb{R}$, which is the diffusion in $[0,1]$ with generator

$$Lf(x) = x(1-x)\tfrac{\partial^2}{\partial x^2} + sx(1-x)\tfrac{\partial}{\partial x}. \tag{1.8}$$

We are immediately cheating here, since L is not a linear operator acting on a finite dimensional space. Ignoring the difficulties associated with infinite dimension, we can write L in terms of simpler "building blocks" as follows. We set

$$A^- f(x) := (1-x)f(x) \quad \text{and} \quad A^+ f(x) := \tfrac{\partial}{\partial x} f(x), \tag{1.9}$$

and we write L in terms of these building blocks as

$$L = A^-(I - A^-)A^+(sI + A^+). \tag{1.10}$$

As our dual space, we choose $\mathbb{N} = \{0, 1, \dots\}$ and as our duality function we choose the function $D : [0,1] \times \mathbb{N} \to \mathbb{R}$ given by

$$D(x,n) := (1-x)^n \qquad (x \in [0,1],\ n \in \mathbb{N}). \tag{1.11}$$

Let $B^\pm$ be operators acting on functions $f : \mathbb{N} \to \mathbb{R}$ as

$$B^- f(n) := f(n+1) \quad \text{and} \quad B^+ f(n) := -nf(n-1). \tag{1.12}$$

Then $B^\pm$ are dual to $A^\pm$ in the sense of (1.3), i.e.,

$$A^\pm D(\,\cdot\,, n)(x) = B^\pm D(x, \,\cdot\,)(n) \qquad (x \in [0,1],\ n \in \mathbb{N}). \tag{1.13}$$

Therefore, in view of Lemma 1.3, the following operator should be dual to L:

$$\hat{L} = (sI + B^+)B^+(I - B^-)B^-. \tag{1.14}$$

(Note that we have replaced $A^\pm$ by $B^\pm$ and reversed the order of the factors.) A little calculation reveals that

$$\hat{L}f(n) = n(n-1)\{f(n-1) - f(n)\} + sn\{f(n+1) - f(n)\}. \tag{1.15}$$

This is not, in general, a Markov generator. For $s \geq 0$, however, it is the generator of a Markov process in $\mathbb{N}$ that jumps from n to $n-1$ with rate $n(n-1)$ and from n to $n+1$ with rate sn.

Recall that the *commutator* of two operators A, B is defined as $[A, B] := AB - BA$. For our operators $A^\pm$, it is easy to check that

$$[A^-, A^+] = I. \tag{1.16}$$

This is similar to the commutation relation between the position and momentum operators in quantum physics. Indeed, the operators $A^{\pm}$ can be used to define a representation of the Heisenberg algebra, which is a particular Lie algebra. The connection to Lie algebras can help us to choose good building blocks and can sometimes also suggest duality functions. To explain this, we need some theory about representations of Lie algebras, which will be presented in the next section.

1.4. *The pathwise approach*

In the remainder of this section, we point out some differences and similarities between the algebraic and pathwise approaches to Markov process duality. A *random mapping representation* of a probability kernel K is a random map M such that

$$K(x, \mathrm{d}y) = \mathbb{P}[M(x) \in \mathrm{d}y]. \tag{1.17}$$

A *stochastic flow* is a collection $(\mathbf{X}_{s,u})_{s\leq u}$ of random maps $\mathbf{X}_{s,u} : \Omega \to \Omega$ such that $\mathbf{X}_{s,s} = I$ and $\mathbf{X}_{t,u} \circ \mathbf{X}_{s,t} = \mathbf{X}_{s,u}$. We say that $(\mathbf{X}_{s,u})_{s\leq u}$ has *independent increments* if

$$\mathbf{X}_{t_1,t_2}, \ldots, \mathbf{X}_{t_{n-1},t_n} \tag{1.18}$$

are independent for any $t_1 < \cdots < t_n$. If $(\mathbf{X}_{s,u})_{s\leq u}$ is a stochastic flow with independent increments such that the law of $\mathbf{X}_{s,u}$ depends only on the difference $u - s$, and X_0 is an independent Ω-valued random variable, then setting

$$X_t := \mathbf{X}_{0,t}(X_0) \qquad (t \geq 0) \tag{1.19}$$

defines a Markov process with transition kernels

$$P_{u-s}(x, \mathrm{d}y) := \mathbb{P}[\mathbf{X}_{s,u}(x) \in \mathrm{d}y] \qquad (s \leq u). \tag{1.20}$$

Note that this formula says that $\mathbf{X}_{s,u}$ is a random mapping representation of P_{u-s}.

Markov processes can often be constructed from stochastic flows. For example, if a stochastic differential equation has unique strong solutions, then these solutions (for different initial states) define a stochastic flow with independent increments that can be used to construct a diffusion process. If L is the generator of a Markov process with finite state space Ω, then L can always be written in the form

$$Lf(x) = \sum_{m\in\mathcal{G}} r_m\{f(m(x)) - f(x)\}, \tag{1.21}$$

where $\mathcal{G}$ is a finite collection of maps $m : \Omega \to \Omega$. We say that two maps $m, \hat{m}$ are *dual* with respect to a duality function D if

$$D\big(m(x), y\big) = D\big(x, \hat{m}(y)\big) \qquad (x \in \Omega,\ y \in \hat{\Omega}). \tag{1.22}$$

Two stochastic flows $(\mathbf{X}_{s,u})_{s\leq u}$ and $(\mathbf{Y}_{s,u})_{s\leq u}$ are dual[c] if for each $s \leq u$, a.s., $\mathbf{Y}_{-u,-s}$ is dual to $\mathbf{X}_{s,u}$. If two stochastic flows are dual, then we say that their associated Markov processes are *pathwise dual*. It is easy to see that this implies Markov process duality.

We recall that in the algebraic approach, there may be many ways in which a given Markov generator can be written in terms of more elementary "building blocks" as in (1.6). Similarly, in the pathwise approach, there are usually many different ways in which a Markov generator can be written in terms of maps as in (1.21). In the algebraic approach we have seen that if all building blocks have duals with respect to a given duality function, then a Markov generator built up from these building blocks also has a dual $\hat{L}$. Similarly, in the pathwise approach, if all maps m occurring in (1.21) have duals $\hat{m}$ with respect to some duality function D, then the process with generator L is pathwise dual to the process with generator

$$\hat{L}f(x) := \sum_{m\in\mathcal{G}} r_m\big\{f\big(\hat{m}(x)\big) - f\big(x\big)\big\}. \tag{1.23}$$

An advantage of the pathwise approach is that an operator $\hat{L}$ of this form is guaranteed to be a Markov generator. On the other hand, not all dualities can be constructed as pathwise dualities, so the algebraic approach is more general. Nevertheless, many known dualities, including the duality for the Wright-Fisher diffusion discussed in the previous subsection, can be obtained in a pathwise way or as limits of such pathwise dualities, see [1, 18].

There are more analogies between the algebraic and pathwise approaches. In Subsection 3.8, we will see that in the algebraic approach, nontrivial dualities can sometimes be found by starting with a "trivial" duality obtained from reversibility and then applying a symmetry transformation. In [14], it is shown that nontrivial pathwise dualities can be found by starting with a "trivial" duality to the inverse image map and then looking for invariant subspaces of the dual process.

---

[c]The definition of duality for stochastic flows that we give here is a weak one. It is often natural to give a somewhat stronger definition, see [14].

2. Representations of Lie algebras

In this section we first introduce Lie algebras and their representations in Sections 2.1 and 2.2. We also state when representations are equivalent by introducing intertwining relationships. The central result of Section 2.2 is Schur's lemma and its corollary stating uniqueness of the intertwiner for equivalent representations.

In Sections 2.3, 2.4 and 2.5 special cases of Lie algebras are considered that are of relevance later on in Section 3: $\mathfrak{su}(2)$, $\mathfrak{su}(1,1)$ and the Heisenberg algebra $\mathfrak{h}$. Finally, in Section 2.6 we define direct sums and tensor products of linear spaces and Lie algebras. As pointed out previously, we tried to keep this section to the necessary essentials that are needed later on. More detail can be found in Appendix A. For a general reference on Lie algebras and their representations we refer to [9].

2.1. *Lie algebras*

A complex[d] (resp. real) *Lie algebra* is a finite dimensional linear space $\mathfrak{g}$ over $\mathbb{C}$ (resp. $\mathbb{R}$) together with a map $[\cdot,\cdot] : \mathfrak{g} \times \mathfrak{g} \to \mathfrak{g}$ called *Lie bracket* such that

(i) $(\mathbf{x},\mathbf{y}) \mapsto [\mathbf{x},\mathbf{y}]$ is bilinear[e],
(ii) $[\mathbf{x},\mathbf{y}] = -[\mathbf{y},\mathbf{x}]$ (skew symmetry),
(iii) $[\mathbf{x},[\mathbf{y},\mathbf{z}]] + [\mathbf{y},[\mathbf{z},\mathbf{x}]] + [\mathbf{z},[\mathbf{x},\mathbf{y}]] = 0$ (Jacobi identity).

An *adjoint operation* on a Lie algebra $\mathfrak{g}$ is a map $\mathbf{x} \mapsto \mathbf{x}^*$ such that

(i) $\mathbf{x} \mapsto \mathbf{x}^*$ is conjugate linear[f],
(ii) $(\mathbf{x}^*)^* = \mathbf{x}$,
(iii) $[\mathbf{x}^*,\mathbf{y}^*] = [\mathbf{y},\mathbf{x}]^*$.

If $\mathfrak{g}$ is a complex Lie algebra, then the space of its skew symmetric elements $\mathfrak{h} := \{\mathbf{x} \in \mathfrak{g} : \mathbf{x}^* = -\mathbf{x}\}$ forms a real Lie algebra. Conversely, starting from a real Lie algebra $\mathfrak{h}$, we can always find a complex Lie algebra $\mathfrak{g}$ equipped with an adjoint operation such that $\mathfrak{h}$ is the space of skew symmetric elements of $\mathfrak{g}$. Then $\mathfrak{g}$ is called the *complexification* of $\mathfrak{h}$.

---

[d] In this section, we mostly focus on complex Lie algebras. Some results stated in the present section (in particular, part (b) of Schur's lemma) are true for complex Lie algebras only. See Appendix A for a more detailed discussion.
[e] That is, both $[\,\cdot\,,\mathbf{x}]$ and $[\mathbf{x},\,\cdot\,]$ are linear maps from $\mathfrak{g}$ to $\mathfrak{g}$ for any $\mathbf{x} \in \mathfrak{g}$.
[f] That is, $(a\mathbf{x} + b\mathbf{y})^* = \bar{a}\mathbf{x}^* + \bar{b}\mathbf{y}^*$ for $a, b \in \mathbb{C}$, $\mathbf{x},\mathbf{y} \in \mathfrak{g}$.

If $\{\mathbf{x}_1, \ldots, \mathbf{x}_n\}$ is a basis for $\mathfrak{g}$, then the Lie bracket on $\mathfrak{g}$ is uniquely characterized by the *commutation relations*

$$[\mathbf{x}_i, \mathbf{x}_j] = \sum_{k=1}^{n} c_{ijk} \mathbf{x}_k \qquad (i < j).$$

The constants c_{ijk} are called the *structure constants.* If $\mathfrak{g}$ is equipped with an adjoint operation, then the latter is uniquely characterized by the *adjoint relations*

$$\mathbf{x}_i^* = \sum_j d_{ij} \mathbf{x}_j.$$

Example Let V be a finite dimensional complex linear space, let $\mathcal{L}(V)$ denote the space of all linear operators $A : V \to V$, and let $\mathrm{tr}(A)$ denote the trace of an operator A. Then

$$\mathfrak{g} := \{A \in \mathcal{L}(V) : \mathrm{tr}(A) = 0\} \quad \text{with} \quad [A, B] := AB - BA$$

is a Lie algebra. Note that $\mathrm{tr}([A, B]) = \mathrm{tr}(AB) - \mathrm{tr}(BA) = 0$ by the basic property of the trace, which shows that $[A, B] \in \mathfrak{g}$ for all $A, B \in \mathfrak{g}$. Note also that $\mathfrak{g}$ is in general not an algebra, i.e., $A, B \in \mathfrak{g}$ does not imply $AB \in \mathfrak{g}$. If V is equipped with an inner product $\langle \cdot | \cdot \rangle$ (which we always take conjugate linear in its first argument and linear in its second argument) and A^* denotes the adjoint of A with respect to this inner product, i.e.,

$$\langle A^* v | w \rangle := \langle v | A w \rangle,$$

then one can check that $A \mapsto A^*$ is an adjoint operation on $\mathfrak{g}$.

By definition, a *Lie algebra homomorphism* is a map $\phi : \mathfrak{g} \to \mathfrak{h}$ from one Lie algebra into another that preserves the structure of the Lie algebra, i.e., ϕ is linear and

$$\phi([A, B]) = [\phi(A), \phi(B)].$$

If ϕ is invertible, then its inverse is also a Lie algebra homomorphism. In this case we call ϕ a *Lie algebra isomorphism.* We say that a Lie algebra homomorphism ϕ is *unitary* if it moreover preserves the structure of the adjoint operation, i.e.,

$$\phi(A^*) = \phi(A)^*.$$

If $\mathfrak{g}$ is a Lie algebra, then we can define a *conjugate* of $\mathfrak{g}$, which is a Lie algebra $\overline{\mathfrak{g}}$ together with a conjugate linear bijection $\mathfrak{g} \ni \mathbf{x} \mapsto \overline{\mathbf{x}} \in \overline{\mathfrak{g}}$ such that

$$[\overline{\mathbf{x}}, \overline{\mathbf{y}}] = [\mathbf{y}, \mathbf{x}].$$

It is easy to see that such a conjugate Lie algebra is unique up to natural isomorphisms, and that $\overline{\overline{\mathfrak{g}}}$ is naturally isomorphic to $\mathfrak{g}$. If $\mathfrak{g}$ is equipped with an adjoint operation, then we can define an adjoint operation on $\overline{\mathfrak{g}}$ by $\overline{\mathbf{x}}^* := \overline{(\mathbf{x}^*)}$.

Example Let V be a complex linear space on which an inner product is defined and let $\mathfrak{g} \subset \mathcal{L}(V)$ be a linear subspace such that $A, B \in \mathfrak{g}$ implies $[A, B] \in \mathfrak{g}$. Then $\mathfrak{g}$ is a sub-Lie-algebra of $\mathcal{L}(V)$. Now $\overline{\mathfrak{g}} := \{A^* : A \in \mathfrak{g}\}$, together with the map $\overline{A} := A^*$ is a realization of the conjugate Lie algebra of $\mathfrak{g}$.

2.2. *Representations*

If V is a finite dimensional linear space, then the space $\mathcal{L}(V)$ of linear operators $A : V \to V$, equipped with the *commutator*

$$[A, B] := AB - BA$$

is a Lie algebra. By definition, a *representation* of a complex Lie algebra $\mathfrak{g}$ is a pair (V, π) where V is a complex linear space of dimension $\dim(V) \geq 1$ and $\pi : \mathfrak{g} \to \mathcal{L}(V)$ is a Lie algebra homomorphism. A representation is *unitary* if this homomorphism is unitary and *faithful* if π is an isomorphism to its image $\pi(\mathfrak{g}) := \{\pi(\mathbf{x}) : \mathbf{x} \in \mathfrak{g}\}$.

There is another way of looking at representations that is often useful. If (V, π) is a representation, then we can define a map

$$\mathfrak{g} \times V \ni (\mathbf{x}, v) \mapsto \mathbf{x}v \in V$$

by $\mathbf{x}v := \pi(\mathbf{x})v$. Such a map satisfies

(i) $(\mathbf{x}, v) \mapsto Av$ is bilinear,
(ii) $[\mathbf{x}, \mathbf{y}]v = \mathbf{x}(\mathbf{y}v) - \mathbf{y}(\mathbf{x}v)$.

Any map with these properties is called a *left action* of $\mathfrak{g}$ on V. It is easy to see that if V is a complex linear space that is equipped with a left action of $\mathfrak{g}$, then setting $\pi(\mathbf{x})v := \mathbf{x}v$ defines a Lie algebra homomorphism from $\mathfrak{g}$ to $\mathcal{L}(V)$. Thus, we can view representations as linear spaces on which a left action of $\mathfrak{g}$ is defined.

Example For any Lie algebra, we may set $V := \mathfrak{g}$. Then, using the Jacobi identity, one can verify that the map $(\mathbf{x}, \mathbf{y}) \mapsto [\mathbf{x}, \mathbf{y}]$ is a left action of $\mathfrak{g}$ on V. (See Lemma A.1 in the appendix.) In this way, every Lie algebra can be represented on itself. This representation is not always faithful, but for many Lie algebras of interest, it is.

Yet another way to look at representations is in terms of commutation relations. Let $\mathfrak{g}$ be a Lie algebra with basis elements $\mathbf{x}_1, \ldots, \mathbf{x}_n$, which satisfy the commutation relations

$$[\mathbf{x}_i, \mathbf{x}_j] = \sum_{k=1}^{n} c_{ijk}\mathbf{x}_k \qquad (i < j).$$

Let V be a complex linear space with $\dim(V) \geq 1$ and let $X_1, \ldots, X_n \in \mathcal{L}(V)$ satisfy

$$[X_i, X_j] = \sum_{k=1}^{n} c_{ijk}X_k \qquad (i < j).$$

Then there exists a unique Lie algebra homomorphism $\pi : \mathfrak{g} \to \mathcal{L}(V)$ such that $\pi(\mathbf{x}_i) = X_i$ $(i = 1, \ldots, n)$. Thus, any collection of linear operators that satisfies the commutation relations of $\mathfrak{g}$ defines a representation of $\mathfrak{g}$. Such a representation is faithful if and only if $X_1, \ldots, X_n$ are linearly independent. If $\mathfrak{g}$ is equipped with an adjoint operation and V is equipped with an inner product, then the representation (V, π) is unitary if and only if $X_1, \ldots, X_n$ satisfy the adjoint relations of $\mathfrak{g}$, i.e.,

$$\mathbf{x}_i^* = \sum_j d_{ij}\mathbf{x}_j \quad \text{and} \quad X_i^* = \sum_j d_{ij}X_j.$$

Let V be a representation of a Lie algebra $\mathfrak{g}$. By definition, an *invariant subspace* of V is a linear subspace $W \subset V$ such that $\mathbf{x}w \in W$ for all $w \in W$ and $\mathbf{x} \in \mathfrak{g}$. A representation is *irreducible* if its only invariant subspaces are $W = \{0\}$ and $W = V$.

Let V, W be two representations of the same Lie algebra $\mathfrak{g}$. By definition, an *intertwiner* of representations is a linear map $\phi : V \to W$ that preserves the structure of a representation, i.e.,

$$\phi(\mathbf{x}v) = \mathbf{x}\phi(v).$$

If ϕ is a bijection then its inverse is also an intertwiner. In this case we call ϕ an *isomorphism* and say that the representations are *equivalent* (or *isomorphic*).

The following result can be found in, e.g., Thm 4.29 of [9]. Below and in what follows, we let $I \in \mathcal{L}(V)$ denote the *identity operator* $Iv := v$.

Proposition 2.1: (Schur's lemma)

(a) *Let V and W be irreducible representations of the same Lie algebra and let $\phi : V \to W$ be an intertwiner. Then either $\phi = 0$ or ϕ is an isomorphism.*

(b) *Let V be an irreducible representation of a Lie algebra and let $\phi : V \to V$ be an intertwiner. Then $\phi = \lambda I$ for some $\lambda \in \mathbb{C}$.*

For us, the following simple consequence of Schur's lemma will be important.

Corollary 2.2: (Unique intertwiner) *Let (V, π_V) and (W, π_W) be equivalent irreducible representations of some Lie algebra. Then there exists an intertwiner $\phi : V \to W$ that is unique up to a multiplicative constant, such that*

$$\phi\pi_V(\mathbf{x}) = \pi_W(\mathbf{x})\phi.$$

Proof: By assumption, V and W are equivalent, so there exists an isomorphism $\phi : V \to W$. Assume that $\psi : V \to W$ is another intertwiner. Then $\phi^{-1} \circ \psi$ is an intertwiner from V into itself, so by part (b) of Schur's lemma, $\phi^{-1} \circ \psi = \lambda I$ and hence $\psi = \lambda\phi$. □

If V is a complex linear space, then we can define a *conjugate* of V, which is a complex linear space $\overline{V}$ together with a conjugate linear bijection $\phi \mapsto \overline{\phi}$.

Example Let V be a complex linear space with inner product $\langle \cdot | \cdot \rangle$. Let V' denote the dual space of V, i.e., the space of all linear forms $l : V \to \mathbb{C}$. For any $v \in V$, we can define a linear form $\langle v| \in V'$ by $\langle v|w := \langle v|w\rangle$. Then V', together with the map $v \mapsto \langle v|$, is a realization of the conjugate of V.

If (V, π) is a representation of a Lie algebra $\mathfrak{g}$, then we can equip the conjugate space $\overline{V}$ with the structure of a representation of the conjugate Lie algebra $\overline{\mathfrak{g}}$ by putting

$$\overline{\mathbf{x}}\,\overline{v} := \overline{\mathbf{x}v}.$$

It is easy to see that this defines a left action of $\overline{\mathfrak{g}}$ on $\overline{V}$. We call $\overline{V}$, equipped with this left action of $\overline{\mathfrak{g}}$, the *conjugate* of the representation V.

There is a close relation between Lie algebras and Lie groups. Roughly speaking, a Lie group is a smooth differentiable manifold that is equipped with a group structure. In particular, a matrix Lie group G is a group whose elements are invertible linear operators acting on some finite dimensional linear space V. The Lie algebra of G is then defined as

$$\mathfrak{h} := \{A \in \mathcal{L}(V) : e^{tA} \in G \ \forall t \geq 0\}.$$

In general, this is a real Lie algebra. More generally, one can associate a Lie algebra to each Lie group (not necessarily a matrix Lie group) and prove that each Lie algebra is the Lie algebra of some Lie group. Under a certain condition (simple connectedness), the Lie algebra determines its associated Lie group uniquely. A finite dimensional representation of a Lie group G is a pair (V, Π) where V is a finite dimensional linear space and $\Pi : G \to \mathcal{L}(V)$ is a group homomorphism. Each representation (V, π) of a real Lie algebra $\mathfrak{h}$ gives rise to a representation (V, Π) of the associated Lie group such that $\Pi(e^{tA}) = e^{t\pi(A)}$. If $\mathfrak{g}$ is the complexification of $\mathfrak{h}$ and (V, π) is a unitary representation of $\mathfrak{g}$, then (V, Π) is a unitary representation of G in the sense that $\Pi(A)$ is a unitary operator for each $A \in G$. All this is explained in more detail in Appendix A.

2.3. *The Lie algebra SU(2)*

The Lie algebra $\mathfrak{su}(2)$ is the three-dimensional complex Lie algebra defined by the commutation relations between its basis elements

$$[\mathbf{s}_{\mathrm{x}}, \mathbf{s}_{\mathrm{y}}] = 2i\mathbf{s}_{\mathrm{z}}, \quad [\mathbf{s}_{\mathrm{y}}, \mathbf{s}_{\mathrm{z}}] = 2i\mathbf{s}_{\mathrm{x}}, \quad [\mathbf{s}_{\mathrm{z}}, \mathbf{s}_{\mathrm{x}}] = 2i\mathbf{s}_{\mathrm{y}}. \tag{2.1}$$

It is customary to equip $\mathfrak{su}(2)$ with an adjoint operation that is defined by

$$\mathbf{s}_{\mathrm{x}}^* = \mathbf{s}_{\mathrm{x}}, \quad \mathbf{s}_{\mathrm{y}}^* = \mathbf{s}_{\mathrm{y}}, \quad \mathbf{s}_{\mathrm{z}}^* = \mathbf{s}_{\mathrm{z}}. \tag{2.2}$$

A faithful unitary representation of $\mathfrak{su}(2)$ is defined by the *Pauli matrices*

$$S_{\mathrm{x}} := \begin{pmatrix} 0 & 1 \\ 1 & 0 \end{pmatrix}, \quad S_{\mathrm{y}} := \begin{pmatrix} 0 & -i \\ i & 0 \end{pmatrix}, \quad \text{and} \quad S_{\mathrm{z}} := \begin{pmatrix} 1 & 0 \\ 0 & -1 \end{pmatrix}. \tag{2.3}$$

It is straightforward to check that these matrices are linearly independent and satisfy the commutation and adjoint relations (2.1) and (2.2). In particular, this shows that $\mathfrak{su}(2)$ is well-defined.[g]

In general, if $S_{\mathrm{x}}, S_{\mathrm{y}}, S_{\mathrm{z}}$ are linear operators on some complex linear space V that satisfy the commutation relations (2.1), and hence define a representation (V, π) of $\mathfrak{su}(2)$, then the so-called *Casimir operator* is defined as

$$C := S_{\mathrm{x}}^2 + S_{\mathrm{y}}^2 + S_{\mathrm{z}}^2.$$

[g]Not every set of commutation relations that one can write down defines a bona fide Lie algebra. By linearity and skew symmetry, specifying $[\mathbf{x}_i, \mathbf{x}_j]$ for all $i < j$ uniquely defines a bilinear map $[\,\cdot\,, \cdot\,]$, but such a map may fail to satisfy the Jacobi identity. Similarly, it is not *a priori* clear that (2.2) defines a bona fide adjoint operation, but the faithful unitary representation defined by the Pauli matrices shows that it does.

The operator C is in general not an element of $\{\pi(\mathbf{x}) : \mathbf{x} \in \mathfrak{su}(2)\}$, i.e., C does not correspond to an element of the Lie algebra $\mathfrak{su}(2)$. It does correspond, however, to an element of the so-called *universal enveloping algebra* of $\mathfrak{su}(2)$; see Appendix A.4 below.

The finite dimensional irreducible representations of $\mathfrak{su}(2)$ are well understood. Part (a) of the following proposition follows from Theorem A.8 in the appendix, using the compactness of the Lie group SU(2). Parts (b) and (c), and also Proposition 2.4 below, follow from Thm 4.32 of [9] and a calculation of the Casimir operator for the representation in Proposition 2.4.

Proposition 2.3: (Irreducible representations of $\mathfrak{su}(2)$) *Let S_x, S_y, S_z be linear operators on a finite dimensional complex linear space V, that satisfy the commutation relations (2.1) and hence define a representation (V, π) of $\mathfrak{su}(2)$. Then:*

(a) *There exists an inner product $\langle \cdot | \cdot \rangle$ on V, which is unique up to a multiplicative constant, such that with respect to this inner product the representation (V, π) is unitary.*

(b) *If the representation (V, π) is irreducible, then there exists an integer $n \geq 1$, which we call the* index *of (V, π), such that the Casimir operator C is given by $C = n(n+2)I$.*

(c) *Two irreducible representations V, W of $\mathfrak{su}(2)$ are equivalent if and only if they have the same index.*

Proposition 2.3 says that the finite dimensional irreducible representations of $\mathfrak{su}(2)$, up to isomorphism, can be labeled by their index n, which is a natural number $n \geq 1$. We next describe what an irreducible representation with index n looks like. In spite of the beautiful symmetry of the commutation relations (2.1), it will be useful to work with a different, less symmetric basis $\{\mathrm{j}^-, \mathrm{j}^+, \mathrm{j}^0\}$ defined as

$$\mathrm{j}^- := \tfrac{1}{2}(\mathrm{s_x} - i\mathrm{s_y}), \quad \mathrm{j}^+ := \tfrac{1}{2}(\mathrm{s_x} + i\mathrm{s_y}), \quad \text{and} \quad \mathrm{j}^0 := \tfrac{1}{2}\mathrm{s_z}, \tag{2.4}$$

which satisfies the commutation and adjoint relations:

$$[\mathrm{j}^0, \mathrm{j}^\pm] = \pm\mathrm{j}^\pm, \quad [\mathrm{j}^-, \mathrm{j}^+] = -2\mathrm{j}^0, \quad (\mathrm{j}^-)^* = \mathrm{j}^+, \quad (\mathrm{j}^0)^* = \mathrm{j}^0. \tag{2.5}$$

The next proposition describes what an irreducible representation of $\mathfrak{su}(2)$ with index n looks like.

Proposition 2.4: (Raising and lowering operators) *Let V be a finite dimensional complex linear space that is equipped with an inner product and let $J^\pm, J^0$ be linear operators on V that satisfy the commutation and adjoint* *relations (2.5) and hence define a unitary representation (V,π) of $\mathfrak{su}(2)$. Assume that (V,π) is irreducible and has index n. Then V has dimension $n+1$ and there exists an orthonormal basis*

$$\{\phi(-n/2),\phi(-n/2+1),\ldots,\phi(n/2)\}$$

such that

$$\begin{aligned} J^0\phi(k) &= k\phi(k),\\ J^-\phi(k) &= \sqrt{(n/2-k+1)(n/2+k)}\phi(k-1),\\ J^+\phi(k) &= \sqrt{(n/2-k)(n/2+k+1)}\phi(k+1) \end{aligned} \tag{2.6}$$

for $k=-n/2,-n/2+1,\ldots,n/2$, with the conventions $J^-\phi(-n/2):=0$ and $J^+\phi(n/2):=0$.

We see from (2.6) that $\phi(k)$ is an eigenvector of J^0 with eigenvalue k, and that the operators $J^\pm$ maps such an eigenvector into an eigenvector with eigenvalue $k\pm1$, respectively. In view of this, $J^\pm$ are called *raising* and *lowering* operators, or also *creation* and *annihilation* operators. It is instructive to see how this property of $J^\pm$ follows rather easily from the commutation relations (2.5). Indeed, if $\phi(k)$ is an eigenvector of J^0 with eigenvalue k, then the commutation relations imply that

$$J^0J^+\phi(k)=\big(J^+J^0+[J^0,J^+]\big)\phi(k)=\big(J^+J^0+J^+\big)\phi(k)=(k+1)J^+\phi(k),$$

which shows that $J^+\phi(k)$ is a (possibly zero) multiple of $\phi(k+1)$. The concept of raising and lowering operators can be generalized to other Lie algebras.

2.4. *The Lie algebra SU(1,1)*

The Lie algebra $\mathfrak{su}(1,1)$ is defined by the commutation relations

$$[\mathbf{t}_{\rm x},\mathbf{t}_{\rm y}]=2i\mathbf{t}_{\rm z},\quad [\mathbf{t}_{\rm y},\mathbf{t}_{\rm z}]=-2i\mathbf{t}_{\rm x},\quad [\mathbf{t}_{\rm z},\mathbf{t}_{\rm x}]=2i\mathbf{t}_{\rm y}. \tag{2.7}$$

Note that this is the same as (2.1) except for the minus sign in the second equality. A faithful representation is defined by the matrices

$$T_{\rm x}:=\begin{pmatrix}0&1\\-1&0\end{pmatrix},\quad T_{\rm y}:=\begin{pmatrix}0&i\\i&0\end{pmatrix},\quad T_{\rm z}:=\begin{pmatrix}1&0\\0&-1\end{pmatrix}. \tag{2.8}$$

It is customary to equip $\mathfrak{su}(1,1)$ with an adjoint operation such that

$$\mathbf{t}_{\rm x}^*=\mathbf{t}_{\rm x},\quad \mathbf{t}_{\rm y}^*=\mathbf{t}_{\rm y},\quad \mathbf{t}_{\rm z}^*=\mathbf{t}_{\rm z}. \tag{2.9}$$

Note however, that the matrices in (2.8) are not self-adjoint and hence do not define a unitary representation of $\mathfrak{su}(1,1)$. In fact, all unitary irreducible representations of $\mathfrak{su}(1,1)$ are infinite dimensional.[h] In a given representation of $\mathfrak{su}(1,1)$, the *Casimir operator* is defined as

$$C := (\tfrac{1}{2}T_{\mathrm{x}})^2 - (\tfrac{1}{2}T_{\mathrm{y}})^2 - (\tfrac{1}{2}T_{\mathrm{z}})^2. \tag{2.10}$$

Again, it is useful to introduce raising and lowering operators, defined as

$$\mathbf{k}^0 := \tfrac{1}{2}\mathbf{t}_{\mathrm{x}} \quad \text{and} \quad \mathbf{k}^\pm := \tfrac{1}{2}(\mathbf{t}_{\mathrm{y}} \pm i\mathbf{t}_{\mathrm{z}}),$$

which satisfy the commutation and adjoint relations

$$[\mathbf{k}^0, \mathbf{k}^\pm] = \pm\mathbf{k}^\pm, \quad [\mathbf{k}^-, \mathbf{k}^+] = 2\mathbf{k}^0, \quad (\mathbf{k}^-)^* = \mathbf{k}^+, \quad (\mathbf{k}^0)^* = \mathbf{k}^0. \tag{2.11}$$

The following proposition is rewritten from formulas (8) and (9) of [12], where this is stated without proof or reference. The constant $r > 0$ below is called the *Bargmann index* [2, 3].

Proposition 2.5: (Representations of $\mathfrak{su}(1,1)$) *For each real constant $r > 0$, there exists an irreducible unitary representation of $\mathfrak{su}(1,1)$ on a Hilbert space with orthonormal basis $\{\phi(0), \phi(1), \dots\}$ on which the operators $K^0, K^\pm$ act as*

$$\begin{aligned} K^0\phi(k) &= (k+r)\phi(k), \\ K^-\phi(k) &= 1_{\{k\geq 1\}}\sqrt{k(k-1+2r)}\phi(k-1), \\ K^+\phi(k) &= \sqrt{(k+1)(k+2r)}\phi(k+1). \end{aligned} \tag{2.12}$$

In this representation, the Casimir operator is given by $C = r(r-1)I$.

In what follows, we will need one more representation of $\mathfrak{su}(1,1)$, as well as a representation of its conjugate Lie algebra. Fix $\alpha > 0$ and consider the following operators acting on smooth functions $f : [0,\infty) \to \mathbb{R}$:

$$\begin{aligned} \mathcal{K}^- f(z) &= z\frac{\partial^2}{\partial z^2}f(z) + \alpha\frac{\partial}{\partial z}f(z), \\ \mathcal{K}^+ f(z) &= zf(z), \\ \mathcal{K}^0 f(z) &= z\frac{\partial}{\partial z}f(z) + \tfrac{1}{2}\alpha f(z). \end{aligned} \tag{2.13}$$

---

[h]Since $\mathfrak{su}(1,1)$ is simple, all representations are faithful. As explained in Subsection A.3, each Lie algebra is the Lie algebra of a unique simply connected Lie group. In the case of $\mathfrak{su}(1,1)$, this is the universal cover H of the Lie group $\mathrm{SU}(1,1)$ (the latter itself not being simply connected). By Theorem A.4 in the appendix, each representation of $\mathfrak{su}(1,1)$ gives rise to a representation of H. Since H is not compact, the existence of a finite dimensional unitary representation would now contradict Lemma A.9 in the appendix.

One can check that these operators satisfy the commutation relations (2.11) of the Lie algebra $\mathfrak{su}(1,1)_{\mathbb{C}}$, i.e.,

$$[\mathcal{K}^0, \mathcal{K}^{\pm}] = \pm\mathcal{K}^{\pm} \quad \text{and} \quad [\mathcal{K}^-, \mathcal{K}^+] = 2\mathcal{K}^0, \tag{2.14}$$

and hence define a representation of $\mathfrak{su}(1,1)$. One can check that the Casimir operator (2.10) for this representation is $C = \frac{\alpha}{2}(\frac{\alpha}{2} - 1)I$ and hence the Bargmann index is $r = \alpha/2$.

Next, fix again $\alpha > 0$ and consider the following operators acting on functions $f : \mathbb{N} \to \mathbb{R}$:

$$\begin{aligned} K^- f(x) &= x f(x-1), \\ K^+ f(x) &= (\alpha + x) f(x+1), \\ K^0 f(x) &= (\tfrac{1}{2}\alpha + x) f(x). \end{aligned} \tag{2.15}$$

One can check that these operators satisfy the commutation relations

$$[K^{\pm}, K^0] = \pm K^{\pm} \quad \text{and} \quad [K^+, K^-] = 2K^0. \tag{2.16}$$

This is similar to (2.11), except that the order of the elements inside the commutator is reversed. In view of the remarks at the end of Section 2.1, this means that the operators $K^0, K^{\pm}$ define a representation of the conjugate Lie algebra associated with $\mathfrak{su}(1,1)$. We will see in Section 3.4 below that the conjugate of the representation in (2.15) is equivalent to the representation in (2.13), provided we choose for both the same α.

A complete classification of all irreducible representations of $\mathfrak{su}(1,1)$, including infinite dimensional ones, is described in the book [21].[i]

2.5. *The Heisenberg algebra*

The *Heisenberg algebra* $\mathfrak{h}$ is the three-dimensional complex Lie algebra defined by the commutation relations

$$[\mathbf{a}^-, \mathbf{a}^+] = \mathbf{a}^0, \quad [\mathbf{a}^-, \mathbf{a}^0] = 0, \quad [\mathbf{a}^+, \mathbf{a}^0] = 0. \tag{2.17}$$

It is customary to equip $\mathfrak{h}$ with an adjoint operation that is defined by

$$(\mathbf{a}^{\pm})^* = \pm\mathbf{a}^{\pm}, \quad (\mathbf{a}^0)^* = \mathbf{a}^0. \tag{2.18}$$

[i]The monumental encyclopedic book [21] is written in a style that some readers may need to get used to, since it does not use the usual theorem-proof layout but rather states an enormous amount of facts in the main text while leaving a lot of detail to be filled in by the reader.

The *Schrödinger representation* of $\mathfrak{h}$ is defined by

$$A^{-}f(x) = \tfrac{\partial}{\partial x} f(x), \quad A^{+}f(x) = xf(x), \quad A^{0}f(x) = f(x), \tag{2.19}$$

which are interpreted as operators on the Hilbert space $L^2(\mathbb{R}, \mathrm{d}x)$ of complex functions on $\mathbb{R}$ that are square integrable with respect to the Lebesgue measure. Note in this representation, A^0 is the identity operator. Any representation of $\mathfrak{h}$ with this property is called a *central* representation.[j] The Schrödinger representation is a unitary representation, i.e., A^- is skew symmetric and A^+ and A^0 are self-adjoint, viewed as linear operators on the Hilbert space $L^2(\mathbb{R}, \mathrm{d}x)$.

Since iA^- and A^+ are self-adjoint, by Stone's theorem, one can define collections of unitary operators $(U^-_t)_{t\in\mathbb{R}}$ and $(U^+_t)_{t\in\mathbb{R}}$ by

$$U^-_s := e^{tA^-} \quad \text{and} \quad U^+_t := e^{itA^+}. \tag{2.20}$$

These operators form one-parameter groups in the sense that $U^\pm_0 = I$ and $U^\pm_s U^\pm_t = U^\pm_{s+t}$ $(s,t \in \mathbb{R})$. Note that we have a factor i in the definition of U^+_t but not in the definition of U^-_s, because A^+ is self-adjoint but A^- is skew symmetric. The commutation relations (2.17) lead, at least formally, to the following commutation relation between U^-_s and U^+_t

$$U^-_s U^+_t = e^{ist} U^+_t U^-_s \qquad (s,t \in \mathbb{R}). \tag{2.21}$$

Indeed, for small ε, we have

$$\begin{aligned}
&U^-_{\varepsilon s} U^+_{\varepsilon t} \\
&= \big(I + \varepsilon s A^- + \tfrac{1}{2}\varepsilon^2 s^2 (A^-)^2 + O(\varepsilon^3)\big)\big(I + i\varepsilon t A^+ - \tfrac{1}{2}\varepsilon^2 t^2 (A^+)^2 + O(\varepsilon^3)\big) \\
&= I + \varepsilon s A^- + \tfrac{1}{2}\varepsilon^2 s^2 (A^-)^2 + i\varepsilon t A^+ - \tfrac{1}{2}\varepsilon^2 t^2 (A^+)^2 + i\varepsilon^2 st A^- A^+ + O(\varepsilon^3) \\
&= I + \varepsilon s A^- + \tfrac{1}{2}\varepsilon^2 s^2 (A^-)^2 + i\varepsilon t A^+ - \tfrac{1}{2}\varepsilon^2 t^2 (A^+)^2 + i\varepsilon^2 st A^+ A^- \\
&\qquad\qquad\qquad\qquad\qquad\qquad + i\varepsilon^2 st [A^-, A^+] + O(\varepsilon^3) \\
&= \big(1 + i\varepsilon^2 st + O(\varepsilon^3)\big) U^+_{\varepsilon t} U^-_{\varepsilon s} + O(\varepsilon^3).
\end{aligned} \tag{2.22}$$

[j]More generally, the *center* of a Lie algebra $\mathfrak{g}$ is the linear space $\mathfrak{c} := \{\mathbf{c} \in \mathfrak{g} : [\mathbf{x}, \mathbf{c}] = 0 \ \forall \mathbf{x} \in \mathfrak{g}\}$. A *central* representation of a Lie algebra is then a representation (V, π) such that for each $\mathbf{c} \in \mathfrak{c}$, there exists a $c \in \mathbb{C}$ such that $\pi(\mathbf{c}) = cI$. Note that with this definition, if (V, π) is a faithful central representation of $\mathfrak{h}$, then we can always "normalize" it by multiplying π with a constant so that $\pi(\mathbf{a}^0) = I$.

The commutation relation (2.21) then follows formally by writing

$$\begin{aligned} U_s^- U_t^+ &= (U_{s/n}^-)^n (U_{t/n}^+)^n \\ &= \big(1 + in^{-2}st + O(n^{-3})\big)^{n^2} (U_{t/n}^+)^n (U_{s/n}^-)^n \underset{n\to\infty}{\longrightarrow} e^{ist} U_t^+ U_s^- . \end{aligned} \tag{2.23}$$

The *Stone-von Neumann theorem* states that all unitary, central representations of the Heisenberg algebra that satisfy (2.21) are equivalent [13].

2.6. *The direct sum and the tensor product*

If V is a linear space and $V_1, \ldots, V_n$ are linear subspaces of V such that every element $v \in V$ can uniquely be written as

$$v = v_1 + \cdots + v_n$$

with $v_i \in V_i$, then we say that V is the *direct sum* of $V_1, \ldots, V_n$ and write $V = V_1 \oplus \cdots \oplus V_n$. If Ω_1, Ω_2 are finite sets and $\mathbb{C}^{\Omega_1}$ denotes the linear space of all functions $f : \Omega_i \to \mathbb{C}$, then we have the natural isomorphism

$$\mathbb{C}^{\Omega_1 \uplus \Omega_2} \cong \mathbb{C}^{\Omega_1} \oplus \mathbb{C}^{\Omega_2},$$

where $\Omega_1 \uplus \Omega_2$ denotes the disjoint union of Ω_1 and Ω_2.

If $\mathfrak{g}_1, \ldots, \mathfrak{g}_n$ are Lie algebras, then we equip $\mathfrak{g}_1 \oplus \cdots \oplus \mathfrak{g}_n$ with the structure of a Lie algebra by putting, for $\mathbf{x}_i, \mathbf{y}_i \in \mathfrak{g}_i$,

$$\big[\mathbf{x}_1 + \cdots + \mathbf{x}_n, \mathbf{y}_1 + \cdots + \mathbf{y}_n\big] := [\mathbf{x}_i, \mathbf{y}_i] + \cdots + [\mathbf{x}_n, \mathbf{y}_n]. \tag{2.24}$$

Note that this has the effect that elements of different Lie algebras $\mathfrak{g}_1, \ldots, \mathfrak{g}_n$ mutually commute. In particular, if $\{\mathbf{x}_1^1, \mathbf{x}_1^2, \mathbf{x}_1^3\}$ and $\{\mathbf{x}_2^1, \mathbf{x}_2^2, \mathbf{x}_2^3\}$ are bases for $\mathfrak{g}_1$ and $\mathfrak{g}_2$, respectively, then

$$\{\mathbf{x}_1^1, \mathbf{x}_1^2, \mathbf{x}_1^3, \mathbf{x}_2^1, \mathbf{x}_2^2, \mathbf{x}_2^3\}$$

is a basis for $\mathfrak{g}_1 \oplus \mathfrak{g}_2$ and $[\mathbf{x}_i^k, \mathbf{x}_j^l] = 0$ whenever $i \neq j$.

By definition, a *bilinear* map of two variables is a function that is linear in each of its arguments. If V and W are finite dimensional linear spaces, then their *tensor product* is a linear space $V \otimes W$ together with a bilinear map

$$V \times W \ni (v, w) \mapsto v \otimes w \in V \otimes W$$

that has the property:

> If F is another linear space and $b : V \times W \to F$ is bilinear, then there exists a unique linear map $\overline{b} : V \otimes W \to F$ such that
>
> $$\overline{b}(v \otimes w) = b(v, w) \qquad (v \in V,\ w \in W).$$

The tensor product of three or more spaces is defined similarly. One can show that all realizations of the tensor product are naturally isomorphic. If $\{e(1),\ldots,e(n)\}$ and $\{f(1),\ldots,f(m)\}$ are bases for V and W, then one can prove that

$$\{e(i)\otimes f(j) : 1\leq i\leq n,\ 1\leq j\leq m\} \tag{2.25}$$

is a basis for $V\otimes W$. In particular, this means that one has the natural isomorphism

$$\mathbb{C}^{\Omega_1\times\Omega_2}\cong\mathbb{C}^{\Omega_1}\otimes\mathbb{C}^{\Omega_2}. \tag{2.26}$$

If $A\in\mathcal{L}(V)$ and $B\in\mathcal{L}(V)$, then one defines $A\otimes B\in\mathcal{L}(V\otimes W)$ by

$$(A\otimes B)(v\otimes w) := (Av)\otimes(Bw). \tag{2.27}$$

We note that not every element of $V\otimes W$ is of the form $v\otimes w$ for some $v\in V$ and $w\in W$. Nevertheless, since the right-hand side of (2.27) is bilinear in v and w, the defining property of the tensor product tells us that this formula unambiguously defines a linear operator on $V\otimes W$.

One can check that the notation $A\otimes B$ is good notation in the sense that the space $\mathcal{L}(V\otimes W)$ together with the bilinear map $(A,B)\mapsto A\otimes B$ is a realization of the tensor product $\mathcal{L}(V)\otimes\mathcal{L}(W)$. Thus, one has the natural isomorphism

$$\mathcal{L}(V\otimes W)\cong\mathcal{L}(V)\otimes\mathcal{L}(W).$$

If V and W are equipped with inner products, then we equip $V\otimes W$ with an inner product by putting

$$\langle v\otimes w|\eta\otimes\xi\rangle := \langle v|\eta\rangle\langle w|\xi\rangle, \tag{2.28}$$

which has the effect that if $\{e(1),\ldots,e(n)\}$ and $\{f(1),\ldots,f(m)\}$ are orthonormal bases for V and W, then the basis for $V\otimes W$ in (2.25) is also orthonormal. Again, one needs the defining property of the tensor product to see that (2.28) is a good definition.

If V,W are representations of Lie algebras $\mathfrak{g},\mathfrak{h}$, respectively, then we can naturally equip the tensor product $V\otimes W$ with the structure of a representation of $\mathfrak{g}\oplus\mathfrak{h}$ by putting

$$(\mathbf{x}+\mathbf{y})(v\otimes w) := (\mathbf{x}v)\otimes(\mathbf{y}w). \tag{2.29}$$

Again, since the right-hand side is bilinear, using the defining property of the tensor product, one can see that this is a good definition.

Let V_1, V_2 be representations of some Lie algebra $\mathfrak{g}$, and let W_1, W_2 be representations of another Lie algebra $\mathfrak{h}$. Let $\phi : V_1 \to V_2$ and $\psi : W_1 \to W_2$ be intertwiners. Then one can check that

$$\phi \otimes \psi : V_1 \otimes W_1 \to V_2 \otimes W_2 \tag{2.30}$$

is also an intertwiner.

If $\mathfrak{h}_1, \ldots, \mathfrak{h}_n$ are n copies of the Heisenberg algebra, and $\mathbf{a}_i^-, \mathbf{a}_i^+, \mathbf{a}_i^0$ are basis elements of $\mathfrak{h}_i$ that satisfy the commutation relations (2.17), then a basis for $\mathfrak{h}_1 \oplus \cdots \oplus \mathfrak{h}_n$ is formed by all elements $\mathbf{a}_i^\pm, \mathbf{a}_i^0$ with $i = 1, \ldots, n$, and these satisfy

$$[\mathbf{a}_i^- \mathbf{a}_j^+] = \delta_{ij}\mathbf{a}_i^0 \quad \text{and} \quad [\mathbf{a}_i^\pm, \mathbf{a}_j^0] = 0.$$

Since the center of $\mathfrak{h}_1 \oplus \cdots \oplus \mathfrak{h}_n$ is spanned by the elements $\mathbf{a}_i^0$ with $i = 1, \ldots, n$, a central representation of $\mathfrak{h}_1 \oplus \cdots \oplus \mathfrak{h}_n$ must map all these elements to multiples of the identity. In particular, a central representation of $\mathfrak{h}_1 \oplus \cdots \oplus \mathfrak{h}_n$ is never faithful (unless $n = 1$). The Lie algebra $\mathfrak{h}(n)$ is the $2n+1$ dimensional Lie algebra with basis elements $\mathbf{a}_i^\pm$ $(i = 1, \ldots, n)$ and $\mathbf{a}^0$, which satisfy the commutation relations

$$[\mathbf{a}_i^- \mathbf{a}_j^+] = \delta_{ij}\mathbf{a}^0 \quad \text{and} \quad [\mathbf{a}_i^\pm, \mathbf{a}^0] = 0.$$

A *central* representation of $\mathfrak{h}(n)$ is a representation (V, π) such that $\pi(\mathbf{a}^0) = I$. The Schrödinger representation of the "n-dimensional" Heisenberg algebra is the central representation of $\mathfrak{h}(n)$ on $L^2(\mathbb{R}^n, \mathrm{d}x)$ given by

$$A^- f(x) = \tfrac{\partial}{\partial x_i} f(x) \quad \text{and} \quad A^+ f(x) := x_i f(x). \tag{2.31}$$

3. The algebraic approach to duality

After our excursion into the theory of Lie algebras, we return to our main topic, which is the algebraic approach to Markov process duality. We recall from Lemma 1.3 that if a Markov generator L can be written in terms of "building blocks" A_i that each have a dual B_i with respect to some duality function D, then also L has a dual $\hat{L}$ with respect to D. As mentioned at the end of Section 1.3, it may be a good idea to choose the A_i's so that they define a representation of some Lie algebra. The next proposition says that in such a situation, other, equivalent representations of the same Lie algebra may lead to dual Markov processes.

Recall the definition of a conjugate Lie algebra $\overline{\mathfrak{g}}$ from Section 2.1. If $Y_1, \ldots, Y_n$ are matrices that define a representation of $\overline{\mathfrak{g}}$, then their adjoints $Y_1^\dagger, \ldots, Y_n^\dagger$ define a representation of the original Lie algebra $\mathfrak{g}$.

Proposition 3.1: (Intertwiners as duality functions) *Let L be the generator of a Markov process with finite state space Ω. Let $X_1, \dots, X_n$ be linear operators on $\mathbb{C}^\Omega$ that form a representation of some Lie algebra $\mathfrak{g}$. Assume that L can be written as a linear combination of products of the operators $X_1, \dots, X_n$*

$$L = \sum_{(i_1,\dots,i_k)\in\mathcal{I}} r_{i_1,\dots,i_k} X_{i_1} \cdots X_{i_k}, \tag{3.1}$$

where $\mathcal{I}$ is some finite set whose elements are sequences $(i_1, \dots, i_k)$ with $k \geq 0$ and $1 \leq i_m \leq n$ for each m. Assume that $Y_1, \dots, Y_n$ are linear operators on $\mathbb{C}^{\hat{\Omega}}$ that define a representation of the conjugate Lie algebra $\overline{\mathfrak{g}}$. Assume that the representations of $\mathfrak{g}$ defined by $Y_1^\dagger, \dots, Y_n^\dagger$ and $X_1, \dots, X_n$ are equivalent. Then there is a bijective intertwiner D, i.e., $X_i D = D Y_i^\dagger$ for each i, and L is dual w.r.t. the duality function D to the operator

$$\hat{L} := \sum_{(i_1,\dots,i_k)\in\mathcal{I}} r_{i_1,\dots,i_k} Y_{i_k} \cdots Y_{i_1}. \tag{3.2}$$

Proof: By definition, two representations are equivalent if and only if there exists a bijective intertwiner. The fact that L is dual to the operator in (3.2) is then immediate from Lemma 1.3. □

At first sight, it may seem unlikely that Proposition 3.1 could be of much use. Even if we can write a generator in terms of a basis of a representation of some Lie algebra $\mathfrak{g}$, and we also find some representation of the conjugate Lie algebra $\overline{\mathfrak{g}}$, we still have to be lucky in the sense that the representations of $\mathfrak{g}$ defined by $Y_1^\dagger, \dots, Y_n^\dagger$ and $X_1, \dots, X_n$ are equivalent, and there is no guarantee that the operator in (3.2) is a Markov generator. Nevertheless, in what follows, we will see that Proposition 3.1 can help us find nontrivial dualities. In the next subsection, we demonstrate this on the operator L from (1.8), which is the generator of a Wright-Fisher diffusion with selection.

3.1. *Self-duality of the Wright-Fisher diffusion*

In Subsection 1.3, we have seen that the operator L from (1.8) can as in (1.10) be written in terms of the "building blocks" $A^\pm$ from (1.9). As we have seen in (1.16), these operators satisfy

$$[A^-, A^+] = I, \tag{3.3}$$

and hence define a central representation of the Heisenberg algebra $\mathfrak{h}$, as defined in Subsection 2.5.

It will be convenient to find a way of writing L in a more symmetric way than in (1.10). To this aim, we change the definitions of $A^\pm$ to

$$A^- f(x) := \frac{-1}{\sqrt{s}} \tfrac{\partial}{\partial x} f(x) \quad \text{and} \quad A^+ f(x) := \sqrt{s} x f(x), \tag{3.4}$$

which again satisfy (3.3), and we write L in terms of these new building blocks as

$$L = -A^+(\sqrt{s} - A^+)A^-(\sqrt{s} - A^-). \tag{3.5}$$

We observe from (3.3) that setting $B^- := A^+$ and $B^+ := A^-$ defines operators such that $[B^-, B^+] = -I$, i.e., B^-, B^+ define a central representation of the conjugate Heisenberg algebra $\overline{\mathfrak{h}}$.

We recall from Section 2.5 that the Stone-von Neumann theorem states that, more or less, all central representations of the Heisenberg algebra are equivalent. In view of this and Proposition 3.1, we may expect that the operator

$$\hat{L} = -(\sqrt{s} - B^-)B^-(\sqrt{s} - B^+)B^+ \tag{3.6}$$

is dual to L with respect to some (so far unknown) duality function D. (Here (3.6) is obtained from (3.5) by replacing $A^\pm$ by $B^\pm$ and reversing the order of the factors.) Since $B^\pm = A^\mp$, we observe that in fact $\hat{L} = L$, so our calculations lead us to suspect that the Wright-Fisher diffusion with selection parameter $s > 0$ should be self-dual.

We still need to find the duality function D. This function must satisfy

$$\frac{-1}{\sqrt{s}} \tfrac{\partial}{\partial x} D(x,y) = A^- D(\,\cdot\,, y)(x) = B^- D(x, \,\cdot\,)(y) = \sqrt{s} y D(x,y), \tag{3.7}$$

which says that $\frac{\partial}{\partial x} D(x,y) = -syD(x,y)$ and leads to the requirement that $D(x,y) = D(0,y)e^{-syx}$. In a similar way, the requirement $A^+D = DB^+$ yields $D(x,y) = D(x,0)e^{-syx}$ and in particular $D(0,y) = D(0,0)$. Thus, we find that up to a multiplicative constant, there is a unique duality function, which is given by

$$D(x,y) = e^{-sxy} \qquad (x, y \in [0,1]), \tag{3.8}$$

and we conclude that the Wright-Fisher diffusion with selection parameter $s > 0$ is self-dual with this duality function.

The argument above was heuristic, but quite smooth. What is remarkable about it is that while usually, the discovery of a duality starts with a clever choice for the duality function, here, the duality function came at the very end. Hidden behind this is the Stone-von Neumann theorem which says that two "good" representations of the Heisenberg algebra must necessarily be equivalent. We did not check the conditions of this theorem in detail (this is why the argument is only heuristic), but rather used it as an inspiration. *A priori*, there was no guarantee that the operator in (3.6) would be a Markov generator, but since $\hat{L} = L$ and L is a Markov generator, this turned out right as well.

Remark It is possible to "discover" the moment dual (1.15) of the Wright-Fisher duality along similar lines as we have discovered its self-duality here, by considering a suitable representation of the conjugate Heisenberg algebra $\overline{\mathfrak{h}}$ on functions $f : \mathbb{N} \to \mathbb{R}$ and applying Propositions 3.1 and 3.2. Such a derivation is less natural, however, since it requires choosing a rather peculiar representation of $\overline{\mathfrak{h}}$ that more or less has the duality function from (1.11) tacitly built into it.

3.2. *Intertwiners and duality functions*

In the previous subsection, just before (3.6) we appealed to Proposition 3.1. In doing so, we cheated in the sense that the operators $A^{\pm}$ from (3.4) do not act on a finite dimensional space. The most obvious consequence of this is that it is not clear how the adjoint operators $B_i^{\dagger}$ from Proposition 3.1 should be defined. Closely related to this is that in the infinite dimensional setting, it is not immediately clear that duality functions define intertwiners and vice versa. In this subsection we show that these difficulties can be resolved by introducing a suitable inner product on the spaces of complex functions on Ω and $\hat{\Omega}$, respectively.

Assume that $X_1, \ldots, X_n$ and $Y_1, \ldots, Y_n$ are linear operators on L^2-spaces $L^2(\Omega, \mu)$ and $L^2(\hat{\Omega}, \nu)$, respectively, that define representations of a Lie algebra $\mathfrak{g}$ and its conjugate $\overline{\mathfrak{g}}$. Let Y_i^* denote the adjoint of Y_i with respect to the inner product on $L^2(\hat{\Omega}, \nu)$. Assume that $\Phi : L^2(\hat{\Omega}, \nu) \to L^2(\Omega, \mu)$ is a linear operator of the form

$$\Phi g(x) = \int g(y) D(x, y) \nu(dy), \tag{3.9}$$

for some function $D : \Omega \times \hat{\Omega} \to \mathbb{C}$ such that the expressions in (3.10) below are well-defined.

Proposition 3.2: (Intertwiners and duality functions) *The operator Φ is an intertwiner of the representations defined by $X_1, \ldots, X_n$ and*

$Y_1^*, \ldots, Y_n^*$, *i.e.*,

$$X_i\Phi = \Phi Y_i^* \quad (i = 1, \ldots, n),$$

if and only if D is a duality function, in the sense that

$$X_i D(\,\cdot\,, y)(x) = Y_i D(x, \,\cdot\,)(y) \qquad (i = 1, \ldots, n) \tag{3.10}$$

for a.e. x, y with respect to the product measure $\mu \otimes \nu$.

Proof: We observe that

$$\begin{aligned}
&\int \overline{f(x)}\mu(\mathrm{d}x) \int g(y)\nu(\mathrm{d}y) X_i D(\,\cdot\,, y)(x) = \int g(y)\nu(\mathrm{d}y)\langle f | X_i D(\,\cdot\,, y)\rangle_\mu \\
&\quad = \int g(y)\nu(\mathrm{d}y)\langle X_i^* f | D(\,\cdot\,, y)\rangle_\mu = \int \overline{X_i^* f(x)}\mu(\mathrm{d}x) \int g(y)\nu(\mathrm{d}y) D(x, y) \\
&\quad = \langle X_i^* f | \Phi g\rangle_\mu = \langle f | X_i \Phi g\rangle_\mu
\end{aligned}$$

and

$$\begin{aligned}
&\int \overline{f(x)}\mu(\mathrm{d}x) \int g(y)\nu(\mathrm{d}y) Y_i D(x, \,\cdot\,)(y) = \int \overline{f(x)}\mu(\mathrm{d}x)\langle g | Y_i D(x, \,\cdot\,)\rangle_\nu \\
&\quad = \int \overline{f(x)}\mu(\mathrm{d}x)\langle Y_i^* g | D(x, \,\cdot\,)\rangle_\nu = \int \overline{f(x)}\mu(\mathrm{d}x) \int Y_i^* g(y)\nu(\mathrm{d}y) D(x, y) \\
&\quad = \langle f | \Phi Y_i^* g\rangle_\mu.
\end{aligned}$$

Since this holds for all f, g, the statement follows. □

Remark Proposition 3.2 allows us to obtain an intertwiner from a duality function. Conversely, if $\Phi : L^2(\hat{\Omega}, \nu) \to L^2(\Omega, \mu)$ is a bounded linear operator, then setting

$$\Delta(f \otimes g) := \int f(x)\Phi g(x)\, \mu(\mathrm{d}x)$$

defines a linear form on the linear span of all functions of the form $f \otimes g$. If Δ is bounded,[k] then it can uniquely be extended to a bounded linear form on

$$L^2(\Omega, \mu) \otimes L^2(\hat{\Omega}, \nu) \cong L^2(\Omega \times \hat{\Omega}, \mu \otimes \nu),$$

[k]Using Cauchy-Schwarz, it is easy to see that $|\Delta(f \otimes g)| \leq \|\Phi\|\, \|f \otimes g\|$, proving that Δ is bounded on functions of the form $f \otimes g$. Nevertheless, Δ may fail to be bounded on the linear span of such functions. A counterexample is $\Omega = \hat{\Omega} = [0, 1]$, $\mu = \nu =$ Lebesgue measure, and Φ the identity map, which gives $\Delta(F) = \int_0^1 F(x, x)\, \mathrm{d}x$. Since the Lebesgue measure on the diagonal $\{(x, y) : x = y\}$ does not have a density w.r.t. $\mu \otimes \nu$, this does not correspond to a bounded linear form on $L^2(\Omega \times \hat{\Omega}, \mu \otimes \nu)$.

so that by the Riesz representation theorem there exists a $D \in L^2(\Omega \times \hat{\Omega}, \mu \otimes \nu)$ such that

$$\Delta(f \otimes g) := \int f(x) D(x, y) g(y)\, \mu(\mathrm{d}x)\nu(\mathrm{d}y),$$

proving that Φ is of the form (3.9) (although there is no guarantee that $D(\,\cdot\,, y)$ and $D(x, \,\cdot\,)$ are in the domains of X_i and Y_i, resp., if these are unbounded operators).

3.3. *The symmetric exclusion process*

In this subsection, we demonstrate Proposition 3.1 on a simple example, which involves the simple exclusion process and the Lie algebra $\mathfrak{su}(2)$. In the end, we find a self-duality that is not entirely trivial, but also not very useful. The present subsection serves mainly as a warm-up for Subsection 3.4 where we will replace $\mathfrak{su}(2)$ by $\mathfrak{su}(1,1)$.

Let S be a finite set and let $r : S \times S \to [0, \infty)$ be a function that is symmetric in the sense that $r(i, j) = r(j, i)$. Consider the Markov process with state space $\Omega = \{0, 1\}^S$ and generator

$$Lf(x) := \sum_{ij} r(i,j) 1_{\{(x_i, x_j)=(1,0)\}} \{f(x - \delta_i + \delta_j) - f(x)\}, \tag{3.11}$$

where $\delta_i \in \Omega$ is defined as $\delta_i(j) := 1_{\{i=j\}}$. Then L is the generator of a *symmetric exclusion process* or *SEP:* If we interpret a 1 at a site as a particle and a 0 as an empty site, then the dynamics of the SEP are such that particles perform independent random walks with kernel r as long as the site they want to jump to is unoccupied; if it is, then nothing happens.

We define operators $J_i^{\pm}$ and J_i^0 by

$$\begin{aligned} J_i^- f(x) &:= 1_{\{x_i=0\}} f(x + \delta_i), \quad J_i^+ f(x) := 1_{\{x_i=1\}} f(x - \delta_i), \\ \text{and} \quad J_i^0 f(x) &:= (x_i - \tfrac{1}{2}) f(x). \end{aligned} \tag{3.12}$$

It is straightforward to check that

$$[J_i^0, J_j^{\pm}] = \pm\delta_{ij} J_i^{\pm} \quad \text{and} \quad [J_i^-, J_j^+] = -2\delta_{ij} J_i^0. \tag{3.13}$$

It follows that the operators $J_i^{\pm}$ and J_i^0 define a representation of a Lie algebra that consists of a direct sum of copies of $\mathfrak{su}(2)$, with one copy for each site $i \in S$. We can write the generator L of the symmetric exclusion process in terms of the operators $J_i^{\pm}$ and J_i^0 as

$$L = \sum_{\{i,j\}} r(i,j) \big[J_i^- J_j^+ + J_j^- J_i^+ + 2J_i^0 J_j^0 - \tfrac{1}{2} I\big], \tag{3.14}$$

where we are summing over all unordered pairs $\{i, j\}$. We observe that the operators

$$K_i^{\pm} := J_i^{\pm}, \quad \text{and} \quad K_i^0 := -J_i^0 \tag{3.15}$$

satisfy the same commutation relations as $J_i^{\pm}$ and J_i^0, except that each commutation relation gets an extra minus sign. This shows that the operators $K_i^{\pm}$ and K_i^0 define a representation of the conjugate Lie algebra $\overline{\mathfrak{su}(2)}$. Moreover, we can alternatively write the generator in (3.14) as

$$L = \sum_{\{i,j\}} r(i,j)\big[K_j^+ K_i^- + K_i^+ K_j^- + 2K_j^0 K_i^0 - \tfrac{1}{2}I\big]. \tag{3.16}$$

We recall from Subsection 2.3 that two irreducible representations of $\mathfrak{su}(2)$ with the same dimension are necessarily equivalent. In view of this, we conjecture that there should exist an intertwiner D, unique up to a multiplicative constant, such that $J_i^{\pm} D = D(K_i^{\pm})^{\dagger}$ and $J_i^0 D = D(K_i^0)^{\dagger}$ for all i. By the general principle in Proposition 3.1, such an intertwiner is a self-duality function for the symmetric exclusion process.

We observe that all our operators act on the space of all complex functions on $\{0,1\}^S$, which in view of (2.26) is given by

$$\mathbb{C}^{\{0,1\}^S} \cong \bigotimes_{i \in S} \mathbb{C}^{\{0,1\}}. \tag{3.17}$$

For example, if $S = \{1, 2, 3\}$ consists of only three sites, then in line with (2.29),

$$J_1^0 = J^0 \otimes I \otimes I, \quad J_2^0 = I \otimes J^0 \otimes I, \quad \text{and} \quad J_3^0 = I \otimes I \otimes J^0,$$

and similarly for $J_1^{\pm}$, $J_2^{\pm}$, and $J_3^{\pm}$. Here

$$\begin{aligned}
J^- f &= \begin{pmatrix} 0 & 0 \\ 1 & 0 \end{pmatrix} \begin{pmatrix} f(1) \\ f(0) \end{pmatrix} = \begin{pmatrix} 0 \\ f(1) \end{pmatrix}, \\
J^+ f &= \begin{pmatrix} 0 & 1 \\ 0 & 0 \end{pmatrix} \begin{pmatrix} f(1) \\ f(0) \end{pmatrix} = \begin{pmatrix} f(0) \\ 0 \end{pmatrix}, \\
J^0 f &= \begin{pmatrix} \frac{1}{2} & 0 \\ 0 & -\frac{1}{2} \end{pmatrix} \begin{pmatrix} f(1) \\ f(0) \end{pmatrix} = \begin{pmatrix} \frac{1}{2} f(1) \\ -\frac{1}{2} f(0) \end{pmatrix}.
\end{aligned} \tag{3.18}$$

We equip $\mathbb{C}^{\{0,1\}}$ and the space in (3.17) with the standard inner product, which has the consequence that $A^* = A^{\dagger}$ and

$$(J_i^-)^* = J_i^+, \quad (J_i^+)^* = J_i^-, \quad \text{and} \quad (J_i^0)^* = J_i^0,$$

showing that the operators $J_i^\pm$ and J_i^0 define a unitary representation of our Lie algebra.

According to the general principle (2.30), to find an intertwiner D which acts on the product space (3.17), it suffices to find an intertwiner for the two-dimensional space corresponding to a single site, and then take the product over all sites. Setting

$$Q := \begin{pmatrix} 0 & 1 \\ 1 & 0 \end{pmatrix},$$

it is straightforward to check that

$$J^\pm Q = QJ^\mp = Q(K^\pm)^\dagger \quad \text{and} \quad J^0 Q = Q(-J^0) = Q(K^0)^\dagger.$$

Now, for example, if $S = \{1,2,3\}$ consists of only three sites, then in view of (2.30)

$$D := Q \otimes Q \otimes Q \quad \text{satisfies} \quad J_i^\pm D = D(K_i^\pm)^\dagger \quad \text{and} \quad J_i^0 D = D(K_i^0)^\dagger$$

$(i = 1,2,3)$. In terms of matrix elements, we have $Q(x_i, y_j) = 1_{\{x_i \neq y_i\}}$ and hence the self-duality function of the symmetric exclusion process that we have found is

$$D(x,y) = \prod_{i\in S} 1_{\{x_i \neq y_i\}} \qquad (x, y \in \{0,1\}^S).$$

3.4. *The symmetric inclusion process*

Let S be a finite set and let $\alpha : S \to (0,\infty)$ and $q : S \times S \to [0,\infty)$ be functions such that $q(i,j) = q(j,i)$ and $q(i,i) = 0$ for each $i \in S$. By definition, the *Brownian energy process* or *BEP* with parameters α, q is a multidimensional diffusion processes $(Z_t)_{t\geq 0}$ with state space $[0,\infty)^S$ and generator

$$L := \tfrac{1}{2} \sum_{i,j\in S} q(i,j)\big[(\alpha_j z_i - \alpha_i z_j)(\tfrac{\partial}{\partial z_j} - \tfrac{\partial}{\partial z_i}) + z_i z_j (\tfrac{\partial}{\partial z_j} - \tfrac{\partial}{\partial z_i})^2\big]. \tag{3.19}$$

This generalizes formula (89) of [7] which arises as the image under a certain map of another model that was introduced in formula (2.4) of [6] with the aim of studying heat conduction. The diffusion in (3.19) has the property that $\sum_i Z_t(i)$ is a preserved quantity. The drift part of the generator is zero if $z_i = \lambda\alpha_i$ for some $\lambda > 0$. If $z_i/\alpha_i > z_j/\alpha_j$, then the drift has the tendency to make z_i smaller and z_j larger.

In analogy with (2.13), we define operators acting on smooth functions $f : [0,\infty)^S \to \mathbb{R}$ by:

$$\begin{aligned} \mathcal{K}_i^- f(z) &= z_i \tfrac{\partial^2}{\partial z_i{}^2} f(z) + \alpha_i \tfrac{\partial}{\partial z_i} f(z), \\ \mathcal{K}_i^+ f(z) &= z_i f(z), \\ \mathcal{K}_i^0 f(z) &= z_i \tfrac{\partial}{\partial z_i} f(z) + \tfrac{1}{2}\alpha_i f(z). \end{aligned} \tag{3.20}$$

By (2.14), these operators satisfy the commutation relations

$$[\mathcal{K}_i^0, \mathcal{K}_j^\pm] = \pm\delta_{ij}\mathcal{K}_i^\pm \quad \text{and} \quad [\mathcal{K}_i^-, \mathcal{K}_j^+] = 2\delta_{ij}\mathcal{K}_i^0.$$

It follows that these operators define a representation of the Lie algebra

$$\bigoplus_{i \in S} \mathfrak{g}_i,$$

where each $\mathfrak{g}_i$ is a copy of $\mathfrak{su}(1,1)$, on the product space

$$\mathbb{C}^{[0,\infty)^S} \cong (\mathbb{C}^{[0,\infty)})^{\otimes S},$$

which is the tensor product of $|S|$ copies of $\mathbb{C}^{[0,\infty)}$.

We can express the generator (3.19) of the Brownian energy process in terms of the operators from (3.20) as

$$L = \tfrac{1}{2} \sum_{i,j \in S} q(i,j)\big[\mathcal{K}_i^+\mathcal{K}_j^- + \mathcal{K}_i^-\mathcal{K}_j^+ - 2\mathcal{K}_i^0\mathcal{K}_j^0 + \tfrac{1}{2}\alpha_i\alpha_j\big]. \tag{3.21}$$

Note that this is very similar to the expression for the symmetric exclusion process in (3.14).

We define operators acting on functions $f : \mathbb{N}^S \to \mathbb{R}$ by

$$\begin{aligned} K_i^- f(x) &= x_i f(x - \delta_i), \\ K_i^+ f(x) &= (\alpha_i + x_i) f(x + \delta_i), \\ K_i^0 f(x) &= (\tfrac{1}{2}\alpha_i + x_i) f(x). \end{aligned} \tag{3.22}$$

In view of (2.16), these operators define a representation of the conjugate of our Lie algebra. It turns out that the conjugate of this representation is equivalent to the representation defined by the operators in (3.20). This is a nontrivial statement that depends crucially on the fact that the parameters α_i are the same in both expressions. Indeed, we have seen in Subsection 2.4 that α is twice the Bargmann index and that representations with a different Bargmann index have a different Casimir operator and hence are not equivalent. Letting Φ denote the intertwiner of $\mathcal{K}_i^\pm$ and $(K_i^\pm)^\dagger$, we can write Φ in the form (3.9), where by Proposition 3.2 D is a duality function.

Similar to what we did at the end of Subsection 3.3, we will choose a duality function of product form:

$$D(z,x) = \prod_{i\in S} Q(z_i,x_i) \qquad (z \in [0,\infty)^S,\ x \in \mathbb{N}^S), \tag{3.23}$$

where Q is a duality function for the single-site operators, i.e.,

$$\mathcal{K}^{\pm}Q(\,\cdot\,,x)(z) = K^{\pm}Q(z,\,\cdot\,)(x), \quad \mathcal{K}^{0}Q(\,\cdot\,,x)(z) = K^{0}Q(z,\,\cdot\,)(x) \tag{3.24}$$

$(z \in [0,\infty),\ x \in \mathbb{N})$. It turns out that

$$Q(z,x) := \frac{\Gamma(\alpha)}{\Gamma(\alpha+x)} z^x = z^x \prod_{k=0}^{x-1} (\alpha+k)^{-1}. \tag{3.25}$$

does the trick. This may look a bit complicated but the form of this duality function can in fact quite easily be guessed from the inductive relation

$$zQ(z,x) = \mathcal{K}^{+}Q(\,\cdot\,,x)(z) = K^{+}Q(z,\,\cdot\,)(x) = (\alpha+x)Q(z,x+1).$$

Our calculations so far imply that the generator in (3.21) is dual with respect to the duality function in (3.23)–(3.25) to the generator

$$\hat{L} = \tfrac{1}{2} \sum_{i,j\in S} q(i,j)\big[K_j^- K_i^+ + K_j^+ K_i^- - 2K_j^0 K_i^0 + \tfrac{1}{2}\alpha_j\alpha_i\big]. \tag{3.26}$$

It turns out that we are lucky in the sense that this is a Markov generator. In view of the similarity with (3.14) (with the role of $\mathfrak{su}(2)$ replaced by $\mathfrak{su}(1,1)$), in [8], the corresponding process has been called the *symmetric inclusion process* or *SIP*. The fact that $\hat{L}$ is a Markov generator can be seen by rewriting it as

$$\begin{aligned} L := \sum_{i,j\in S} q(i,j)\Big[&\alpha_j x_i\{f(x-\delta_i+\delta_j) - f(x)\} \\ &+ x_i x_j\{f(x-\delta_i+\delta_j) - f(x)\}\Big]. \end{aligned} \tag{3.27}$$

The Markov process $(X_t)_{t\geq 0}$ with generator $\hat{L}$ has the property that $\sum_i X_t(i)$ is a preserved quantity. The terms in the generator involving the constants α_j describe a system of independent random walks, where each particle at i jumps with rate α_j to the site j. A reversible law for this part of the dynamics is a Poisson field with local intensity $\lambda\alpha_i$ for some $\lambda > 0$. The remaining terms in the generator describe a dynamics where particles at i jump to j with a rate that is proportional to the number $x(j)$ of particles at j. This part of the dynamics causes an attraction between particles.

3.5. *Duality functions of product form*

In the previous two subsections, we have seen that for a Markov process whose state space is a Carthesian product of other spaces, it is often natural to choose duality functions of *product form* as in (3.23). This idea does not depend on Lie algebras and is in fact older than the use of Lie algebras in duality theory.

In a series of papers [15, 16, 17], Lloyd and Sudbury have systematically searched for dualities in a large class of interacting particle systems, which contains many well-known systems such as the voter model, contact process, and symmetric exclusion process. Let S be a finite set and let $q : S^2 \to [0, \infty)$ be a function such that $q(i,j) = q(j,i)$ and $q(i,i) = 0$ for all $i \in S$. On the state space $\{0,1\}^S$,[1] let $L = L(a,b,c,d,e)$ be the Markov generator, acting on functions $f : \{0,1\}^S \to \mathbb{R}$, as

$$\begin{aligned} Lf(x) = \sum_{i,j \in S} q(i,j) \Big[&\tfrac{1}{2} a 1_{\{(x(i),x(j))=(1,1)\}} \{f(x - \delta_i - \delta_j) - f(x)\} \\ &b 1_{\{(x(i),x(j))=(0,1)\}} \{f(x + \delta_i) - f(x)\} \\ &c 1_{\{(x(i),x(j))=(1,1)\}} \{f(x - \delta_i) - f(x)\} \\ &d 1_{\{(x(i),x(j))=(0,1)\}} \{f(x - \delta_j) - f(x)\} \\ &e 1_{\{(x(i),x(j))=(0,1)\}} \{f(x + \delta_i - \delta_j) - f(x)\} \Big]. \end{aligned} \tag{3.28}$$

The dynamics of the Markov process with generator L can be described by saying that for each pair of sites i, j, the configuration of the process at these sites makes the following transitions with the following rates:

$$\begin{array}{llll} 11 \mapsto 00 & \text{with rate} & aq(i,j) & \text{(annihilation)}, \\ 01 \mapsto 11 & \text{with rate} & bq(i,j) & \text{(branching)}, \\ 11 \mapsto 01 & \text{with rate} & cq(i,j) & \text{(coalecence)}, \\ 01 \mapsto 00 & \text{with rate} & dq(i,j) & \text{(death)}, \\ 01 \mapsto 10 & \text{with rate} & eq(i,j) & \text{(exclusion dynamics)}. \end{array}$$

Note that the factor $\frac{1}{2}$ in front of a disappears since the total rate of this transition is $\frac{1}{2}a(q(i,j) + q(j,i)) = aq(i,j)$. A lot of well-known interacting

[1]In principle the set $\{0,1\}$ could be replaced by a finite set A. However, the combinatorial effort quickly explodes. For example, the duality matrix Q in (3.29) would change from a 2×2 matrix to an $|A| \times |A|$ matrix, making the set of potential duality matrices vastly larger.

particle systems fall into this class. For example

$$\begin{aligned} &\text{voter model} \quad b = d = 1, \text{ other parameters } 0, \\ &\text{contact process} \quad b = \lambda, \ c = d = 1, \text{ other parameters } 0, \\ &\text{symmetric exclusion} \quad e = 1, \text{ other parameters } 0. \end{aligned}$$

As we have already seen in (3.17), the class of all functions $f : \{0,1\}^S \to \mathbb{R}$ can be written as the tensor product

$$\mathbb{R}^{\{0,1\}^S} \cong \bigotimes_{i \in S} \mathbb{R}^{\{0,1\}},$$

with one 'factor' $\mathbb{R}^{\{0,1\}}$ for each site $i \in S$. Moreover, duality functions D on the space $\{0,1\}^S \times \{0,1\}^S$ can be viewed as matrices corresponding to linear operators that act on $\mathbb{R}^{\{0,1\}^S}$. Based on various arguments that are not very important at this point, Lloyd and Sudbury decided to look for duality functions of *product form*

$$D(x, y) = \prod_{i \in S} Q(x_i, y_i), \tag{3.29}$$

where Q is a 2×2 matrix. After a more or less systematic search for suitable matrices Q, Lloyd and Sudbury found a rich class of dualities for matrices of the form

$$\begin{pmatrix} Q_q(0,0) \ Q_q(0,1) \\ Q_q(1,0) \ Q_q(1,1) \end{pmatrix} = \begin{pmatrix} 1 \ 1 \\ 1 \ q \end{pmatrix}, \tag{3.30}$$

where $q \in \mathbb{R} \backslash \{1\}$ is a constant. This choice of Q yields the duality function

$$D_q(x, y) := \prod_{i \in S} Q_q(x_i, y_i) = q^{\sum_{i \in S} x_i y_j} \qquad (x, y \in \{0,1\}^S). \tag{3.31}$$

In particular, setting $q = 0$ yields

$$D_0(x, y) = 1_{\{\sum_{i \in S} x_i y_j = 0\}},$$

which corresponds to the well-known *additive systems duality*, while $q = -1$ is known as *cancellative systems duality*. For these special values of q, the duality can in fact be upgraded to a pathwise duality as in Subsection 1.4, using a construction in terms of open paths in a graphical representation; see for example [1] and [10] as well as [14] for more detail. Interestingly, for other values of q, there seems to be no pathwise interpretation of the duality with duality function D_q.

We cite the following theorem from [15, 17]. A somewhat more general version of this theorem which drops the symmetry assumption $q(i, j) =$

$q(j,i)$ at the cost of replacing (3.32) by a somewhat more complicated set of conditions can be found in Appendix A of the version on the ArXiv of [18].

Theorem 3.3: (q-duality) *The generators* $L(a,b,c,d,e)$ *and* $L(a',b',c',d',e')$ *from (3.28) are dual with respect to the duality function* D_q *from (3.31) if and only if*

$$a' = a + 2q\gamma, \quad b' = b + \gamma, \quad c' = c - (1+q)\gamma, \quad d' = d + \gamma, \quad e' = e - \gamma, \tag{3.32}$$

where $\gamma := (a + c - d + qb)/(1-q)$.

3.6. *Intertwining and thinning*

In Subsection 1.2, when we introduced Markov process duality, we also defined the similar concept of intertwining of Markov processes. So far, we have not discussed this second concept very much, but it turns out that the two are closely related. In particular, as Lloyd and Sudbury already observed [15, 17], there is a close connection between q-duality and thinning relations. To explain this, we start with a general principle, that says that if two Markov processes are both dual to a third Markov process, then we can expect an intertwining relation between the first two processes.

Lemma 3.4: (Duality and intertwining) *Let* Ω *and* $\hat{\Omega}$ *be finite sets, and let* $L_i : \mathbb{R}^\Omega \to \mathbb{R}^\Omega$, $\hat{L} : \mathbb{R}^{\hat{\Omega}} \to \mathbb{R}^{\hat{\Omega}}$, *and* $D_i : \mathbb{R}^{\hat{\Omega}} \to \mathbb{R}^\Omega$ *be linear operators such that*

$$L_i D_i = D_i \hat{L}^\dagger \qquad (i = 1, 2). \tag{3.33}$$

Assume that D_1 *and* D_2 *are invertible. Then*

$$L_1(D_1 D_2^{-1}) = (D_1 D_2^{-1}) L_2. \tag{3.34}$$

Proof: This follows by writing $D_1^{-1} L_1 D_1 = \hat{L}^\dagger = D_2^{-1} L_2 D_2$. □

We have seen that for interacting particle systems, there are good reasons to look for duality functions of product form as in (3.29). Likewise, it is natural to look for intertwining probability kernels of product form. If the state space is of the form $\{0,1\}^S$, this means that we are looking for kernels of the form

$$K(x,y) = \prod_{i \in S} M(x_i, y_i) \qquad (x, y \in \{0,1\}^S),$$

where M is a probability kernel on $\{0,1\}$. If we moreover require that $M(0,0)=1$ (which is natural for interacting particle systems for which the all zero state is a trap), then there is only a one-parameter family of such kernels. For $p\in[0,1]$, let M_p be the probability kernel on $\{0,1\}$ given by

$$M_p = \begin{pmatrix} M_p(0,0) & M_p(0,1) \\ M_p(1,0) & M_p(1,1) \end{pmatrix} := \begin{pmatrix} 1 & 0 \\ 1-p & p \end{pmatrix}, \tag{3.35}$$

and let

$$K_p(x,y) := \prod_{i\in S} M_p(x_i,y_i) \qquad (x,y\in\{0,1\}^S) \tag{3.36}$$

be the corresponding kernel on $\{0,1\}^S$ of product form. We can interpret $x\in\{0,1\}^S$ as a configuration of particles on S, where $x_i=1$ if the site $i\in S$ is occupied by a particle, and $x_i=0$ otherwise. Then K_p is a *thinning* kernel that independently for each site throws away particles with probability $1-p$ or keeps them with probability p. It is easy to see that

$$K_pK_{p'} = K_{pp'},$$

i.e., first thinning with p and then with p' is the same as thinning with pp'. There is a close relation between Lloyd and Sudbury's duality function D_q from (3.31) and thinning kernels of the form (3.36). We claim that

$$D_qD_{q'}^{-1} = K_p \quad \text{with} \quad p = \frac{1-q}{1-q'} \qquad (q,q'\in\mathbb{R},\ q'\neq 1). \tag{3.37}$$

Since both D_q and K_p are of product form, i.e.,

$$D_q = \bigotimes_{i\in S} Q_q \quad \text{and} \quad K_p = \bigotimes_{i\in S} M_p$$

with Q_q and M_p as in (3.30) and (3.35), it suffices to check that

$$Q_qQ_{q'}^{-1} = M_p \quad \text{with} \quad p = \frac{1-q}{1-q'}.$$

Indeed, one can check that

$$Q_q^{-1} = \begin{pmatrix} 1 & 1 \\ 1 & q \end{pmatrix}^{-1} = (1-q)^{-1}\begin{pmatrix} -q & 1 \\ 1 & -1 \end{pmatrix} \qquad (q\neq 1),$$

and that

$$Q_qQ_{q'}^{-1} = (1-q')^{-1}\begin{pmatrix} 1 & 1 \\ 1 & q \end{pmatrix}\begin{pmatrix} -q' & 1 \\ 1 & -1 \end{pmatrix} = \begin{pmatrix} 1 & 0 \\ \frac{q-q'}{1-q'} & \frac{1-q}{1-q'} \end{pmatrix} = M_p,$$

as claimed.

Proposition 3.5: (Thinning and q-duality) *Let L_1 and L_2 be generators of Markov processes with state space $\{0,1\}^S$. Assume that there exists an operator $\hat{L}$ such that*

$$L_i D_{q_i} = D_{q_i} \hat{L}^\dagger \qquad (i = 1, 2) \tag{3.38}$$

for some $q_1, q_2 \in \mathbb{R}$ such that $q_2 \neq 1$ and $p := (1 - q_1)/(1 - q_2) \in [0, 1]$. Then

$$L_1 K_p = K_p L_2. \tag{3.39}$$

Proof: This follows from (3.37) and Lemma 3.4. Note that in general, there is no guarantee that the operator $D_1 D_2^{-1}$ from Lemma 3.4 is a probability kernel. In a way, Proposition 3.5 explains why the q-duality function D_q is natural, because it is closely linked to the natural concept of thinning. □

3.7. *The biased voter model*

In this section, we demonstrate Lloyd-Sudbury theory on the example of the *biased voter model* with *selection parameter* $s > 0$, which is the interacting particle system with generator

$$L(a, b, c, d, e) = L(0, 1 + s, 0, 1, 0) =: L_{\text{bias}}.$$

We apply Theorem 3.3 to find q-duals of the biased voter model. For simplicity, we restrict ourselves here to dual generators of the form $L(a', b', c', d', e')$ with $a' = 0$, which means that we must choose the parameter q as

$$q = 0 \quad \text{or} \quad q = (1 + s)^{-1}.$$

For $q = 0$ we find the dual generator

$$L(a', b', c', d', e') = L(0, s, 1, 0, 1) =: L_{\text{braco}},$$

which describes a system of branching and coalescing random walks with branching parameter s. For $q = (1 + s)^{-1}$, we find a self-duality, i.e., in this case $L(a', b', c', d', e') = L(a, b, c, d, e) = L_{\text{bias}}$.

Since L_{bias} and L_{braco} are both q-dual to $\hat{L} = L_{\text{bias}}$, Proposition 3.5 tells us that there is a thinning relation between biased voter models and systems of branching and coalescing random walks of the form

$$L_{\text{bias}} K_p = K_p L_{\text{braco}} \quad \text{with} \quad p = \frac{1 - (1 + s)^{-1}}{1 - 0} = \frac{s}{1 + s}.$$

As explained in Subsection 3.6, this implies that if we start a biased voter model $(X_t)_{t\geq 0}$ and a system of branching and coalescing random walks $(Y_t)_{t\geq 0}$ in initial states $\mu_t^{\rm bias}$ and $\mu_t^{\rm braco}$ denote the laws of X_t and Y_t, then

$$\mu_0^{\rm bias} K_p = \mu_0^{\rm braco} \quad \text{implies} \quad \mu_t^{\rm bias} K_p = \mu_t^{\rm braco} \quad (t \geq 0).$$

In other words, the following two procedures are equivalent:

(i) Evolve a particle configuration for time t according to biased voter model dynamics, then thin with p.
(ii) Thin a particle configuration with p, then evolve for time t according to branching coalescing random walk dynamics.

In particular, if we start X in the initial state $X_0(i) = 1$ for all $i \in S$, then because of the nature of the voter model, we will have $X_t(i) = 1$ for all $i \in S$ and $t \geq 0$. Applying the thinning relation now shows that product measure with intensity p is an invariant law for branching coalescing random walk dynamics. Thus, there is a close connection between:

I. q-duality,
II. thinning relations,
III. invariant laws of product form.

Although Lloyd-Sudbury theory is restricted to Markov processes with state space of the form $\{0,1\}^S$, many other dualities, including the self-duality of the Wright-Fisher diffusion from Section 3.1, can be derived from Lloyd-Sudbury duals by taking a suitable limit [18].

3.8. *Time-reversal and symmetry*

In this subsection we present an idea from [7], which says that nontrivial dualities can sometimes be found by starting from a "trivial" duality which is based on time reversal, and then using a symmetry of the model to transform such a duality into a nontrivial one. Although Lie algebras are not strictly needed in this approach, writing generators in terms of the basis elements of a representation of a Lie algebra can help finding suitable symmetries.

Each irreducible Markov process with finite state space Ω has a unique invariant measure, i.e., a probability measure μ such that

$$\mu L = 0 \quad \text{or equivalently} \quad \mu P_t = \mu \quad (t \geq 0),$$

where L denotes the generator and $(P_t)_{t\geq 0}$ the semigroup of the Markov process. Irreducibility implies that $\mu(x) > 0$ for all $x \in \Omega$. Letting $(X_t)_{t\in\mathbb{R}}$

denote the stationary process, we see that the semigroup $(\tilde{P}_t)_{t\geq 0}$ of the time-reversed process is given by

$$\begin{aligned}\tilde{P}_t(x,y) &= \frac{\mathbb{P}[X_0 = y,\ X_t = x]}{\mathbb{P}[X_t = x]}\\ &= \frac{\mu(y)P_t(y,x)}{\mu(x)} = \mu(y)P_t(y,x)\mu(x)^{-1} \quad (t \geq 0).\end{aligned}$$

Differentiating shows that the generator $\tilde{L}$ of the time-reversed process is given by[m]

$$\tilde{L}(x,y) = \mu(y)L(y,x)\mu(x)^{-1}.$$

Let R denote the diagonal matrix

$$R(x,y) := \delta_{x,y}\mu(x)^{-1}.$$

Then $L(y,x)\mu(x)^{-1} = \tilde{L}(x,y)\mu(y)^{-1} = \mu(y)^{-1}\tilde{L}^{\dagger}(y,x)$ can be rewritten as

$$LR = R\tilde{L}^{\dagger},$$

which shows that $\tilde{L}$ is dual to L with duality function R. In particular, reversible processes (for which $\tilde{L} = L$) are always *self-dual* with duality function $R(x,y)$. Note that since R is diagonal, it is invertible with

$$R^{-1}(x,y) := \delta_{x,y}\mu(x) \qquad (x,y \in \Omega).$$

Let V be a finite dimensional complex linear space and let $L : V \to V$ be any linear operator (not necessarily a Markov generator). Then it is known that there exists an invertible matrix $Q \in \mathcal{L}(V)$ such that

$$LQ = QL^{\dagger} \quad \text{or equivalently} \quad L^{\dagger}Q^{-1} = Q^{-1}L. \tag{3.40}$$

Thus, *every* finite dimensional linear operator is self-dual and the self-duality function Q can be chosen such that it is invertible, viewed as a matrix. Let

$$\mathcal{C}_L := \{A \in \mathcal{L}(V) : AL = LA\}$$

be the algebra of all elements of $\mathcal{L}(V)$ that commute with L. We call this the space of *symmetries* of L. In Thm 2.6 of [7], the following simple observation is made.

Lemma 3.6: (Self-duality functions) *Let L be a linear operator on some finite dimensional linear space V. Fix some Q as in (3.40). Then the set of all self-duality functions of L is given by*

$$\{SQ : S \in \mathcal{C}_L\}.$$

[m]This formula is wrong below (12) in [7].

Proof: Clearly, if $S \in \mathcal{C}_L$, then

$$LSQ = SLQ = SQL^\dagger,$$

showing that SQ is a self-duality function. Conversely, if D is a self-duality function, then we can write $D = SQ$ with $S = DQ^{-1}$. Now, since D is a self-duality function,

$$SL = DQ^{-1}L = DL^\dagger Q^{-1} = LDQ^{-1} = LS,$$

which shows that $S \in \mathcal{C}_L$. □

For dualities, we can play a similar game. Once we have two operators $L, \hat{L}$ that are dual with duality function D, i.e.,

$$LD = D\hat{L}^\dagger,$$

we have that for any $S \in \mathcal{C}_L$, the operators $L, \hat{L}$ are also dual with duality function SD, as follows by writing

$$LSD = SLD = SD\hat{L}^\dagger.$$

If D is invertible, then every duality function of L and $\hat{L}$ is of this form. Indeed, if $\tilde{D}$ is any duality function, then we can write $\tilde{D} = SD$ with $S = \tilde{D}D^{-1}$. Now

$$SL = \tilde{D}D^{-1}L = \tilde{D}L^\dagger D^{-1} = L\tilde{D}D^{-1} = LS,$$

proving that $S \in \mathcal{C}_L$. See also Thm 2.10 of [7].

3.9. *The symmetric exclusion process revisited*

Following Sect. 3.1 of [7], we demonstrate the principles explained in the previous subsections to derive a self-duality of the symmetric exclusion process. Our starting point is formula (3.14), which expresses the generator L in terms of operators $J_i^\pm, J_i^0$ that define a representation (V, π) of a Lie algebra $\mathfrak{g}$ that is the direct sum of finitely many copies of the Lie algebra $\mathfrak{su}(2)$, with one copy for each site $i \in S$. Since $r(i,j) = r(j,i)$, we can rewrite this formula as

$$L = \tfrac{1}{2}\sum_{i,j} r(i,j)\big[J_i^- J_j^+ + J_j^- J_i^+ + 2J_i^0 J_j^0 - \tfrac{1}{2}I\big]. \qquad (3.41)$$

A straightforward calculation shows that

$$\sum_k [J_k^\pm, L] = 0 \quad \text{and} \quad \sum_k [J_k^0, L] = 0 \qquad (k \in S). \qquad (3.42)$$

We need a bit of general theory. If U, V, W are representations of the same Lie algebra $\mathfrak{g}$, then we can equip their tensor product $U \otimes V \otimes W$ with the structure of a representation of $\mathfrak{g}$ by putting

$$A(u \otimes v \otimes w) := Au \otimes v \otimes w + u \otimes Av \otimes w + u \otimes v \otimes Aw \qquad (A \in \mathfrak{g}), \tag{3.43}$$

and similar for the tensor product of any finite number of representations, see formula (A.13) in the appendix. This definition also naturally equips $U \otimes V \otimes W$ with the structure of a representation of the Lie group G associated with $\mathfrak{g}$, in such a way that

$$e^{tA}(u \otimes v \otimes w) = e^{tA}u \otimes e^{tA}v \otimes e^{tA}w \qquad (A \in \mathfrak{g},\ t \geq 0),$$

where for each $A \in \mathfrak{g}$, the operator e^{tA} is an element of the Lie group G associated with $\mathfrak{g}$. Thus, the representation (3.43) corresponds to letting the Lie group act in the same way on each space in the tensor product.

In our specific set-up, this means that the operators K^-, K^+, K^0 defined by

$$K^- := \sum_k J_k^-, \quad K^+ := \sum_k J_k^+, \quad K^0 := \sum_k J_k^0 \tag{3.44}$$

define a representation of $\mathfrak{su}(2)$ on the product space

$$\mathbb{C}^{\{0,1\}^S} \cong \bigotimes_{i \in S} \mathbb{C}^{\{0,1\}}.$$

(Indeed, one can check that K^-, K^+, K^0 satisfy the commutation relations of $\mathfrak{su}(2)$.) Let $c_- K^- + c_+ K^+ + c_0 K^0$ be an operator in the linear space spanned by K^-, K^+, K^0. Then

$$e^{t(c_- K^- + c_+ K^+ + c_0 K^0)} = \bigotimes_{i \in S} e^{t(c_- J^- + c_+ J^+ + c_0 J^0)} \qquad (t \geq 0), \tag{3.45}$$

i.e., a natural group of symmetries of the generator L is formed by all operators of the form (3.45) and their products, and this actually corresponds to a representation of the Lie group SU(2).

We take this as our motivation to look at one specific operator of the form (3.45), which is e^{K^+}. One can check that the uniform distribution is an invariant law for the exclusion process, so by the principle of Subsection 3.8, the function

$$D(x, y) = 1_{\{x=y\}} = \prod_{i \in S} 1_{\{x_i = y_i\}}$$

is a trivial self-duality function. Applying Lemma 3.6 to the symmetry $S = e^{K^+}$, we see that $SD = SI = S$ is also a self-duality function. Since S

factorizes over the sites, it suffices to calculate S for a single site, and then take the product. We recall from (3.18) that

$$J^+ f \begin{pmatrix} 0 & 1 \\ 0 & 0 \end{pmatrix} \begin{pmatrix} f(1) \\ f(0) \end{pmatrix} = \begin{pmatrix} f(0) \\ 0 \end{pmatrix},$$

which gives

$$e^{J^+} = \sum_{n=0}^{\infty} \frac{1}{n!}(J^+)^n = I + J^+ = \begin{pmatrix} 1 & 1 \\ 0 & 1 \end{pmatrix}$$

and finally yields the duality function

$$S(x, y) = \prod_{i \in S} 1_{\{x_i \geq y_i\}} \qquad (x, y \in \{0, 1\}^S).$$

A. A crash course in Lie algebras

A.1. *Lie groups*

In the present appendix, we give a bit more background on Lie algebras. In particular, we explain how Lie algebras are closely linked to Lie groups, and how every Lie algebra can naturally be embedded in an algebra, called the universal enveloping algebra. We also explain how properties of the Lie group (in particular, compactness) are related to representations of its associated Lie algebra.

A *group* is a set G which contains a special element I, called the *identity*, and on which a *group product* $(A, B) \mapsto AB$ and *inverse operation* $A \mapsto A^{-1}$ are defined such that

(i) $IA = AI = A$
(ii) $(AB)C = A(BC)$
(iii) $A^{-1}A = AA^{-1} = I$.

A group is *abelian* (also called *commutative*) if $AB = BA$ for all $A, B \in G$. A *group homomorphism* is a map Φ from one group G into another group H that preserves the group structure, i.e.,

(i) $\Phi(I) = I$,
(ii) $\Phi(AB) = \Phi(A)\Phi(B)$,
(iii) $\Phi(A^{-1}) = \Phi(A)^{-1}$.

If Φ is a bijection, then Φ^{-1} is also a group homomorphism. In this case, we call Φ a *group isomorphism*. A *subgroup* of a group G is a subset $H \subset G$ such that $I \in H$ and H is closed under the product and inverse, i.e., $A, B \in H$

imply $AB \in H$ and $A \in H$ implies $A^{-1} \in H$. A subgroup is in a natural way itself a group.

A *Lie group* is a smooth manifold G which is also a group such that the group product and inverse functions

$$G \times G \ni (A, B) \mapsto AB \in G \quad \text{and} \quad G \ni A \mapsto A^{-1} \in G$$

are smooth. A *finite dimensional representation* of G is a finite dimensional linear space V over $\mathbb{R}$ or $\mathbb{C}$ together with a map

$$G \times V \ni (A, v) \mapsto Av \in V$$

such that

(i) $v \mapsto Av$ is linear,
(ii) $Iv = v$,
(iii) $A(Bv) = (AB)v$.

Letting $\mathcal{L}(V)$ denote the space of all linear operators $A : V \to V$, these conditions are equivalent to saying that the map $\Pi : G \to \mathcal{L}(V)$ defined by

$$\Pi(A)v := Av$$

is a group homomorphism from G into the *general linear group* $\mathrm{GL}(V)$ of all invertible linear maps $A : V \to V$. A representation is *faithful* if Π is one-to-one, i.e., if $A \mapsto \Pi(A)$ is a group isomorphism between G and the subgroup $\Pi(G) := \{\Pi(A) : A \in G\}$ of $\mathrm{GL}(V)$.

One can prove that if G is a Lie group and V is a faithful finite dimensional representation, then $\Pi(G)$ is a closed subset of $\mathrm{GL}(V)$ and $\Pi : G \to \Pi(G)$ is a homeomorphism. Conversely, each closed subgroup of $\mathrm{GL}(V)$ is a Lie group. Such Lie groups are called *matrix Lie groups*. Not every Lie group has a finite dimensional faithful representation, so not every Lie group is a matrix Lie group, but many important Lie groups are matrix Lie groups and following [9] we will mostly focus on them from now on.

A.2. *Lie algebras*

An *algebra* is a finite dimensional linear space $\mathfrak{a}$ over $\mathbb{R}$ or $\mathbb{C}$ with a special element I called *unit element* and on which there is defined a product

$$\mathfrak{a} \times \mathfrak{a} \ni (A, B) \mapsto AB \in \mathfrak{a}$$

such that

(i) $(A, B) \mapsto AB$ is bilinear,
(ii) $IA = AI = A$,
(iii) $(AB)C = A(BC)$.

In some textbooks, algebras are not required to contain a unit element. We speak of a *real* resp. *complex* algebra depending on whether $\mathfrak{a}$ is a linear space over $\mathbb{R}$ or $\mathbb{C}$. An algebra is *abelian* if $AB = BA$ for all $A, B \in G$. In any algebra, the *commutator* of two elements A, B is defined as $[A, B] = AB - BA$. If V is a linear space, then $\mathcal{L}(V)$ is an algebra.

An *algebra homomorphism* is a map $\phi : \mathfrak{a} \to \mathfrak{b}$ from one algebra into another that preserves the structure, i.e.,

(i) ϕ is linear,
(ii) $\phi(I) = I$,
(iii) $\phi(AB) = \phi(A)\phi(B)$.

Algebra homomorphisms that are bijections have the property that ϕ^{-1} is also a homomorphism; these are called algebra isomorphisms. A *subalgebra* of an algebra $\mathfrak{a}$ is a linear subspace $\mathfrak{b} \subset \mathfrak{a}$ that contains I and is closed under the product.

Lie algebras, *Lie algebra homomorphisms*, and *isomorphisms* have already been defined in Section 2.1. A *sub-Lie-algebra* is a linear subspace $\mathfrak{h} \subset \mathfrak{g}$ such that

$$A, B \in \mathfrak{h} \quad \text{implies} \quad [A, B] \in \mathfrak{h}.$$

If $\mathfrak{g}$ is an algebra, then $\mathfrak{g}$, equipped with the commutator map $[\cdot, \cdot]$, is a Lie algebra. As the example in Section 2.1 shows, Lie algebras need not be algebras.

A *representation* of an algebra $\mathfrak{a}$ is a linear space V together with a map $\mathfrak{a} \times V \to V$ that satisfies

(i) $(A, v) \mapsto Av$ is bilinear,
(ii) $Iv = v$,
(iii) $A(Bv) = (AB)v$.

If $\mathfrak{a}$ is a complex algebra, then we require V to be a linear space over $\mathbb{C}$, but even when $\mathfrak{a}$ is a real algebra, it is often useful to allow for the case that V is a linear space over $\mathbb{C}$. In this case, bilinearity means real linearity in the first argument and complex linearity in the second argument. We speak of *real* or *complex* representations depending on whether V is a linear space over $\mathbb{R}$ or $\mathbb{C}$.

A representation V of an algebra $\mathfrak{a}$ gives in a natural way rise to an algebra homomorphism $\pi : \mathfrak{a} \to \mathcal{L}(V)$ defined as

$$\pi(A)v := Av \qquad (A \in \mathfrak{a},\ v \in V).$$

Conversely, given an algebra homomorphism $\pi : \mathfrak{a} \to \mathcal{L}(V)$ we can equip V with the structure of a representation by defining $Av := \pi(A)v$. Thus, a representation V of an algebra $\mathfrak{a}$ is equivalent to a pair (V, π) where V is a linear space and $\pi : \mathfrak{a} \to \mathcal{L}(V)$ is an algebra homomorphism. A representation (V, π) is *faithful* if π is an isomorphism between $\mathfrak{a}$ and the subalgebra $\pi(\mathfrak{a}) = \{\pi(A) : A \in \mathfrak{a}\}$ of $\mathcal{L}(V)$.

Representations of Lie algebras have already been defined in Section 2.2. If V is a complex representation of a real algebra or Lie algebra $\mathfrak{a}$, then the image of $\mathfrak{a}$ under π is only a real subspace of $\mathcal{L}(V)$. We can define a complex algebra or Lie algebra $\mathfrak{a}_{\mathbb{C}}$ whose elements can formally be written as $A+iB$ with $A, B \in \mathfrak{a}$; this is called the *complexification* of $\mathfrak{a}$. Then π extends uniquely to a homomorphism from $\mathfrak{a}_{\mathbb{C}}$ to $\mathcal{L}(V)$, see Prop. 3.39 of [9], so V is also a representation of $\mathfrak{a}_{\mathbb{C}}$.

Every algebra has a faithful representation. Indeed, $\mathfrak{a}$ together with the map $(A, B) \mapsto AB$ is a representation of itself, and it is not hard to see (using our assumption that $I \in \mathfrak{a}$) that this representation is faithful. Lie algebras can be represented on themselves in a construction that is very similar to the one for algebras.

Lemma A.1: (Lie algebra represented on itself) *A Lie algebra $\mathfrak{g}$, equipped with the map $(A, B) \mapsto [A, B]$, is a representation of itself.*

Proof: It will be convenient to use somewhat different notation for the Lie bracket. If $\mathfrak{g}$ is a Lie algebra and $X \in \mathfrak{g}$, then we define $\mathrm{ad}_X : \mathfrak{g} \to \mathfrak{g}$ by

$$\mathrm{ad}_X(A) := [X, A].$$

We need to show that $\mathfrak{g} \ni X \mapsto \mathrm{ad}_X \in \mathcal{L}(\mathfrak{g})$ is a Lie algebra homomorphism. Bilinearity follows immediately from the bilinear property (i) of the Lie bracket, so it remains to show that

$$\mathrm{ad}_{[X,Y]}(Z) = \mathrm{ad}_X(\mathrm{ad}_Y(Z)) - \mathrm{ad}_Y(\mathrm{ad}_X(Z)).$$

This can be rewritten as

$$[[X, Y], Z] = [X, [Y, Z]] - [Y, [X, Z]].$$

Using also the skew symmetric property (ii) of the Lie bracket, this can be rewritten as

$$0 = [Z, [X, Y]] + [X, [Y, Z]] + [Y, [Z, X]],$$

which is the Jacobi identity. □

In general, representing a Lie algebra on itself as in Lemma A.1 need not yield a faithful representation. (For example, any abelian algebra is also a Lie algebra and for such Lie algebras $\mathrm{ad}_X = 0$ for each X.) By definition, the *center* of a Lie algebra $\mathfrak{g}$ is the set

$$\{X \in \mathfrak{g} : [X, A] = 0 \ \forall A \in \mathfrak{g}\}. \tag{A.1}$$

We say that the center is *trivial* if it contains only the zero element. If $\mathfrak{g}$ has a trivial center, then the representation $X \mapsto \mathrm{ad}_X$ of $\mathfrak{g}$ on itself is faithful. Indeed, $\mathrm{ad}_X = \mathrm{ad}_Y$ implies $[X, A] = [Y, A]$ for all $A \in \mathfrak{g}$ and hence $X - Y$ is an element of the center of $\mathfrak{g}$. If the center is trivial, this implies $X = Y$.

A.3. *Relation between Lie groups and Lie algebras*

Let V be a linear space and let $G \subset \mathrm{GL}(V)$ be a matrix Lie group. By definition, the *Lie algebra* $\mathfrak{g}$ of G is the space of all matrices A such that there exists a smooth curve γ in G with

$$\gamma(0) = I \quad \text{and} \quad \tfrac{\partial}{\partial t}\gamma(t)\big|_{t=0} = A.$$

In manifold terminology, this says that $\mathfrak{g}$ is the *tangent space* to G at I. For any matrix A, we define

$$e^A := \sum_{n=0}^{\infty} \frac{1}{n!} A^n. \tag{A.2}$$

The following lemma follows from Cor. 3.46 of [9]. The main idea behind this lemma is that the elements of the Lie algebra act as "infinitesimal generators" of the Lie group.

Lemma A.2: (Exponential formula) *Let $\mathfrak{g}$ be the Lie algebra of a Lie group $G \subset \mathrm{GL}(V)$. Then the following conditions are equivalent.*

(i) $A \in \mathfrak{g}$,
(ii) $e^{tA} \in G$ *for all* $t \in \mathbb{R}$.

The following lemma (a precise proof of which can be found in Thm 3.20 of [9]) says that our terminology is justified.

Lemma A.3: (Lie algebra property) *The Lie algebra of any matrix Lie group is a real Lie algebra.*

there exists a unique Lie group homomorphism $\tilde{\Psi} : H_2 \to H_1$ such that $\tilde{\Psi}(e^A) = e^{\psi^{-1}(A)}$ $(A \in \mathfrak{h}_2)$. Now

$$\tilde{\Psi}(\Psi(e^A)) = \tilde{\Psi}(e^{\psi(A)}) = e^{\psi^{-1} \circ \psi(A)} = e^A \qquad (A \in \mathfrak{h}_1)$$

and similarly $\Psi(\tilde{\Psi}(e^A))$ $(A \in \mathfrak{h}_2)$, which (using the fact that elements of the form e^A with $A \in \mathfrak{h}_i$ generate H_i) proves that Ψ is invertible and $\tilde{\Psi} = \Psi^{-1}$. □

Informally, the universal cover H of G is the unique simply connected Lie group that has the same Lie algebra as G. The universal cover of a matrix Lie group need in general not be a matrix Lie group. Lie's third theorem (Thm 5.25 of [9]) says:

Theorem A.6: (Lie's third theorem) *Every real Lie algebra $\mathfrak{g}$ is the Lie algebra of some connected Lie group G.*

By Conclusion 5.26 of [9], we can even take G to be a matrix Lie group, and by restricting to the identity component we can take G to be connected. By going to the universal cover, we can also take G to be simply connected, but in this case we may loose the property that G is a *matrix* Lie group. Anyway, we can conclude:

> There is a one-to-one correspondence between Lie algebras and simply connected Lie groups. Every Lie group has a unique universal cover, which is a simply connected Lie group with the same Lie algebra.

Let G be a Lie group with Lie algebra $\mathfrak{g}$ and let (V, Π) be a representation of G. Then, by (A.4), there exists a unique Lie algebra homomorphism $\pi : \mathfrak{g} \to \mathcal{L}(V)$ such that

$$\Pi(e^A) = e^{\pi(A)} \qquad (A \in \mathfrak{g}). \tag{A.5}$$

More concretely, one has (see Prop. 4.4 of [9])

$$\pi(A)v = \frac{\partial}{\partial t}\Pi(e^{tA})v\Big|_{t=0} \qquad (A \in \mathfrak{g},\ v \in V). \tag{A.6}$$

We say that (V, π) is the representation of $\mathfrak{g}$ *associated* with the representation (V, Π) of G. Conversely, if G is simply connected, then by grace of Theorem A.4, through (A.5), each representation (V, π) of $\mathfrak{g}$ gives rise to a unique associated representation (V, Π) of G.

A.4. *Relation between algebras and Lie algebras*

If $\mathfrak{a}$ is an algebra and $\mathfrak{c} \subset \mathfrak{a}$ is any subset of $\mathfrak{a}$, then there exists a smallest subalgebra $\mathfrak{b} \subset \mathfrak{a}$ such that $\mathfrak{b}$ contains $\mathfrak{c}$. This algebra consists of the linear span of the unit element I and all finite products of elements of $\mathfrak{c}$. We call $\mathfrak{b}$ the algebra *generated* by $\mathfrak{c}$. If $\mathfrak{b} = \mathfrak{a}$, then we say that $\mathfrak{c}$ *generates* $\mathfrak{a}$.

Let $\mathfrak{g}$ be a Lie algebra. By definition, an *enveloping algebra* for $\mathfrak{g}$ is a pair $(\mathfrak{a}, \mathfrak{i})$ such that

(i) $\mathfrak{a}$ is an algebra and $\mathfrak{i} : \mathfrak{g} \to \mathfrak{a}$ is a Lie algebra homomorphism.
(ii) The image $\mathfrak{i}(\mathfrak{g})$ of $\mathfrak{g}$ under $\mathfrak{i}$ generates $\mathfrak{a}$.

We cite the following theorem from Thms 9.7 and 9.9 of [9].

Theorem A.7: (Universal enveloping algebra) *For every Lie algebra $\mathfrak{g}$, there exists an enveloping algebra $(\mathfrak{a}, \mathfrak{i})$ with the following properties.*

(i) *If $(\mathfrak{b}, \mathfrak{j})$ is an enveloping algebra of $\mathfrak{g}$, then there exists a unique algebra homomorphism $\phi : \mathfrak{a} \to \mathfrak{b}$ such that $\phi(\mathfrak{i}(A)) = \mathfrak{j}(A)$ for all $A \in \mathfrak{g}$.*
(ii) *If $\{X_1, \ldots, X_n\}$ is a basis for $\mathfrak{g}$, then a basis for $\mathfrak{a}$ is formed by all elements of the form*

$$\mathfrak{i}(X_1)^{k_1} \cdots \mathfrak{i}(X_n)^{k_n},$$

where $k_1, \ldots, k_n \geq 0$ are integers. In particular, these elements are linearly independent.

An argument similar to the proof of Lemma A.5 shows that the pair $(\mathfrak{a}, \mathfrak{i})$ from Theorem A.7 is unique up to natural isomorphisms. We call $(\mathfrak{a}, \mathfrak{i})$ the *universal enveloping algebra* of $\mathfrak{g}$ and use the notation $U(\mathfrak{g}) := \mathfrak{a}$. By property (ii), the map $\mathfrak{i}$ is one-to-one, so we often identify $\mathfrak{g}$ with its image under $\mathfrak{i}$ and pretend $\mathfrak{g}$ is a sub-Lie-algebra of $U(\mathfrak{g})$.

As an immediate consequence of property (i) of Theorem A.7, we see that if V is a representation of a Lie algebra $\mathfrak{g}$ and $\pi : \mathfrak{g} \to \mathcal{L}(V)$ is the associated Lie algebra homomorphism, then there exists a unique algebra homomorphism $\overline{\pi} : U(\mathfrak{g}) \to \mathcal{L}(V)$ such that $\overline{\pi}(A) = \pi(A)$ $(A \in \mathfrak{g})$. (Here we view $\mathfrak{g}$ as a sub-Lie-algebra of $U(\mathfrak{g})$.) Conversely, of course, every representation of $U(\mathfrak{g})$ is also a representation of $\mathfrak{g}$.

If (V, π) is a representation of a Lie algebra $\mathfrak{g}$, then we usually denote the associated representation of $U(\mathfrak{g})$ also by (V, π), i.e., we identify the map π with its extension $\overline{\pi}$. Note, however, that a representation (V, π) of a Lie algebra $\mathfrak{g}$ can be faithful even when the associated representation (V, π) of $U(\mathfrak{g})$ is not. Indeed, by property (ii) of Theorem A.7, $U(\mathfrak{g})$ is always infinite dimensional, even though $\mathfrak{g}$ is finite dimensional, so finite dimensional faithful representations of $\mathfrak{g}$ are not faithful when viewed as a representation of $U(\mathfrak{g})$.

A.5. *Adjoints and unitary representations*

Let V be a finite dimensional linear space equipped with an inner product $\langle\cdot|\cdot\rangle$, which for linear spaces over $\mathbb{C}$ is conjugate linear in its first argument and linear in its second argument. Each $A \in \mathcal{L}(V)$ has a unique *adjoint* $A^* \in \mathcal{L}(V)$ such that

$$\langle A^*v|w\rangle = \langle v|Aw\rangle \qquad (v, w \in V). \tag{A.7}$$

An operator A is *self-adjoint* (also called *hermitian*) if $A^* = A$ and *skew symmetric* if $A^* = -A$. A *positive operator* is an operator such that $\langle v|Av\rangle \geq 0$ for all v. If V, W are linear spaces equipped with inner products, then an operator $U \in \mathcal{L}(V, W)$ is called *unitary* if it preserves the inner product, i.e.,

$$\langle Uv|Uw\rangle = \langle v|w\rangle \qquad (v, w \in V). \tag{A.8}$$

In particular, an operator $U \in \mathcal{L}(V)$ is unitary if and only if it is invertible and $U^{-1} = U$. If V is a finite dimensional linear space over $\mathbb{C}$, then for $v \in V$ we define operators $\langle v| \in \mathcal{L}(V, \mathbb{C})$ and $|v\rangle \in \mathcal{L}(\mathbb{C}, V)$ by

$$\langle v|w := \langle v|w\rangle \quad \text{and} \quad |v\rangle c := cv.$$

Then $\langle v||w\rangle$ is an operator in $\mathcal{L}(\mathbb{C}, \mathbb{C})$ which we can identify with the complex number $\langle v|w\rangle$. Moreover, $|v\rangle\langle w|$ is an operator in $\mathcal{L}(V)$. An *orthonormal* basis $\{e(1), \ldots, e(n)\}$ of V is a basis such that $\langle e(i)|e(j)\rangle = \delta_{ij}$. Then

$$A = \sum_{ij} A_{ij}|e(i)\rangle\langle e(j)|,$$

where A_{ij} denotes the matrix of A with respect to the orthonormal basis $\{e(1), \ldots, e(n)\}$. An operator $A \in \mathcal{L}(V)$ is *normal* if $[A, A^*] = 0$. An operator is normal if and only if it is diagonal w.r.t. some orthonormal basis, i.e., if it can be written as

$$A = \sum_{i} \lambda_i|e(i)\rangle\langle e(i)|,$$

where the λ_i are the eigenvalues of A. For operators, the following properties are equivalent.

$$\begin{aligned}
A \text{ is hermitian} &\quad\Leftrightarrow\quad A \text{ is normal with real eigenvalues,}\\
A \text{ is skew symmetric} &\quad\Leftrightarrow\quad A \text{ is normal with imaginary eigenvalues,}\\
A \text{ is positive} &\quad\Leftrightarrow\quad A \text{ is normal with nonnegative eigenvalues,}\\
A \text{ is unitary} &\quad\Leftrightarrow\quad A \text{ is normal with eigenvalues of norm 1.}
\end{aligned}$$

By definition, a *unitary* representation of a Lie group G is a complex representation (V, Π) where V is equipped with an inner product such that $\Pi(A)$ is a unitary operator for each $A \in G$. A *unitary* representation of a real Lie algebra $\mathfrak{g}$ is a complex representation V that is equipped with an inner product such that

$$\pi(A) \quad \text{is skew symmetric for all } A \in \mathfrak{g}.$$

Since $e^{\pi(A)}$ is unitary if and only if $\pi(A)$ is skew symmetric, our definitions imply that a representation (V, Π) of a Lie group G is unitary if and only if the associated representation (V, π) of the real Lie algebra $\mathfrak{g}$ of G is unitary.

Theorem A.8: (Compact Lie groups) *Let K be a compact Lie group and let V be a representation of K. Then it is possible to equip V with an inner product so that V becomes a unitary representation of K.*

Proof: (sketch) Choose an arbitrary inner product $\langle\cdot|\cdot\rangle$ on V and define

$$\langle v|w\rangle_K := \int \langle\Pi(A)v|\Pi(A)w\rangle \mathrm{d}A,$$

where $\mathrm{d}A$ denotes the *Haar measure* on K, which is finite by the assumption that K is compact. It is easy to check that $\langle\cdot|\cdot\rangle_K$ is an inner product. In particular, since $\Pi(A)$ is invertible for each $A \in K$, we have $\Pi(A)v \neq 0$ and hence $\langle\Pi(A)v|\Pi(A)v\rangle > 0$ for all $v \in V$ and $A \in K$. Now by the fact that the Haar measure is invariant under the action of the group

$$\begin{aligned}
\langle\Pi(B)v|\Pi(B)w\rangle_K &= \int \langle\Pi(A)\Pi(B)v|\Pi(A)\Pi(B)w\rangle \mathrm{d}A\\
&= \int \langle\Pi(AB)v|\Pi(AB)w\rangle \mathrm{d}A = \int \langle\Pi(C)v|\Pi(C)w\rangle \mathrm{d}C = \langle v|w\rangle_K,
\end{aligned}$$

which proves that V, equipped with the inner product $\langle\cdot|\cdot\rangle_K$, is a unitary representation of K. □

The following lemma is a sort of converse to Theorem A.8 since it says that noncompact Lie groups do not have faithful unitary representations, at least when we restrict ourselves to finite dimensional representations, as we do here.

Lemma A.9: (Noncompact Lie groups) *Let K be a noncompact Lie group and let V be a faithful (finite dimensional) representation of K. Then it is not possible to equip V with an inner product so that V becomes a unitary representation of K.*

Proof: Equip V with an inner product and let $\mathrm{U}(V)$ denote the group of all unitary maps $A : V \to V$. If (V, Π) is a faithful representation of K, then the image $\Pi(K)$ of K under Π is a closed subset of $\mathrm{GL}(V)$ and $\Pi : K \to \Pi(K)$ is a homeomorphism. If (V, Π) is a unitary representation, then $\Pi(K) \subset \mathrm{U}(V)$ and hence by the compactness of the latter, $\Pi(K)$ is compact. Since $\Pi : K \to \Pi(K)$ is a homeomorphism, it follows that K is compact. □

A $*$-*algebra* is a complex algebra on which there is defined an *adjoint* operation $A \mapsto A^*$ such that

(i) $A \mapsto A^*$ is conjugate linear,
(ii) $(A^*)^* = A$,
(iii) $(AB)^* = B^* A^*$.

If V is a complex finite dimensional linear space equipped with an inner product, then $\mathcal{L}(V)$, equipped with the adjoint operation (A.7), is a $*$-algebra.

A $*$-*algebra homomorphism* is an algebra homomorphism that satisfies

$$\phi(A^*) = \phi(A)^*.$$

A *sub-$*$-algebra* of a $*$-algebra is a subalgebra that is closed under the adjoint operation. By definition, a $*$-*representation* of a $*$-algebra $\mathfrak{a}$ is a representation (V, π) such that V is equipped with an inner product and π is a $*$-algebra homomorphism.

In general, a $*$-algebra may fail to have a faithful $*$-representation. For finite dimensional $*$-algebras, a necessary and sufficient condition for the existence of a faithful representation is that

$$A^* A = 0 \quad \text{implies} \quad A = 0,$$

but it is rather difficult to prove this; see [20] and references therein. In infinite dimensions, one needs the theory of C$*$-algebras, which are $*$-algebras equipped with a norm that in faithful representations corresponds to the operator norm $\|A\| = \sup_{\|v\| \leq 1} \|Av\|$.

Recall the definition of an *adjoint* operation on a complex Lie algebra $\mathfrak{g}$ from Section 2.1. Recall also that we called a Lie algebra homomorphism *unitary* if $\phi(A^*) = \phi(A)^*$, and that a *unitary* representation is a representation (V, π) such that V is equipped with an inner product and π is a unitary Lie algebra homomorphism.

Lemma A.10: (Universal enveloping $*$-algebra) *Let $\mathfrak{g}$ be a Lie-$*$-algebra. Then there exists a unique adjoint operation on its universal enveloping algebra $U(\mathfrak{g})$ that coincides with the adjoint operation on $\mathfrak{g}$.*

Proof: Recall from Sections 2.2 that every complex linear space V has a *conjugate* space which is a linear space $\overline{V}$ together with a conjugate linear bijection $V \ni v \mapsto \overline{v} \in \overline{V}$. If $\mathfrak{a}$ is a complex algebra, then we can equip $\overline{\mathfrak{a}}$ with the structure of an algebra by putting

$$\overline{A}\,\overline{B} := \overline{BA}.$$

It is not hard to see that a map $A \mapsto A^*$ defined on some algebra $\mathfrak{a}$ is an adjoint operation if and only if the map $A \mapsto \overline{A^*}$ from $\mathfrak{a}$ into $\overline{\mathfrak{a}}$ is an algebra homomorphism. By the definition of an adjoint operation on a Lie algebra, $[A^*, B^*] = -[A, B]^*$ for all $A, B \in \mathfrak{g}$. It follows that the map

$$\mathfrak{g} \ni X \mapsto \overline{X^*} \in \overline{U(\mathfrak{g})}$$

is a Lie algebra homomorphism, which by the defining property of the universal enveloping algebra (Theorem A.7 (i)) extends to a unique algebra homomorphism from $U(\mathfrak{g})$ to $\overline{U(\mathfrak{g})}$. □

A.6. *Dual, quotient, sum, and product spaces*

Dual spaces

The *dual* V' of a finite dimensional linear space V over $\mathbb{K} = \mathbb{R}$ or $= \mathbb{C}$ is the space of all linear forms $l : V \to \mathbb{K}$. Each element $v \in V$ naturally defines a linear form L_v on V' by $L_v(l) := l(v)$ and each linear form on V arises in this way, so we can identify $V'' \cong V$. If $\{e(1), \ldots, e(n)\}$ is a basis for V, then setting $f(i)(e(j)) := 1_{\{i=j\}}$ defines a basis $\{f(1), \ldots, f(n)\}$ for V' called the *dual basis*. If V is equipped with an inner product, then setting

$$\langle v|w := \langle v|w\rangle$$

defines a linear form on V and $V' := \{\langle v| : v \in V\}$. Through this identification, we also equip V' with an inner product. Then if $\{e(1), \ldots, e(n)\}$ is an orthonormal basis for V, the dual basis is an orthonormal basis for V'. Each linear map $A : V \to W$ gives naturally rise to a *dual map* $A' : W' \to V'$ defined by

$$A'(l) := l \circ A,$$

and indeed every linear map from W' to V' arises in this way, i.e., $\mathcal{L}(W', V') = \{A' : A \in \mathcal{L}(V, W)\}$. If V, W are equipped with inner products and $A \in \mathcal{L}(V, W)$, then

$$A'(\langle\phi|) = \langle A^*\phi|,$$

where A^* denotes the adjoint of A, i.e., this is the linear map $A^* \in \mathcal{L}(W, V)$ defined by

$$\langle\phi|A\psi\rangle = \langle A^*\phi|\psi\rangle \qquad (\phi \in W, \ \psi \in V).$$

If (V, Π) is a representation of a Lie group G, then we can define group homomorphism $\Pi' : G \to \mathcal{L}(V')$ by

$$\Pi'(A)l := \Pi(A^{-1})'l = l \circ \Pi(A^{-1}).$$

In this way, the dual space V' naturally obtains the structure of a representation of G. Note that

$$\Pi'(AB)l = l \circ \Pi((AB)^{-1}) = l \circ \Pi(A^{-1})\Pi(B^{-1}) = \Pi'(A)(\Pi'(B)l),$$

proving that Π' is indeed a group homomorphism. Similarly, if (V, π) is a representation of a Lie algebra $\mathfrak{g}$, then we can equip the dual space V' with the structure of a representation of $\mathfrak{g}$ by putting

$$\pi'(A)l := -\pi(A)'(l) = -l \circ \pi(A),$$

where in this case the minus sign guarantees that

$$\begin{aligned}\pi'([A, B])l &= -l \circ \pi([A, B]) = -l \circ \big(\pi(A)\pi(B) - \pi(B)\pi(A)\big)\\ &= -\big(\pi'(B)(\pi'(A)l) - \pi'(A)(\pi'(B)l)\big) = \pi'(A)(\pi'(B)l) - \pi'(B)(\pi'(A)l).\end{aligned}$$

This is called the *dual representation* or *contragredient representation* of G or $\mathfrak{g}$, respectively, associated with V, see Def. 4.21 of [9]. If two representations of G and $\mathfrak{g}$ are associated as in (A.6), then their dual representations are also associated.

Quotient spaces

By definition, a *normal subgroup* of a group $\mathcal{G}$ is a subgroup $\mathcal{H}$ such that

$$A\mathcal{H} := \{AB : B \in \mathcal{H}\} = \{BA : B \in \mathcal{H}\} =: \mathcal{H}A \qquad \forall A \in \mathcal{G}, \tag{A.9}$$

or equivalently, if $B \in \mathcal{H}$ implies $ABA^{-1} \in \mathcal{H}$ for all $A \in \mathcal{G}$. Sets of the form $A\mathcal{H}$ and $\mathcal{H}A$ are called *left* and *right cosets*, respectively. If $\mathcal{H}$ is a normal subgroup, then left cosets are right cosets and vice versa, and we can equip the set

$$\mathcal{G}/\mathcal{H} := \{A\mathcal{H} : A \in \mathcal{G}\} = \{\mathcal{H}A : A \in \mathcal{G}\}$$

of all cosets with a group structure such that

$$(A\mathcal{H})(B\mathcal{H}) = (AB)\mathcal{H}.$$

We call $\mathcal{G}/\mathcal{H}$ the *quotient group* of $\mathcal{G}$ and $\mathcal{H}$. Note that as a set this is obtained from $\mathcal{G}$ by dividing out the equivalence relation

$$A \sim B \quad \Leftrightarrow \quad A = BC \quad \text{for some} \quad C \in \mathcal{H}.$$

If V is a linear space and $W \subset V$ is a linear subspace, then we can define an equivalence relation on V by setting

$$v_1 \sim v_2 \quad \Leftrightarrow \quad v_1 = v_2 + w \quad \text{for some} \quad w \in W.$$

The equivalence classes with respect to this equivalence relation are the sets of the form

$$v + W := \{v + w : w \in W\}$$

and we can equip the space

$$V/W := \{v + W : v \in V\}$$

with the structure of a linear space by setting

$$a_1(v_1 + W) + a_2(v_2 + W) := \big(a_1v_1 + a_2v_2\big) + W.$$

An *invariant subspace* of a representation V of a Lie group G, Lie algebra $\mathfrak{g}$, or algebra $\mathfrak{a}$ is a linear space $W \subset V$ such that $Aw \in W$ for all $w \in W$ and A from G, $\mathfrak{g}$, or $\mathfrak{a}$, respectively. If W is an invariant subspace, then we can equip the quotient space V/W with the structure of a representation by setting

$$A(v + W) := (Av) + W.$$

Note that this is a good definition since $v_1 = v_2 + w$ for some $w \in W$ implies $Av_1 = Av_2 + Aw$ where $Aw \in W$ by the assumption that W is invariant.

A *left ideal* (resp. *right ideal*) of an algebra $\mathfrak{a}$ is a linear subspace $\mathfrak{i} \subset \mathfrak{a}$ such that $AB \in \mathfrak{i}$ (resp. $BA \in \mathfrak{i}$) for all $A \in \mathfrak{a}$ and $B \in \mathfrak{i}$. An *ideal* is a linear subspace that is both a left and right ideal. If $\mathfrak{i}$ is an ideal of $\mathfrak{a}$, then we can equip the quotient space $\mathfrak{a}/\mathfrak{i}$ with the structure of an algebra by putting

$$(A + \mathfrak{i})(B + \mathfrak{i}) := (AB) + \mathfrak{i}.$$

To see that this is a good definition, write $A_1 \sim A_2$ if $A_1 = A_2 + B$ for some $B \in \mathfrak{i}$. Then $A_1 \sim A_2$ and $B_1 \sim B_2$ imply that $A_1 = A_2 + C$ and $B_1 = B_2 + D$ for some $C, D \in \mathfrak{i}$ and hence

$$A_1B_1 = (A_2 + C)(B_2 + D) = A_2B_2 + \big(CB_2 + A_2D + CD\big)$$

with $CB_2 + A_2D + CD \in \mathfrak{i}$, so $A_1B_1 \sim A_2B_2$. If $\mathfrak{a}$ is a $*$-algebra, then a $*$-*ideal* of $\mathfrak{a}$ is an ideal $\mathfrak{i}$ such that $A \in \mathfrak{i}$ implies $A^* \in \mathfrak{i}$. If $\mathfrak{i}$ is a $*$-ideal, then we can equip the quotient algebra $\mathfrak{a}/\mathfrak{i}$ with an adjoint operation by putting

$$(A + \mathfrak{i})^* := A^* + \mathfrak{i}.$$

A linear subspace $\mathfrak{h}$ of a Lie algebra $\mathfrak{g}$ is said to be an *ideal* if $[A, B] \in \mathfrak{h}$ for all $A \in \mathfrak{g}$ and $B \in \mathfrak{h}$. Note that this automatically implies that also $[B, A] = -[A, B] \in \mathfrak{h}$. If $\mathfrak{h}$ is an ideal of a Lie algebra, then we can equip the quotient space $\mathfrak{g}/\mathfrak{h}$ with the structure of a Lie algebra by putting

$$[A + \mathfrak{h}, B + \mathfrak{h}] := [A, B] + \mathfrak{h}.$$

The proof that this is a good definition is the same as for algebras.

The direct sum

The direct sum $V_1 \oplus \cdots \oplus V_n$ of linear spaces $V_1, \ldots, V_n$ has already been defined in Section 2.6. There is a natural isomorphism between $V_1 \oplus \cdots \oplus V_n$ and the Carthesian product

$$V_1 \times \cdots \times V_n = \big\{\big(\phi(1), \ldots, \phi(n)\big) : \phi(i) \in V_i \ \forall i\big\},$$

which we equip with a linear structure by defining

$$a\big(\phi(1), \ldots, \phi(n)\big) + b\big(\psi(1), \ldots, \psi(n)\big) := \big(a\phi(1) + b\phi(1), \ldots, a\phi(n) + b\phi(n)\big).$$

If $V_1, \ldots, V_n$ are equipped with inner products, then we require that the inner product on $V_1 \oplus \cdots \oplus V_n$ is given by

$$\langle \phi(1) + \cdots + \phi(n) | \psi(1) + \cdots + \psi(n) \rangle := \sum_{k=1}^{n} \langle \phi(k) | \psi(k) \rangle, \tag{A.10}$$

which has the effect that $V_1, \ldots, V_n$ are (mutually) orthogonal. One has the natural isomorphism

$$(V_1 \oplus V_2)/V_2 \cong V_1.$$

In general, given a subspace V_1 of some larger linear space W, there are many possible ways to choose another subspace V_2 such that $W = V_1 \oplus V_2$ and hence $W \cong (W/V_1) \oplus V_1$.

If V is a linear subspace of some larger linear space W, and W is equipped with an inner product, then we define the *orthogonal complement* of V as

$$V^{\perp} := \{w \in W : \langle v|w\rangle = 0 \ \forall v \in V\}.$$

Then one has the natural isomorphisms

$$W/V \cong V^{\perp} \quad \text{and} \quad W \cong V \oplus V^{\perp},$$

where the inner product $V \oplus V^{\perp}$ is given in terms of the inner products on V and $V^{\perp}$ as in (A.10). Thus, given a linear subspace V_1 of a linear space W that is equipped with an inner product, there is a canonical way to choose another subspace V_2 such that $W = V_1 \oplus V_2$.

If $V_1, \ldots, V_n$ are representations of the same Lie group, Lie algebra, or algebra, then we equip $V_1 \oplus \cdots \oplus V_n$ with the structure of a representation by putting

$$A\big(\phi(1) + \cdots + \phi(n)\big) := A\phi(1) + \cdots + A\phi(n).$$

If V, W are representations, then W is an invariant subspace of $V \oplus W$ and one has the natural isomorphism of representations $(V \oplus W)/W \cong V$.

If $\mathfrak{a}_1, \ldots, \mathfrak{a}_n$ are algebras, then we equip their direct sum $\mathfrak{a}_1 \oplus \cdots \oplus \mathfrak{a}_n$ with the structure of an algebra by putting

$$\big(A(1) + \cdots + A(n)\big)\big(B(1) + \cdots + B(n)\big) := A(1)B(1) + \cdots + A(n)B(n). \tag{A.11}$$

If $\mathfrak{a}, \mathfrak{b}$ are algebras, then $\mathfrak{b}$ is an ideal of $\mathfrak{a} \oplus \mathfrak{b}$ and one has the natural isomorphism $(\mathfrak{a} \oplus \mathfrak{b})/\mathfrak{b} \cong \mathfrak{a}$. Note that $\mathfrak{b}$ is not a subalgebra of $\mathfrak{a} \oplus \mathfrak{b}$ since $I \notin \mathfrak{b}$ (unless $\mathfrak{a} = \{0\}$). For $*$-algebras, we also put

$$\big(A(1) + \cdots + A(n)\big)^* := \big(A(1)^* + \cdots + A(n)^*\big).$$

The direct sum of Lie algebras has already been defined in Section 2.6. It is easy to see that this is consistent with the definition of the direct sum of algebras.

The tensor product

The *tensor product* of two (or more) linear spaces has already been defined in Section 2.6. A proof similar to the proof of Lemma A.5 shows that the tensor product is unique up to natural isomorphisms, i.e., if $V\tilde{\otimes}W$ and $(\phi,\psi)\mapsto\phi\tilde{\otimes}\psi$ is another linear space together with a bilinear map which satisfies the defining property of the tensor product, then there exists a unique linear bijection $\Psi: V\otimes W\to V\tilde{\otimes}W$ such that $\Psi(V\otimes W)=V\tilde{\otimes}W$.

If V,W are representations of the same Lie group, then we equip $V\otimes W$ with the structure of a representation by putting

$$A(\phi\otimes\psi):=A\phi\otimes A\psi. \tag{A.12}$$

If V,W are representations of the same Lie algebra or algebra, then we equip $V\otimes W$ with the structure of a representation by putting

$$A(\phi\otimes\psi):=A\phi\otimes\psi+\phi\otimes A\psi. \tag{A.13}$$

The reason why we define things in this way is that in view of (A.6), if $\mathfrak{g}$ is the Lie algebra of G, then the representation of $\mathfrak{g}$ defined in (A.13) is the representation of $\mathfrak{g}$ that is associated with the representation of G defined in (A.12). Note that (A.13) is bilinear in ϕ and ψ and hence by the defining property of the tensor product uniquely defines a linear operator on $V\otimes W$.

If $\mathfrak{a},\mathfrak{b}$ are algebras, then we equip their tensor product $\mathfrak{a}\otimes\mathfrak{b}$ with the structure of an algebra by putting

$$\big(A(1)\otimes B(1)\big)\big(A(2)\otimes B(2)\big):=\big(A(1)A(2)\otimes B(1)B(2)\big).$$

Using the defining property of the tensor product, one can show that this unambiguously defines a linear map

$$(\mathfrak{a}\otimes\mathfrak{b})^2\ni(A,B)\mapsto AB\in\mathfrak{a}\otimes\mathfrak{b}.$$

We can identify $\mathfrak{a}$ and $\mathfrak{b}$ with the subalgebras of $\mathfrak{a}\otimes\mathfrak{b}$ given by

$$\mathfrak{a}\cong\{A\otimes I: A\in\mathfrak{a}\}\quad\text{and}\quad\mathfrak{b}\cong\{I\otimes B: B\in\mathfrak{b}\}.$$

Note that if we identify $\mathfrak{a}$ and $\mathfrak{b}$ with subalgebras of $\mathfrak{a}\otimes\mathfrak{b}$, then every element of $\mathfrak{a}$ commutes with every element of $\mathfrak{b}$. If $\mathfrak{a},\mathfrak{b}$ are $*$-algebras, then we equip the algebra $\mathfrak{a}\otimes\mathfrak{b}$ with an adjoint operation by setting

$$(A\otimes B)^*:=(A^*\otimes B^*).$$

If $\mathfrak{g}$ and $\mathfrak{h}$ are Lie algebras, then the universal enveloping algebra of their direct sum is naturally isomorphic to the tensor product of their universal enveloping algebras:

$$U(\mathfrak{g}\oplus\mathfrak{h})\cong U(\mathfrak{g})\otimes U(\mathfrak{h}). \tag{A.14}$$

Indeed, if $\{X_1,\ldots,X_n\}$ is a basis for $\mathfrak{g}$ and $\{Y_1,\ldots,Y_m\}$ is a basis for $\mathfrak{h}$, then we can define a bilinear map $(A,B)\mapsto A\otimes B$ from $U(\mathfrak{g})\times U(\mathfrak{h})$ into $U(\mathfrak{g}\oplus\mathfrak{h})$ by

$$\begin{aligned}&\big(X_1^{k_1}\cdots X_n^{k_n},Y_1^{l_1}\cdots Y_m^{l_m}\big)\\&\quad\mapsto X_1^{k_1}\cdots X_n^{k_n}\otimes Y_1^{l_1}\cdots Y_m^{l_m}:=X_1^{k_1}\cdots X_n^{k_n}Y_1^{l_1}\cdots Y_m^{l_m}.\end{aligned}$$

where we view $\mathfrak{g}$ and $\mathfrak{h}$ as sub-Lie-algebras of $\mathfrak{g}\oplus\mathfrak{h}$ such that $[X,Y]=0$ for each $X\in\mathfrak{g}$ and $Y\in\mathfrak{h}$. In view of Theorem A.7, the space $U(\mathfrak{g}\oplus\mathfrak{h})$ together with this bilinear map is a realization of the tensor product $U(\mathfrak{g})\otimes U(\mathfrak{h})$.

On a philosophical note, recall that elements of a Lie algebra are related to elements of a matrix Lie group via an exponential map. We can view (A.14) as a reflection of the property of the exponential map that converts sums into products.

If V and W are representations of algebras $\mathfrak{a}$ and $\mathfrak{b}$, respectively, then we can make $V\otimes W$ into a representation of $\mathfrak{a}\otimes\mathfrak{b}$ by setting

$$(A\otimes B)(\phi\otimes\psi):=(A\phi)\otimes(B\psi). \tag{A.15}$$

Again, by bilinearity and the defining property of the tensor product, this is a good definition. Note that this is consistent with (A.14) and our definition in (2.29) where we showed that if V and W are representations of Lie algebras $\mathfrak{g}$ and $\mathfrak{h}$, then $V\otimes W$ is naturally a representation of $\mathfrak{g}\oplus\mathfrak{h}$. On the other hand, one should observe that in the special case that $\mathfrak{a}=\mathfrak{b}$, our present construction differs from our earlier construction in (A.13).

A.7. *Irreducible representations*

Let $\mathfrak{g}$ be a Lie algebra on which an adjoint operation is defined, and let $\mathfrak{h}:=\{\mathbf{a}\in\mathfrak{g}:\mathbf{a}^*=-\mathbf{a}\}$ denote the real sub-Lie-algebra[p] consisting of all skew-symmetric elements of $\mathfrak{g}$. It is not hard to see that $\mathfrak{g}$ is the complexification of $\mathfrak{h}$, i.e., each $\mathbf{a}\in\mathfrak{g}$ can uniquely be written as $\mathbf{a}=\mathbf{a}_1+i\mathbf{a}_2$ with $\mathbf{a}_1,\mathbf{a}_2\in\mathfrak{h}$.[q] Let $\{\mathbf{x}_1,\ldots,\mathbf{x}_n\}$ be a basis for $\mathfrak{g}$. The Lie bracket on $\mathfrak{g}$ is uniquely characterized by the commutation relations

$$[\mathbf{x}_i,\mathbf{x}_j]=\sum_{j=1}^n c_{ijk}\mathbf{x}_k, \tag{A.16}$$

[p]To see that this is a sub-Lie-algebra, note that $\mathbf{a},\mathbf{b}\in\mathfrak{h}$ imply $[\mathbf{a},\mathbf{b}]^*=-[\mathbf{a}^*,\mathbf{b}^*]$ and hence $[\mathbf{a},\mathbf{b}]\in\mathfrak{h}$.

[q]Equivalently, we may show that each $\mathbf{a}\in\mathfrak{g}$ can uniquely be written as $\mathbf{a}=\mathrm{Re}(\mathbf{a})+i\mathrm{Im}(\mathbf{a})$ with $\mathrm{Re}(\mathbf{a}),\mathrm{Im}(\mathbf{a})$ self-adjoint. This follows easily by putting $\mathrm{Re}(\mathbf{a}):=\frac{1}{2}(\mathbf{a}+\mathbf{a}^*)$ and $\mathrm{Im}(\mathbf{a}):=\frac{1}{2}i(\mathbf{a}^*-\mathbf{a})$.

where c_{ijk} are the structure constants. Likewise, the adjoint operation on $\mathfrak{g}$ is uniquely characterized by its action on basis elements

$$\mathbf{x}_i^* = \sum_j d_{ij}\mathbf{x}_j, \tag{A.17}$$

where d_{ij} is another set of constants.

By Theorem A.6, the real Lie algebra $\mathfrak{h}$ is the Lie algebra of some Lie group G. By going to the universal cover, we can take G to be simply connected, in which case it is uniquely determined by $\mathfrak{h}$. Conversely, if G is a simply connected Lie group, $\mathfrak{h}$ is its real Lie algebra, and $\mathfrak{g} := \mathfrak{h}_\mathbb{C}$ is the complexification of $\mathfrak{h}$, then we can equip $\mathfrak{g}$ with an adjoint operation such that the set of skew symmetric elements is exactly $\mathfrak{h}$, by putting $(\mathbf{a}_1 + i\mathbf{a}_2)^* := -\mathbf{a}_1 + i\mathbf{a}_2$ for each $\mathbf{a}_1, \mathbf{a}_2 \in \mathfrak{h}$.

If V is a linear space and $X_1, \ldots, X_n \in \mathcal{L}(V)$ satisfy (A.16), then there exists a unique Lie algebra homomorphism $\pi : \mathfrak{g} \to \mathcal{L}(V)$ such that $\pi(\mathbf{x}_i) = X_i$ $(i = 1, \ldots, n)$. If V is equipped with an inner product and the operators $X_1, \ldots, X_n$ moreover satisfy (A.17), then π is a unitary representation. By Theorem A.7 (i) and Lemma A.10, π can in a unique way be extended to a $*$-algebra homomorphism $\overline{\pi} : U(\mathfrak{g}) \to \mathcal{L}(V)$. Moreover, if G is the simply connected Lie group associated with $\mathfrak{h}$, then by Theorem A.4, there exists a unique Lie group homomorphism $\Pi : G \to \mathcal{L}(V)$ such that (A.5) holds, so (V, Π) is a representation of G. Since every element of $\mathfrak{h}$ is skew symmetric, (V, π) and hence also (V, Π) are unitary representations of $\mathfrak{h}$ and G, respectively.

Let $W \subset V$ be a linear subspace. It is not hard to see that

$$\begin{aligned} & W \text{ is an invariant subspace of } (V, \Pi) \\ \Leftrightarrow \quad & W \text{ is an invariant subspace of } (V, \pi) \\ \Leftrightarrow \quad & W \text{ is an invariant subspace of } (V, \overline{\pi}). \end{aligned}$$

We say that V is *irreducible* if its only invariant subspaces are $\{0\}$ and V.

Let V, W be two representations of the same Lie group G, Lie algebra $\mathfrak{g}$, or algebra $\mathfrak{a}$. Generalizing our earlier definition for ie algebras, a *homomorphism* of representations (of any kind) is a linear map $\phi : V \to W$ such that

$$\phi(\mathbf{a}v) = \mathbf{a}\phi(v) \tag{A.18}$$

for all $\mathbf{a} \in G$, $\mathbf{a} \in \mathfrak{g}$, or $\mathbf{a} \in \mathfrak{a}$, respectively. Homomorphisms of representations are called *intertwiners* of representations. If ϕ is a bijection, then its inverse is also an intertwining map. In this case we call ϕ an *isomorphism* and say that the representations are *equivalent* (or *isomorphic*). If G is a simply connected Lie group, $\mathfrak{g}$ its associated complexified Lie algebra, and $U(\mathfrak{g})$ its universal enveloping algebra, then it is not hard to see that

$$(A.18) \text{ holds } \forall \mathbf{a} \in G \;\Leftrightarrow\; (A.18) \text{ holds } \forall \mathbf{a} \in \mathfrak{g} \;\Leftrightarrow\; (A.18) \text{ holds } \forall \mathbf{a} \in U(\mathfrak{g}).$$

The following result can be found in, e.g., Thm 4.29 of [9]. In the special case of complex Lie algebras, we have already stated this in Proposition 2.1.

Proposition A.11: (Schur's lemma)

(a) *Let V and W be irreducible representations of a Lie group, Lie algebra, or algebra, and let $\phi : V \to W$ be an intertwiner. Then either $\phi = 0$ or ϕ is an isomorphism.*
(b) *Let V be an irreducible complex representation of a Lie group, Lie algebra, or algebra, and let $\phi : V \to V$ be an intertwiner. Then $\phi = \lambda I$ for some $\lambda \in \mathbb{C}$.*

By definition, the *center* of an algebra is the subalgebra $\mathcal{C}(\mathfrak{a}) := \{C \in \mathfrak{a} : [A, C] = 0 \ \forall A \in \mathfrak{a}\}$. The center is *trivial* if $\mathcal{C}(\mathfrak{a}) = \{\lambda I : \lambda \in \mathbb{K}\}$. The following is adapted from Cor. 4.30 of [9].

Corollary A.12: (Center) *Let (V, π) be an irreducible complex representation of an algebra $\mathfrak{a}$ and let $C \in \mathcal{C}(\mathfrak{a})$. Then $\pi(C) = \lambda I$ for some $\lambda \in \mathbb{C}$.*

Proof: Define $\phi : V \to V$ by $\phi v := \pi(C)v$. Then $\phi(Av) = \pi(C)\pi(A)v = \pi(CA)v = \pi(AC)v = \pi(A)\pi(C)v = A(\phi v)$ for all $A \in \mathfrak{a}$, so $\phi : V \to V$ is an intertwiner. By part (b) of Schur's lemma, $\phi = \lambda I$ for some $\lambda \in \mathbb{C}$. □

A.8. *Semisimple Lie algebras*

A Lie algebra $\mathfrak{g}$ is called *irreducible* (see Def. 3.11 of [9]) if its only ideals are $\{0\}$ and $\mathfrak{g}$, and *simple* if it is irreducible and has dimension $\dim(\mathfrak{g}) \geq 2$. A Lie algebra is called *semisimple* if it can be written as the direct sum of simple Lie algebras. Recall the definition of the center of a Lie algebra in (A.1).

Lemma A.13: (Trivial center) *The center of a semisimple Lie algebra is trivial.*

Proof: If $\mathfrak{g}$ is simple and A is an element of its center, then the linear space spanned by A is an ideal. Since $\dim(\mathfrak{g}) \geq 2$ and its only ideals are $\{0\}$ and $\mathfrak{g}$, this implies that $A = 0$. If $\mathfrak{g} = \mathfrak{g}_1 \oplus \cdots \oplus \mathfrak{g}_n$ is the direct sum of simple Lie algebras, then we can write any element A of the center of $\mathfrak{g}$ as

$A = A(1) + \cdots + A(n)$ with $A(k) \in \mathfrak{g}$. By the definition of the Lie bracket on $\mathfrak{g}$ (see (2.24)), $A(k)$ lies in the center of $\mathfrak{g}$ for each k, and hence $A = 0$ by what we have already proved. □

The following proposition is similar to Prop. 7.4 of [9].

Proposition A.14: (Inner product on Lie algebra) *Let $\mathfrak{g}$ be a Lie algebra on which an adjoint operation is defined, let $\mathfrak{h} := \{\mathbf{a} \in \mathfrak{g} : \mathbf{a}^* = -\mathbf{a}\}$, and let G be the simply connected Lie group with Lie algebra $\mathfrak{h}$. Assume that G is compact. Then the Lie algebra $\mathfrak{g}$, equipped with the map*

$$\mathfrak{g} \ni \mathbf{x} \mapsto \mathrm{ad}_{\mathbf{x}} \in \mathcal{L}(\mathfrak{g}),$$

is a faithful representation of itself. It is possible to equip $\mathfrak{g}$ with an inner product such that this is a unitary representation, i.e., $\mathrm{ad}_{\mathbf{x}^*} = (\mathrm{ad}_{\mathbf{x}})^*$ $(\mathbf{x} \in \mathfrak{g})$.

Proof: By Prop. 7.7 of [9], the center of $\mathfrak{g}$ is trivial. By Lemma A.1 and the remarks below it, it follows that $\mathfrak{g}$, equipped with the map $\mathfrak{g} \ni \mathrm{ad}_X \in \mathcal{L}(\mathfrak{g})$, is a faithful representation of itself. This representation naturally gives rise to a representation of G. By assumption, G is compact, so by Theorem A.8, we can equip $\mathfrak{g}$ with an inner product so that this representation is unitary. It follows that the representation of $\mathfrak{h}$ on $\mathfrak{g}$ is also unitary and hence the representation of $\mathfrak{g}$ on itself is a unitary representation. □

The following theorem follows from Thm 7.8 of [9].

Theorem A.15: (Semisimple algebras) *Let G be a compact simply connected Lie group and let $\mathfrak{g}$ be the complexification of its Lie algebra. Then $\mathfrak{g}$ is semisimple.*

Proof: (main idea) If $\mathfrak{g}$ is not simple, then it has some ideal $\mathfrak{i}$ that is neither $\{0\}$ nor $\mathfrak{g}$. Let $\mathfrak{i}^\perp$ denote the orthogonal complement of $\mathfrak{i}$ with respect to the inner product on $\mathfrak{g}$ defined in Proposition A.14. It is shown in Prop. 7.5 of [9] that $\mathfrak{i}^\perp$ is an ideal of $\mathfrak{g}$ and one has $\mathfrak{g} \cong \mathfrak{i} \oplus \mathfrak{i}^\perp$, where $\oplus$ denotes the direct sum of Lie algebras. Continuing this process, one arrives at a decomposition of $\mathfrak{g}$ as a direct sum of simple Lie algebras. □

In fact, the converse statement to Theorem A.15 also holds: if $\mathfrak{g}$ is a semisimple complex Lie algebra, then it is the complexification of the Lie algebra of a compact simply connected Lie group. This is stated (with references for a proof) in Sect. 10.3 of [9].

Let G be a compact simply connected Lie group, let $\mathfrak{h}$ be its real Lie algebra, let $\mathfrak{g} := \mathfrak{h}_{\mathbb{C}}$ be the complexification of $\mathfrak{h}$, and let $U(\mathfrak{g})$ denote the universal enveloping algebra of $\mathfrak{g}$. The *Casimir element* is the element $C \in U(\mathfrak{g})$ defined as

$$\mathbf{c} := -\sum_j \mathbf{x}_j^2,$$

where $\{\mathbf{x}_1, \dots, \mathbf{x}_n\}$ is a basis for $\mathfrak{h}$ that is orthonormal with respect to the inner product from Proposition A.14.[r] We cite the following result from Prop. 10.5 of [9].

Proposition A.16: (Casimir element) *The definition of the Casimir element does not depend on the choice of the orthonormal basis* $\{\mathbf{x}_1, \dots, \mathbf{x}_n\}$ *of* $\mathfrak{h}$. *Moreover* $\mathbf{c}$ *lies in the center of* $U(\mathfrak{g})$.

In irreducible representations, the Casimir element has a simple form.

Lemma A.17: (Representations of Casimir element) *For each irreducible representation* (V, π) *of* $\mathfrak{g}$, *there exists a constant* $\lambda_V \geq 0$ *such that* $\pi(\mathbf{c}) = \lambda_V I$.

Proof: Proposition A.16 and Corollary A.12 imply that for each irreducible representation (V, π) of $U(\mathfrak{g})$, there exists a constant $\lambda \in \mathbb{C}$ such that $\pi(\mathbf{c}) = \lambda I$. By Theorem A.8, we can equip V with an inner product such that it is a unitary representation of $\mathfrak{h}$. This means that $\mathbf{x}_j$ is skew symmetric and hence $i\mathbf{x}_j$ is hermitian, so $\mathbf{c} = \sum_j (i\mathbf{x}_j)^2$ is a positive operator. In particular, its eigenvalues are ≥ 0. □

A.9. *Some basic matrix Lie groups*

For any finite dimensional linear space V over $V = \mathbb{R}$ or $= \mathbb{C}$, we let $\mathrm{GL}(V)$ denote the *general linear group* of all invertible linear maps $A : V \to V$. In particular, we write $\mathrm{GL}(n; \mathbb{R}) = \mathrm{GL}(\mathbb{R}^n)$ and $\mathrm{GL}(n; \mathbb{C}) = \mathrm{GL}(\mathbb{C}^n)$.

The *special linear group* $\mathrm{SL}(V)$ is defined as

$$\mathrm{SL}(V) := \{A \in \mathrm{GL}(V) : \det(A) = 1\}.$$

[r]The inner product from Proposition A.14 is not completely unique; at best it is only determined up to a multiplicative constant. So the Casimir operator depends on the choice of the inner product, but once this is fixed, it does not depend on the choice of the orthonormal basis.

Again, we write $\mathrm{SL}(n;\mathbb{R}) = \mathrm{SL}(\mathbb{R}^n)$ and $\mathrm{SL}(n;\mathbb{C}) = \mathrm{SL}(\mathbb{C}^n)$. If V is a finite dimensional linear space over $\mathbb{C}$ and V is equipped with an inner product $\langle\,\cdot\,|\,\cdot\,\rangle$, then we call

$$\mathrm{U}(V) := \{A \in \mathcal{L}(V) : A \text{ is unitary}\}$$

the *unitary group* and

$$\mathrm{SU}(V) := \{A \in \mathrm{U}(V) : \det(A) = 1\}$$

the *special unitary group*, and write $\mathrm{U}(n) := \mathrm{U}(\mathbb{C}^n)$ and $\mathrm{SU}(n) := \mathrm{SU}(\mathbb{C}^n)$.

If V is a finite dimensional linear space over $\mathbb{R}$ and V is equipped with an inner product $\langle\,\cdot\,|\,\cdot\,\rangle$, then an operator $O \in \mathcal{L}(V)$ that preserves the inner product as in (A.8) is called *orthogonal.* (This is the equivalent of unitarity in the real setting.) We call

$$\mathrm{O}(V) := \{A \in \mathcal{L}(V) : A \text{ is orthogonal}\}$$

the *orthogonal group* and

$$\mathrm{SO}(V) := \{A \in \mathrm{O}(V) : \det(A) = 1\}$$

the *special orthogonal group*, and write $\mathrm{O}(n) := \mathrm{O}(\mathbb{R}^n)$ and $\mathrm{SO}(n) := \mathrm{SO}(\mathbb{R}^n)$. There also exists a group $\mathrm{O}(n;\mathbb{C})$, which consists of all complex matrices that preserve the bilinear form $(v, w) := \sum_i v_i w_i$. Note that this is *not* the inner product on $\mathbb{C}^n$; as a result $\mathrm{O}(n;\mathbb{C})$ is not the same as $\mathrm{U}(n)$.

Unitary operators satisfy $|\det(A)| = 1$ and orthogonal operators satisfy $\det(A) = \pm 1$. The group $\mathrm{O}(3)$ consists of rotations and reflections (and combinations thereof) while $\mathrm{SO}(3)$ consists only of rotations.

By Prop. 3.23 of [9], for $\mathbb{K} = \mathbb{R}$ or $= \mathbb{C}$, the Lie algebra of $\mathrm{GL}(n,\mathbb{K})$ is the space $M_n(\mathbb{K})$ of all $\mathbb{K}$-valued $n \times n$ matrices, and the Lie algebra of $\mathrm{SL}(n,\mathbb{K})$ is given by

$$\mathfrak{sl}(n,\mathbb{K}) = \{A \in M_n(\mathbb{K}) : \mathrm{tr}(A) = 0\}.$$

By Prop. 3.24 of [9], the Lie algebras of $\mathrm{U}(n)$ and $\mathrm{O}(n)$ are given by

$$\mathfrak{u}(n) = \{A \in M_n(\mathbb{C}) : A^* = -A\} \quad \text{and} \quad \mathfrak{o}(n) = \{A \in M_n(\mathbb{R}) : A^* = -A\}.$$

Moreover, again by Prop. 3.24 of [9], the Lie algebras of $\mathrm{SU}(n)$ and $\mathrm{SO}(n)$ are given by

$$\mathfrak{su}(n) = \{A \in M_n(\mathbb{C}) : A^* = -A,\ \mathrm{tr}(A) = 0\} \quad \text{and} \quad \mathfrak{so}(n) = \mathfrak{o}(n).$$

By formula (3.17) of [9], the complexifications of the real Lie algebras introduced above are given by

$$\begin{aligned}\mathfrak{gl}(n,\mathbb{R})_{\mathbb{C}} &\cong \mathfrak{gl}(n,\mathbb{C}),\\ \mathfrak{u}(n)_{\mathbb{C}} &\cong \mathfrak{gl}(n,\mathbb{C}),\\ \mathfrak{su}(n)_{\mathbb{C}} &\cong \mathfrak{sl}(n,\mathbb{C}),\\ \mathfrak{sl}(n,\mathbb{R})_{\mathbb{C}} &\cong \mathfrak{sl}(n,\mathbb{C}),\\ \mathfrak{so}(n,\mathbb{R})_{\mathbb{C}} &\cong \mathfrak{so}(n,\mathbb{C}).\end{aligned}$$

As mentioned in Sect. 1.3.1 of [9], the following Lie groups are compact:

$$\mathrm{O}(n), \quad \mathrm{SO}(n), \quad \mathrm{U}(n), \quad \text{and} \quad \mathrm{SU}(n).$$

By Prop 1.11, 1.12, and 1.13 and Exercise 1.13 of [9], the following Lie groups are connected:

$$\mathrm{GL}(n;\mathbb{C}) \quad \mathrm{SL}(n;\mathbb{C}) \quad \mathrm{U}(n) \quad \mathrm{SU}(n), \quad \text{and} \quad \mathrm{SO}(n).$$

By Prop. 13.11 of [9], the group $\mathrm{SU}(n)$ is simply connected. By Example 5.15 of [9], $\mathrm{SU}(2)$ is the universal cover of $\mathrm{SO}(3)$.

Of further interest are the real and complex *symplectic groups* $\mathrm{SP}(n,\mathbb{R})$ and $\mathrm{SP}(n,\mathbb{C})$, and the *compact symplectic group* $\mathrm{SP}(n)$; for their definitions we refer to Sect. 1.2.4 of [9].

A.10. *The Lie group SU(1,1)*

Let us define a Minkowski form $\{\,\cdot\,,\,\cdot\,\}:\mathbb{C}^2\to\mathbb{C}$ by

$$\{v,w\} := v_1^* w_1 - v_2^* w_2.$$

Note that this is almost identical to the usual definition of the inner product on $\mathbb{C}^2$ (in particular, it is conjugate linear in its first argument and linear in its second argument), except for the minus sign in front of the second term. Letting

$$M := \begin{pmatrix} 1 & 0 \\ 0 & -1 \end{pmatrix},$$

we can write

$$\{v,w\} = \langle v|M|w\rangle,$$

where $\langle\,\cdot\,,\,\cdot\,\rangle$ is the usual inner product. The Lie group $\mathrm{SU}(1,1)$ is the matrix Lie group consisting of all matrices $Y\in\mathcal{L}(\mathbb{C}^2)$ with determinant 1 that preserve this Minkowski form, i.e.,

$$\det(Y) = 1 \quad \text{and} \quad \{Yv,Yw\} = \{v,w\} \quad (v,w\in\mathbb{C}^2).$$

The second condition can be rewritten as $\langle Yv|M|Yw\rangle = \langle v|M|w\rangle$ which holds for all v, w if and only if

$$Y^*MY = M, \tag{A.19}$$

where Y^* denotes the usual adjoint of a matrix. Since

$$(e^{tA})^*Me^{tA} = M + t(A^*M + MA) + O(t^2),$$

it is not hard to see that a matrix of the form $Y = e^{tA}$ satisfies (A.19) if and only if

$$A^*M + MA = 0 \quad \Leftrightarrow \quad MA^*M = -A,$$

and the Lie algebra $\mathfrak{su}(1,1)$ associated with $\mathrm{SU}(1,1)$ is given by

$$\mathfrak{su}(1,1) = \big\{A \in M_2(\mathbb{C}) : MA^*M = -A,\ \mathrm{tr}(A) = 0\big\}.$$

It is easy to see that

$$A = \begin{pmatrix} A_{11} & A_{12} \\ A_{21} & A_{22} \end{pmatrix} \quad \Rightarrow \quad MA^*M = \begin{pmatrix} A_{11} & -(A_{21})^* \\ -(A_{12})^* & A_{22} \end{pmatrix}$$

and in fact the map $A \mapsto MA^*M$ satisfies the axioms of an adjoint operation. Let $\mathfrak{su}(1,1)_{\mathbb{C}}$ denote the Lie algebra

$$\mathfrak{su}(1,1)_{\mathbb{C}} := \big\{A \in M_2(\mathbb{C}) : \mathrm{tr}(A) = 0\big\},$$

equipped with the adjoint operation $A \mapsto MA^*M$. Then $\mathfrak{su}(1,1)$ is the real sub-Lie-algebra of $\mathfrak{su}(1,1)_{\mathbb{C}}$ consisting of all elements that are skew symmetric with respect to the adjoint operation $A \mapsto MA^*M$.

A basis for $\mathfrak{su}(1,1)_{\mathbb{C}}$ is formed by the matrices in (2.8), which satisfy the commutation relations (2.7). The adjoint operation $A \mapsto MA^*M$ leads to the adjoint relations (2.9). Some elementary facts about the Lie algebra $\mathfrak{su}(1,1)_{\mathbb{C}}$ are already stated in Section 2.4. Note that the definition of the "Casimir operator" in (2.10) does not follow the general definition for compact Lie groups in Proposition A.16, but is instead defined in an analogous way, replacing the inner product by a Minkowski form.

A.11. *The Heisenberg group*

Consider the matrices

$$X := \begin{pmatrix} 0&1&0 \\ 0&0&0 \\ 0&0&0 \end{pmatrix}, \quad Y := \begin{pmatrix} 0&0&0 \\ 0&0&1 \\ 0&0&0 \end{pmatrix}, \quad Z := \begin{pmatrix} 0&0&1 \\ 0&0&0 \\ 0&0&0 \end{pmatrix}.$$

We observe that

$$\begin{aligned} XX &= 0, & XY &= Z, & XZ &= 0, \\ YX &= 0, & YY &= 0, & YZ &= 0, \\ ZX &= 0, & ZY &= 0, & ZZ &= 0. \end{aligned}$$

The *Heisenberg group* H (Sect. 1.2.6 of [9]) is the matrix Lie group consisting of all 3×3 real matrices of the form

$$B = I + xX + yY + zZ \qquad (x, y, z \in \mathbb{R}).$$

To see that this is really a group, we note that if B is as above, then its inverse B^{-1} is given by

$$B^{-1} = -xX - yY + (xy - z)Z.$$

It is easy to see that $\{X, Y, Z\}$ is a basis for the Lie algebra $\mathfrak{h}$ of H. In fact, the expansion formula for $e^{t(xX+yY+zZ)}$ terminates and

$$e^{t(xX+yY+zZ)} = I + t(xX + yY + zZ) + \tfrac{1}{2}t^2xyZ \qquad (t \geq 0).$$

The basis elements X, Y, Z satisfy the commutation relations

$$[X, Y] = Z, \quad [X, Z] = 0, \quad [Y, Z] = 0.$$

Thus, we can abstractly define the *Heisenberg Lie algebra* as the real Lie algebra $\mathfrak{h}$ with basis elements $\mathbf{x}, \mathbf{y}, \mathbf{z}$ that satisfy the commutation relations

$$[\mathbf{x}, \mathbf{y}] = \mathbf{z}, \quad [\mathbf{x}, \mathbf{z}] = 0, \quad [\mathbf{y}, \mathbf{z}] = 0. \tag{A.20}$$

Representations of the Heisenberg algebra have already been discussed in Subsection 2.5.

References

1. R. Alkemper and M. Hutzenthaler, "Graphical representation of some duality relations in stochastic population models", *Electron. Commun. Probab.*, **12** (2007), 206–220.
2. V. Bargmann, "Irreducible unitary representations of the Lorentz group", *Ann. Math.*, **48** (1947), 568–640.
3. V. Bargmann, "On a Hilbert space of analytic functions and an associated integral transform", *Commun. Pure Appl. Math.*, **14** (1961), 187–214.
4. G. Carinci, C. Giardinà, C. Giberti, and F. Redig, "Dualities in population genetics: a fresh look with new dualities", *Stochastic Processes Appl.*, **125(3)** (2015), 941–969.
5. J. A. Fill, "Strong stationary duality for continuous-time Markov chains. I. Theory", *J. Theor. Probab.*, **5(1)** (1992), 45–70.
6. C. Giardinà, J. Kurchan, and F. Redig, "Duality and exact correlations for a model of heat conduction", *J. Math. Phys.*, **48** (2007), 033301, 1–15.

7. C. Giardinà, J. Kurchan, F. Redig, and K. Vafayi, "Duality and hidden symmetries in interacting particle systems", *J. Stat. Phys.*, **135(1)** (2009), 25–55.
8. C. Giardinà, F. Redig, and K. Vafayi, "Correlation Inequalities for Interacting Particle Systems with Duality", *J. Stat. Phys.*, **141(2)** (2010), 242–263.
9. B. C. Hall, *Lie Groups, Lie Algebras, and Representations: An Elementary Introduction,* Graduate Texts in Mathematics, vol. 222, Springer, 2003.
10. S. Jansen and N. Kurt, "Pathwise construction of certain moment dualities and application to population models with balancing selection", *Electron. Comm. Probab.*, **18** (14) (2013).
11. S. Jansen and N. Kurt, "On the notion(s) of duality for Markov processes", *Prob. Surveys,* **11** (2014), 59–120.
12. M. Novaes, "Some basics of $\mathfrak{su}(1,1)$", *Revista Brasileira de Ensino de Física,* **26(4)** (2014), 351–357.
13. J. Rosenberg, "A selective history of the Stone-von Neumann theorem", *Contemp. Math.*, **365** (2004), 331–353.
14. A. Sturm and J. M. Swart, "Pathwise duals of monotone and additive Markov processes", *J. Theor. Probab.* (2016). 52 pages. doi:10.1007/s10959-016-0721-5.
15. A. Sudbury and P. Lloyd, "Quantum operators in classical probability theory. II: The concept of duality in interacting particle systems", *Ann. Probab.*, **23(4)** (1995), 1816–1830.
16. A. Sudbury and P. Lloyd, "Quantum operators in classical probability theory. IV: Quasi-duality and thinnings of interacting particle systems", *Ann. Probab.*, **25(1)** (1997), 96–114.
17. A. Sudbury, "Dual families of interacting particle systems on graphs", *J. Theor. Probab.*, **13(3)** (2000), 695–716.
18. J. M. Swart, "Duals and thinnings of some relatives of the contact process", Pages 203-214 in: *Prague Stochastics 2006*, M. Hušková and M. Janžura (eds.), Matfyzpress, Prague, 2006, ArXiv:math.PR/0604335.
19. J. M. Swart, *Duality and Intertwining of Markov Chains,* Lecture notes for the ALEA in Europe School October 21-25 2013, Luminy (Marseille), available at: `http://staff.utia.cas.cz/swart/tea_index.html`.
20. J. M. Swart, *Introduction to Quantum Probability,* Lecture notes (2017) available at: `http://staff.utia.cas.cz/swart/tea_index.html`.
21. N. Ya. Vilenkin and A.U. Klimyk. *Representation of Lie groups and special functions. Volume 1: Simplest Lie groups, special functions and integral transforms,* Kluwer, Dordrecht, 1991.

The Genealogy of Extremal Particles of Branching Brownian Motion

Sandra Kliem* and Kumarjit Saha†

*Institute of Mathematics
Goethe University Frankfurt
Robert-Mayer-Str. 10, Room 914
60325 Frankfurt am Main, Germany
kliem@math.uni-frankfurt.de

†Ashoka University
Rajiv Gandhi Education City
Rai 131029, Haryana, India
kumarjit.saha@ashoka.edu.in

Branching Brownian motion (BBM) is used as a population model where the location of a particle on the real line represents his or her fitness which evolves according to a Brownian motion due to mutations. In this context, the genealogical tree of the particles is important to understand the evolution of the population over time.

The main goal of this article is to give an overview and motivations for some of the results on the genealogy of extremal particles of BBM.

In the first part of this article, we discuss the results of L.-P. Arguin, A. Bovier and N. Kistler from [Genealogy of Extremal Particles of BBM, 2011] and [Poissonian Statistics in the Extremal Process of BBM, 2012]. Their articles give a partial answer to what the distribution of the limiting extremal process of BBM looks like. These two articles highlight, how genetical information can successfully be used to obtain rather extensive information for the limiting particle picture.

In the second part, we discuss the results of J. Berestycki, N. Berestycki and J. Schweinsberg from [The Genealogy of BBM with Absorption, 2013]. BBM with absorption is related to a population model with selection. This article shows that the genealogy of the surviving particles (with a selective advantage), which are 'local' extremal particles, is then governed by a Bolthausen-Sznitman coalescent process. This result is in sharp contrast to the standard population models without selection.

Contents

1. Introduction

Branching Brownian motion (BBM) is a stochastic process that describes the evolution of a set of particles over time, where particles split at random times and move in space $\mathbb{R}$. To be more precise, define BBM as follows.

Definition 1.1: (Branching Brownian motion (BBM)). At time $t = 0$, start with a single particle, say $x_1(0)$, at the origin $0 \in \mathbb{R}$. The path of the particle is modeled as a Brownian motion (BM). At a (random) point in time $\tau \sim \text{Exp}(1)$, that is, τ has exponential distribution with parameter 1, the particle splits into two identical particles that start (both) at $x_1(\tau)$ and continue to move as two independent BMs each. Repeat this procedure for each of the ensuing particles independently.

In the first part of this article, Section 2, we shortly discuss BBM as a stochastic process. The distribution of BBM can be described in terms of solutions of the so-called F-KPP equation. For the latter, an implicit solution is given in Bramson, [14] and the convergence of solutions to travelling waves is established. His techniques are of fundamental use in Arguin, Bovier and Kistler [2], [3] and we therefore provide a short motivation of the techniques involved in the first part. This is followed by an overview of results on the distribution of the *maximum* of BBM. In particular, a deterministic centering term $m(t)$ is introduced that is of order-1-distance to the *median* of the maximum. In their two articles, Arguin, Bovier and Kistler give a partial answer to what the distribution of the limiting extremal process of BBM looks like, that is, the distribution around the "tip", centered by $m(t)$. In [2] they show that the genetic distance of extremal particles at time t is either 0 or t up to order 1 with high probability. In [3], they use this result to deduce much more. Namely, they conclude that the limiting extremal process takes the form of a randomly shifted cluster point process. Although the resulting description of the limiting process is not complete, it provides a very instructive and illustrative example of how information on genealogies can be of use in understanding the evolution of a particle system over time. It is for this reason, that we chose to focus on these two articles. Let us point out that Arguin, Bovier and Kistler [4] and Aïdékon, Berestycki, Brunet and Shi [1] established simultaneously the complete descriptions of the limit of the extremal process.

In [7], Brownian particles whose fitness becomes too low, that is, comes below a linear absorption barrier, are eliminated. This model is the content of the second part of this article, Section 3. Consider a family of $N \in \mathbb{N}$ independent BBMs with initial positions distributed on the positive real line. Fitness of a particle is measured in terms of position. If the fitness of a particle becomes too low, it gets killed. The parameters of the absorption barrier and the initial positions in the Nth model are chosen such that the population size does not fluctuate much and remains of the same order as that of the initial size. Time is measured in units of $(\log N)^3$. Mostly, the cloud of the (surviving) particles has diameter of logarithmic order and moves with a linear speed. Most of the particles stay inside this region throughout and the number of such particles essentially behaves like a deterministic process. In addition to this, occasionally a particle travels too far to the right and the descendants of such particle quickly build up to a positive proportion of the population. Proper time scaling to observe such particles and proper estimates on the contribution of such particles give the required result in [7]. In particular, in the limit of $N \to \infty$, the genealogy of the surviving particles, which are 'local' extremal particles, is then governed by a so-called Bolthausen-Sznitman coalescent, see Definition 3.5 below for details.

The results of Section 2 are very different from the results of Section 3. Indeed, as we will see in Section 2 for BBM without selection, following the path of two extremal particles at time t, t big, backwards in time, only pairwise merging (coalescence) of two separate lineages occurs by construction, almost surely. The coalescence takes place either very early (that is, at time 0 up to order 1) or very late (that is, at time t up to order 1). By killing off particles if their fitness becomes too low, that is under selection in the form of an absorption barrier, the behavior at the front changes drastically. Note that in contrast to BBM without selection, for N big, the extremal particles of BBM under selection coincide with the whole cloud of particles. Tracing back the lineages of particles backwards on a time scale of units $(\log N)^3$, their genealogy is now close to a Bolthausen-Sznitman coalescent process. The latter allows for the coalescence of blocks of particles of any size at random times. These coalescence events occur at constant rates backwards over time. Thus the behavior of the extremal particles in BBM under selection is in sharp contrast with the behavior of the extremal particles in BBM without selection.

It is our goal to give an overview of the results obtained and to present the main ideas and motivations of the proofs.

2. Extremal particles in branching Brownian motion

2.1. *Branching Brownian motion*

Recall Definition 1.1 where we introduced BBM as a stochastic process that describes the evolution of a set of particles over time, where particles split at random times and move in space $\mathbb{R}$. We use binary branching for the sake of clarity. There are two ways of interpreting a splitting-event. Either a particle gives birth to *one* new particle or a particle dies and at the time of death gives birth to *two* new particles. We will choose the latter interpretation and call the particle before the splitting-event, respectively

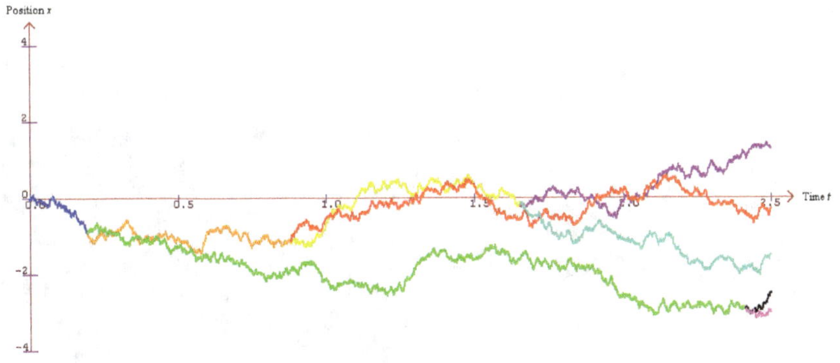

Fig. 1. Realisation of BBM over a short time-interval, cf. homepage of Matt Roberts, University of Bath for the colored version.

the two particles after the splitting-event, the "parent", respectively the "children".

Note the *recursive structure* of the process. After a split, each child and its progeny can be thought of as constituting again a BBM, independent of everything else, starting with a single particle (the child) shifted (in space) to the position of the parent at the time of its death (the time of the splitting-event).

The resulting process is a collection of a (random) number $n(t)$ of particles

$$\big(x_1(t), x_2(t), \ldots, x_{n(t)}(t)\big)_{t\geq 0} = \big\{x_k(t) : k \leq n(t)\big\}_{t\geq 0}.$$

Note the following slight abuse of notation. When investigating the path of a particle alive at time t, that is $(x_k(s))_{s\in[0,t]}$ for some $k \leq n(t)$, the position of the particle (or its parent or its grandparent etc.) at time $0 \leq s \leq t$ will also be denoted by $x_k(s)$ when the context is clear.

Two realisations of BBM are given in Figures 1 and 2.

Remark 2.1: There are connections of BBM to spin glass theory, in particular, to the Generalised Random Energy models (GREM). There are also connections to extremes of the free Gaussian random field in $d = 2$. Many results can be extended to branching random walk and variable speed BBM. The interested reader may have a look at Bovier's lecture notes [12] for some of these results and for pointers to relevant literature.

The genealogy of the particles of BBM can be read off from its construction (recall the notion of "parent" and "children").

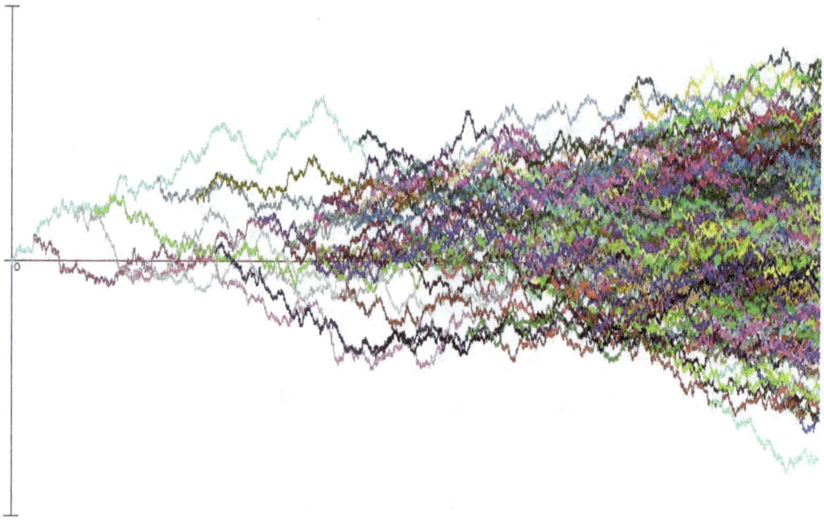

Fig. 2. Realisation of BBM over a long time-interval, cf. homepage of Matt Roberts, University of Bath for the colored version.

Definition 2.2: (Genealogies of BBM). Let the genealogical distance between two particles be the *time of death of the longest surviving ancestor of both particles*, that is, the *unique time where the most recent common ancestor (MRCA) splits*. That is,

$$\begin{aligned} d(x_k(t), x_\ell(t)) &\equiv \inf\{0 \le s \le t : x_k(s) \ne x_\ell(s)\} \\ &= \text{time (from 0) to } MRCA. \end{aligned}$$

The average number of particles at time $t \ge 0$ is given explicitly below. We only make use of (2.1) in what follows. A proof of the results can be found in [12]. Equation (2.1) is deduced from the recursive structure (cf. comment below Definition 1.1) and the strong Markov property of the process.

Lemma 2.3: *The average of the number of particles* $n(t)$ *at time* $t \ge 0$ *is*

$$\mathbb{E}[n(t)] = e^t. \tag{2.1}$$

Moreover, $e^{-t}n(t)$ *is a martingale that converges, almost surely and in* L^1, *to an exponential random variable of parameter* 1.

A proof can be found in Lemmas 3.2, 3.3 in [12]. Note that equation (2.1) follows directly from the many-to-one lemma below using $g_t(v) \equiv 1$.

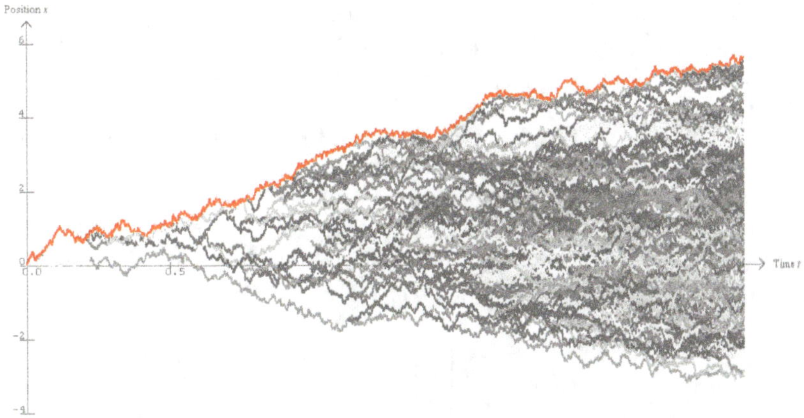

Fig. 3. The maximum of BBM, see J. Berestycki (Lecture Notes, "Topics on BBM", http://www.stats.ox.ac.uk/~berestyc/Articles/EBP18_v2.pdf, Figure 3.1), image by Matt Roberts.

The latter is a useful tool in branching processes (see for instance Harris and Roberts [19] for background and extensions).

Lemma 2.4: *(Many-to-one lemma, cf. [33], Subsection 3.1). Let $g_t(x)$ be a measurable functional of t and the path of a particle $x = (x(s))_{s\in[0,t]}$ up to time t. Then*

$$\mathbb{E}\left[\sum_{k\le n(t)} g_t\big((x_k(s))_{s\in[0,t]}\big)\right] = e^t E[g_t(\xi)],$$

where $\xi_t, t \ge 0$ is standard Brownian motion under P.

2.2. *The F-KPP equation*

Solutions to the following partial differential equation (PDE) turn out to be a crucial tool in the description of the law of the maximum of BBM ("the tip", see Figure 3) and the distribution of BBM in general as we will see in the examples given below.

Consider the partial differential equation

$$u_t = \frac{1}{2}u_{xx} + u^2 - u, \qquad u = u(t,x) \in [0,1],\ t \ge 0,\ x \in \mathbb{R}, \tag{2.2}$$
$$u(0,x) = f(x).$$

Remark 2.5: Set $v = 1 - u$. Then this PDE is a special case of the Kolmogorov-Petrovskii-Piskunov (KPP) equation (also known as the Kolmogorov- or Fisher-equation).

The crucial connection between BBM and the F-KPP equation is given by the following lemma.

Lemma 2.6: *(cf. Lemma 3.4 of [12]). Let* $\{x_k(t) : k \leq n(t)\}_{t\geq 0}$ *be a BBM starting at* 0. *Let* $f : \mathbb{R} \to [0,1]$ *and* $u(t,x)$ *the solution to the F-KPP equation (2.2) with* $u(0,x) = f(x)$. *Then*

$$u(t,x) = \mathbb{E}\left[\prod_{k=1}^{n(t)} f(x - x_k(t))\right]. \tag{2.3}$$

Idea of proof. Branching property: Let $p_t(x) = \frac{1}{\sqrt{2\pi t}} e^{-x^2/2t}$ denote the Heat kernel. Recall that the time to the first split is $\tau \sim \text{Exp}(1)$ with density $f_\tau(s) = \mathbb{1}_{\{s\geq 0\}} e^{-s}$ and consequently $\mathbb{P}(\tau > t) = e^{-t}$. Use $u(t,x)$ to denote the right hand side in (2.3). According to if the split happens after time t or before, one obtains

$$u(t,x) = e^{-t} \int p_t(z) f(x-z) dz + \int_0^t e^{-s} \int p_s(z) u^2(t-s, x-z) dz ds.$$

Indeed, for the second term, use that if the first split occurs at time $s \leq t$ at position z, the progeny of the two children are independent BBM starting at $z \in \mathbb{R}$ each.

Now differentiate with respect to t, use integration by parts and $\frac{\partial}{\partial t} p_t(x) = \frac{1}{2} \frac{\partial^2}{\partial_x \partial_x} p_t(x)$ to see that $u(t,x)$ solves indeed the F-KPP equation with initial condition $f(x)$. By the uniqueness of solutions to (2.2), the claim follows. □

Example 2.7: Let $M(t) = \max_{k\leq n(t)} x_k(t)$. With $f(x) = \mathbb{1}_{[0,\infty)}(x)$ one obtains

$$\begin{aligned} u(t,x) &= \mathbb{E}\left[\prod_{k=1}^{n(t)} \mathbb{1}_{[0,\infty)}(x - x_k(t))\right] = \mathbb{E}\left[\prod_{k=1}^{n(t)} \mathbb{1}_{\{x_k(t)\leq x\}}\right] \\ &= \mathbb{P}\left(\max_{k\leq n(t)} x_k(t) \leq x\right) = \mathbb{P}(M(t) \leq x) = F_{M(t)}(x), \end{aligned}$$

where we use F_X to denote the cumulative distribution function of a real-valued random variable X. For an example of the evolution of $M(t)$, see Figure 3.

Example 2.8: Let $f(x) = e^{-\phi(x)}, \phi \in \mathcal{C}_c^+$, the set of non-negative continuous functions with compact support. Set $\mathcal{P}_t = \sum_{k=1}^{n(t)} \delta_{x_k(t)}$, the point process associated to BBM at time t. Then

$$u(t,x) = \mathbb{E}\left[\prod_{k=1}^{n(t)} e^{-\phi(x-x_k(t))}\right] = \mathbb{E}\left[e^{-\int \phi(x-z)\mathcal{P}_t(dz)}\right].$$

The right hand side gives the *Laplace functional* of the point process $\mathcal{P}_t$. Recall that the Laplace functional of a point process uniquely characterizes its distribution (cf. [12], Proposition 6.6 and note that it suffices to consider $\phi \in \mathcal{C}_c^+$ by dominated convergence).

2.2.1. *An implicit representation of the solution of the F-KPP equation*

In what follows, consider $v(t,x) \equiv 1-u(t,x)$ instead of $u(t,x)$. Then $v(t,x)$ solves

$$v_t = \frac{1}{2}v_{xx} - v^2 + v = \frac{1}{2}v_{xx} + (1-v)v, \qquad v(0,x) = 1-u(0,x). \tag{2.4}$$

We next give two implicit probabilistic representations of the solution of the F-KPP equation. Several observations follow to give the interested reader a flavor of the way this representation can be used to derive analytical properties of $v(t,x)$ respectively $u(t,x)$.

The starting point is the celebrated Feynman-Kac formula (cf. Bramson [14], (1.31)–(1.32)). It gives a representation of solutions to the *linear* PDE

$$v_t = \frac{1}{2}v_{xx} + k(t,x)v, \qquad v_0(x) = v(0,x). \tag{2.5}$$

Namely,

$$v(t,x) = \mathbb{E}\left[\exp\left(\int_0^t k(t-s, B^x(s))ds\right) v(0, B^x(t))\right],$$

where $(B^x(t))_{t\geq 0}$ denotes BM, starting in $x \in \mathbb{R}$. Bramson [14] sets $k(t,x) = 1-v(t,x)$ with v solving (2.4) and obtains as an *implicit probabilistic description of* v,

$$v(t,x) = \mathbb{E}\left[\exp\left(\int_0^t (1-v(t-s, B^x(s)))ds\right) v(0, B^x(t))\right]. \tag{2.6}$$

Bramson's interpretation is that of $v(t,x)$ as "the weighted average of the different sample paths of BM". His interpretation becomes clearer in the context of the following remark.

Remark 2.9: (Observations). Recall that $k(t,x) = 1 - v(t,x) = u(t,x)$ and consider $u(0,x) = \mathbb{1}_{[0,\infty)}(x)$ as in Example 2.7. Then one can show (cf. [14], (1.34)–(1.36) and note that "his" u is "our" v) that

(a) $0 < k(r,y) = u(r,y) < 1$ for all $r > 0, y \in \mathbb{R}$.
(b) For all $0 < \epsilon < 1$,

$$\limsup_{y\to-\infty} k(r,y) \le \epsilon \text{ and } \lim_{y\to\infty} k(r,y) = 1 \text{ for } r \ge r(\epsilon),$$

that is, for y large, the weighting of a path is nearly maximal and for y small, insignificant. Bramson [14] thus distinguishes paths $y(s), 0 \le s \le t$ according to

$$\exp\left(\int_0^t k(t-s, y(s))ds\right) \sim e^t \text{ or } \exp\left(\int_0^t k(t-s, y(s))ds\right) \ll e^t.$$

(c) There is a transition near $k(r,y) = u(r,y) = 1/2$. Note that this can be taken as a motivation for considering $m(t) = \sup\{x : u(t,x) \le 1/2\} + O(1)$ for the *centering term* in (2.8).

To describe, before weighting, which path $(B^x(s))_{0\le s\le t}$ follows (cf. for instance part (b) in the remark above), it is convenient to rewrite (2.6) into *another implicit probabilistic description* of v by conditioning on the final (at time t) position of the path of the Brownian motion, starting in x.

$$\begin{aligned} v(t,x) &= \mathbb{E}\left[\mathbb{E}\left[\exp\left(\int_0^t (1 - v(t-s, B^x(s)))ds\right) v(0, B^x(t)) \,\Big|\, B^x(t)\right]\right] \\ &= \int_{-\infty}^{\infty} \frac{e^{-(x-y)^2/2t}}{\sqrt{2\pi t}} \mathbb{E}\left[\exp\left(\int_0^t (1 - v(t-s, \mathfrak{z}^t_{x,y}(s)))ds\right)\right] v(0,y)dy, \end{aligned}$$

where $\mathfrak{z}^t_{x,y}$ denotes a *Brownian bridge* starting at x (at time 0) and ending at y (at time t). For the reader unfamiliar with Brownian bridges, we shortly recall some basic facts (cf. [14], beginning of Chapter 2 until including Lemma 2.1) that turn out to be useful for the analysis of solutions to the F-KPP equation.

Remark 2.10: (Brownian bridges).

(a) A Brownian bridge $(\mathfrak{z}^t_{x,y})_{t\ge 0}$ has the distribution of a BM starting at x, conditional on being in y at time t.
(b) Starting in $x = 0$, it can be defined as $\mathfrak{z}^t_{0,0}(s) = B^0(s) - \frac{s}{t}B^0(t)$, $0 \le s \le t$. It is independent of $B^0(t)$.
(c) For arbitrary initial condition $x \in \mathbb{R}$, $\mathfrak{z}^t_{x,y}(s)$ is equal in distribution to $\mathfrak{z}^t_{0,0}(s) + \frac{s}{t}y + \frac{t-s}{t}x$, $0 \le s \le t$.

(d) A Brownian bridge $(\mathfrak{z}^t_{x,y})_{t\geq 0}$ is a Gaussian process, almost surely continuous on $[0,t]$ and it has the strong Markov property.
(e) Finally, $\mathrm{Var}(\mathfrak{z}^t_{0,0}(s)) = \frac{s(t-s)}{t}$ attains its maximum in the middle of the interval $[0,t]$, namely $\mathrm{Var}(\mathfrak{z}^t_{0,0}(t/2)) = t/4$.

2.3. *The distribution of the maximum of BBM*

In this subsection we collect first results on the distribution of the maximum of BBM $M(t) = \max_{k\leq n(t)} x_k(t)$. The interested reader is referred to the references given below. Due to the dependence structure of particles (between time 0 up to their MRCA they follow the same path), particles tend to stay closer together than in the i.i.d. case, a fact heuristically confirmed in Remark 2.11 below. As a result, the behavior of the extremas of the process also changes.

Let us return to Example 2.7. Then $u(t,x) = \mathbb{P}(M(t) \leq x) = \mathbb{P}(\max_{k\leq n(t)} x_k(t) \leq x)$ solves (2.2) with $u(0,x) = \mathbb{1}_{[0,\infty)}(x)$. In particular, $0 < u(t,x) < 1$ for all $x \in \mathbb{R}, t > 0$.

Take as *centering term*,

$$m(t) \equiv \sqrt{2}t - \frac{3}{2\sqrt{2}}\log(t). \tag{2.7}$$

Then

$$m(t) = \sup\{x : u(t,x) \leq 1/2\} + O(1), \tag{2.8}$$

see Bramson [13], Theorem 1 (note that here, "his" u is "our" u) or Roberts [33], the latter including a short proof of the former result. Also recall Remark 2.9(c). Note that $m(t)$ is deterministic. Moreover,

$$\begin{aligned}\mathbb{P}(M(t) - m(t) \leq x) &= u(t, m(t) + x) \\ &\to w(x) \ \text{ uniformly in } x \text{ as } t \to \infty.\end{aligned} \tag{2.9}$$

Here, $w(x)$ is the unique (up to translation) solution of the equation

$$\frac{1}{2}w_{xx} + \sqrt{2}w_x + w^2 - w = 0 \tag{2.10}$$

satisfying $0 < w(x) < 1$ for all $x \in \mathbb{R}$ and $w(x) \to 0$ as $x \to -\infty$, $w(x) \to 1$ as $x \to \infty$. Thus, $M(t)$ is indeed in the vicinity of the centering term $m(t)$ with high probability.

Remark 2.11: (The i.i.d. case). For a BBM, $\mathbb{E}[n(t)] = e^t$ by Lemma 2.3. Set $n = e^t$ and consider $B_k, 1 \leq k \leq n$ independent standard BMs. In this case, convergence in distribution as in (2.9) follows with $m(t) = \sqrt{2}t -$

$\frac{1}{2\sqrt{2}}\log(t)$. Indeed, let $X_1(t),\dots,X_n(t)$ be $n \in \mathbb{N}$ i.i.d. standard normal random variables and let

$$b_n = \sqrt{2\log(n)} - \frac{\log(\log(n)) + \log(4\pi)}{2\sqrt{2\log(n)}} \quad \text{and} \quad a_n = \sqrt{2\log(n)}.$$

Then, by Theorem 1.5.3 in Leadbetter, Lindgren and Rootzén [28], for all $x \in \mathbb{R}$,

$$\lim_{n\to\infty} \mathbb{P}\Big(\max_{k\le n} X_k(t) \le b_n + x/a_n\Big) = \exp(-e^{-x}).$$

Set $n = e^t$ and consider $B_k, 1 \le k \le n$ independent standard BMs starting in 0. Set $X_k(t) = B_k(t)/\sqrt{t}$. Then, for $y \equiv x/\sqrt{2}$,

$$\begin{aligned}&\lim_{t\to\infty} \mathbb{P}\Big(\max_{k\le e^t} \frac{B_k(t)}{\sqrt{t}} \le \sqrt{2t} - \frac{\log(t)}{2\sqrt{2t}} + \frac{\log(4\pi)+2x}{2\sqrt{2t}}\Big)\\ &= \lim_{t\to\infty} \mathbb{P}\Big(\max_{k\le e^t} B_k(t) \le \sqrt{2}t - \frac{\log(t)}{2\sqrt{2}} + O(1) + y\Big) = \exp\Big(-e^{-\sqrt{2}y}\Big).\end{aligned}$$

Lalley and Sellke [24], Theorem 1 give an explicit description of $w(x)$ (cf. (2.9)–(2.10)). Let

$$Z(t) = \sum_{k=1}^{n(t)} (\sqrt{2}t - x_k(t))e^{-\sqrt{2}(\sqrt{2}t - x_k(t))} \tag{2.11}$$

be the so-called *derivative martingale*, then

$$Z = \lim_{t\to\infty} Z(t) \tag{2.12}$$

exists, is finite and is strictly positive almost surely. Moreover, for some $C > 0$,

$$w(x) = \mathbb{E}\Big[\exp\Big(-CZe^{-\sqrt{2}x}\Big)\Big] = \mathbb{E}\Big[\exp\Big(-e^{-\sqrt{2}\left(x - \frac{\log(CZ)}{\sqrt{2}}\right)}\Big)\Big]. \tag{2.13}$$

Note that the definition of $m(t)$ in (6) of [24] differs from (2.7) by a constant $+o(1)$ term that only changes the value of the constant C in the above statement. The expectation is taken with respect to Z. Observe in particular that

$$\text{for } x \to \infty, \quad 1 - w(x) \sim Cxe^{-\sqrt{2}x} \tag{2.14}$$

(cf. [24], (11)).

Remark 2.12: The so-called *Gumbel distribution* has cumulative distribution function (cf. Johnson, Kotz and Balakrishnan [20], pages 2–4)

$$F_G(x) = \mathbb{P}(G \le x) = \exp\big(-e^{-(x-\mu)/\beta}\big) = \exp\big(-e^{\mu/\beta}e^{-x/\beta}\big)$$

with parameters $\mu \in \mathbb{R}, \beta > 0$. It appears as one of three extreme value distributions (cf. [28], page 4 and Theorem 1.4.2). Hence, $w(x)$ represents a *random shift* of "the" Gumbel distribution with $\mu = 0$ and $\beta = 1/\sqrt{2}$. Note that in Remark 2.11, the limiting cumulative distribution function is "the" Gumbel distribution.

To conclude, let us mention Theorem 2 of [24] and the Corollary that follows it.

Theorem 2.13: *Suppose two independent BBMs* $\left(X_1^A(t), \ldots, X_{n^A(t)}^A(t)\right)$ *and* $\left(X_1^B(t), \ldots, X_{n^B(t)}^B(t)\right)$ *are started at* 0 *and* x*, respectively, where* $x < 0$*. Then with probability* 1 *there exist random times* $t_n, n \in \mathbb{N}, t_n \to \infty$ *such that*

$$M^A(t_n) < M^B(t_n)$$

for all $n \in \mathbb{N}$.

Idea of proof. Use (2.9), that is, $\lim_{t\to\infty} \mathbb{P}(M(t) - m(t) \leq x) = w(x)$. □

Corollary 2.14: *Every particle born in a BBM has a descendant particle in the "lead" at some future time.*

We note that this result stands in sharp contrast to the behavior of BBM under absorption, where particles are killed upon hitting a space time line, see Section 3.

2.4. *Extremal particles in BBM*

This subsection gives an introduction to the results of Arguin, Bovier and Kistler [2] and Arguin, Bovier and Kistler [3]. We are interested in the limit $(t \to \infty)$ of the *extremal process*

$$\mathcal{E}_t \equiv \sum_{k \leq n(t)} \delta_{x_k(t) - m(t)} \equiv \sum_{k \leq n(t)} \delta_{\overline{x_k}(t)}.$$

Recall: The *centering term* $m(t)$ is deterministic and satisfies

$$m(t) = \sqrt{2}t - \frac{3}{2\sqrt{2}} \log(t) = \sup\{x : u(t,x) \leq 1/2\} + O(1).$$

By (2.9), we can choose $D \subset \mathbb{R}$ big enough such that $M(t) = \max_{k \leq n(t)} x_k(t) \in m(t) + D \equiv \{x \in \mathbb{R} : \exists d \in D : x = m(t) + d\}$ with high probability.

Note that the description of the limit ($t \to \infty$) of $\mathcal{E}_t$ by means of [2] and [3] is not complete although it gives a good description of the general evolution of the extremal process over time. A complete description was later obtained simultaneously by [4] respectively [1], see Subsection 2.7.

2.4.1. *Genealogy of extremal particles of BBM*

We start with the main result of [2]. Although this result might not look very revealing at a first glance, we will see in what follows that a lot on the "distribution at the front" can be deduced from it. Recall the definition of genealogical distances from Definition 2.2.

Theorem 2.15: *(Genealogy of Extremal Particles, Theorem 2.1 of [2]). For any compact set $D \subset \mathbb{R}$,*

$$\lim_{r\to\infty} \sup_{t>3r} \mathbb{P}\big(\exists\, 1 \le k, \ell \le n(t) : \overline{x_k}(t), \overline{x_\ell}(t) \in D \quad \text{and } d(x_k(t), x_\ell(t)) \in (r, t-r)\big) = 0. \tag{2.15}$$

Remark 2.16: We conclude that the MRCA of two extremal particles at time t splits / branches off with high probability at a time

(1) in the interval $(0, r)$ ("very early branching") or
(2) in the interval $(t - r, t)$ ("very late branching").

Have a look at Figure 4 to get a general idea of this result. The idea of the proof is given below in Subsection 2.6.

2.4.2. *Poissonian statistics in the extremal process of BBM*

By taking $r = \epsilon t$, we obtain the following Corollary to Theorem 2.15.

Corollary 2.17: *For any compact set $D \subset \mathbb{R}$ and $0 < \epsilon < 1/3$,*

$$\lim_{t\to\infty} \mathbb{P}\left(\exists\, 1 \le k, \ell \le n(t) : \overline{x_k}(t), \overline{x_\ell}(t) \in D \text{ and } \frac{d(x_k(t), x_\ell(t))}{t} \in (\epsilon, 1-\epsilon)\right) = 0.$$

Without loss of generality, order the (centered) particles in decreasing order, that is, let

$$\overline{x_1}(t) \ge \overline{x_2}(t) \ge \cdots \ge \overline{x_{n(t)}}(t). \tag{2.16}$$

Set

$$\overline{Q}(t) = \{\overline{Q}_{k\ell}(t)\}_{k,\ell \le n(t)} = \left\{\frac{d(x_k(t), x_\ell(t))}{t}\right\}_{k,\ell \le n(t)}.$$

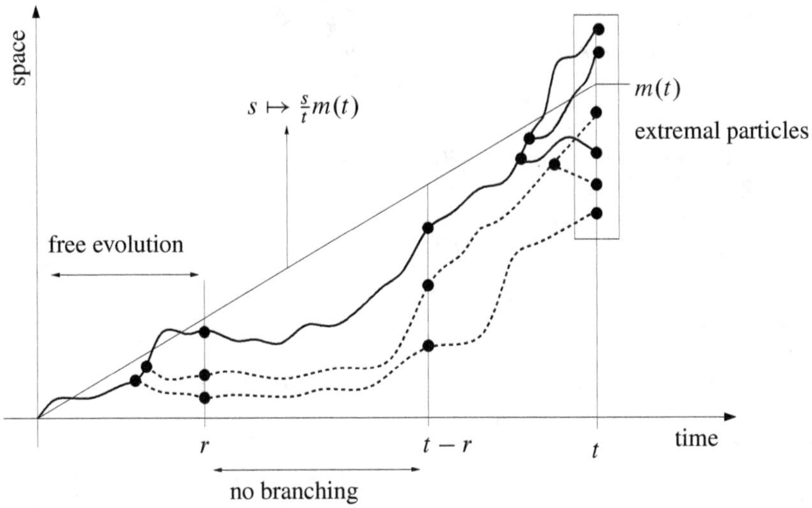

Fig. 4. Evolution of the system (Figure 2.4 of [2]).

Now partition extremal particles into disjoint families. Two particles are in the same family, or equivalence class, if their MRCA split "very late" (cf. (2) in Remark 2.16). By contrast, the MRCA of two disjoint families has to have split "very early". The following definition allows for a choice of a representative for each family with the additional property, that it is maximal (according to (2.16)) in its class.

Definition 2.18: Let $0 < q < 1$. The *q-thinning* $\mathcal{E}_t^{(q)}$ of the pair $(\mathcal{E}_t, \overline{Q}(t))$ is defined as follows:

(i) Consider equivalence classes of particles alive at time t, where two particles are equivalent if their MRCA split at a time later than $q \cdot t$, that is,

$$k \sim_q \ell \iff \overline{Q}_{k\ell}(t) \geq q.$$

(ii) Select the maximal (according to (2.16)) particle within each class.
(iii) Then $\mathcal{E}_t^{(q)}$ is the point process of these representatives.

Remark 2.19: The thinning map $(\mathcal{E}_t, \overline{Q}(t)) \mapsto \mathcal{E}^{(q)}(t)$ is continuous on an appropriate space, cf. (2.6) in [3]. The construction extends to $q = q(t) \in (0, 1)$.

Theorem 2.20: *(Theorem 2, [3]). For any $0 < q < 1$, the processes $\mathcal{E}_t^{(q)}$ converge in law to the same limit, $\mathcal{E}^0$. Also,*

$$\lim_{r\to\infty}\lim_{t\to\infty}\mathcal{E}_t^{(1-r/t)} = \mathcal{E}^0.$$

Moreover, conditionally on Z, (the limit of the derivative martingale, cf. (2.12)),

$$\mathcal{E}^0 = PPP\big(CZ\sqrt{2}e^{-\sqrt{2}x}dx\big) \tag{2.17}$$

(PPP stands for "Poisson Point Process"), where $C > 0$ is the constant appearing in (2.13).

The idea of the proof is given below in Subsection 2.5.

Remark 2.21: This result can be visualized as in Figures 5 and 6. The particles at the frontier of BBM for large times can thus be constructed as follows:

(1) set down so-called *cluster-extrema* according to $\mathcal{E}^0$, that is, according to a randomly shifted PPP with "exponential" ($x \in \mathbb{R}$) intensity;
(2) attach to each cluster-extrema a so-called *cluster*. (The distribution of these clusters was later determined in [4] respectively [1], cf. Subsection 2.7. As it turns out, the clusters are independent identically distributed.)
(3) Note that particles in one cluster lie to the left (in space) of their corresponding cluster-extrema (Poissonian particles).

Remark 2.22: The PPP-structure from (2.17) may not come as a big surprise to those acquainted with the theory of extreme values. See for instance Corollary 4.19(i) in Resnick [32] (for the definition of Λ, see Proposition 0.3(iii), PRM stands for "Poisson random measure"). Recall Figure 4. With high probability, the ancestors of extremal particles from different clusters evolve independently for the time-interval $[r, t-r]$ thus suggesting the PPP-structure.

Finally, recall Remark 2.12. The random shift can be seen as a result of the free evolution up to time r (cf. paragraph below (1.11) in [3]).

We conclude this subsection with an invariance property for the limiting extremal process.

Remark 2.23: (Invariance property, cf. [4], (3.9)). The law of the *limiting extremal process* $\mathcal{E} \equiv \sum_{i\in\mathbb{N}}\delta_{e_i}$ satisfies the following invariance property:

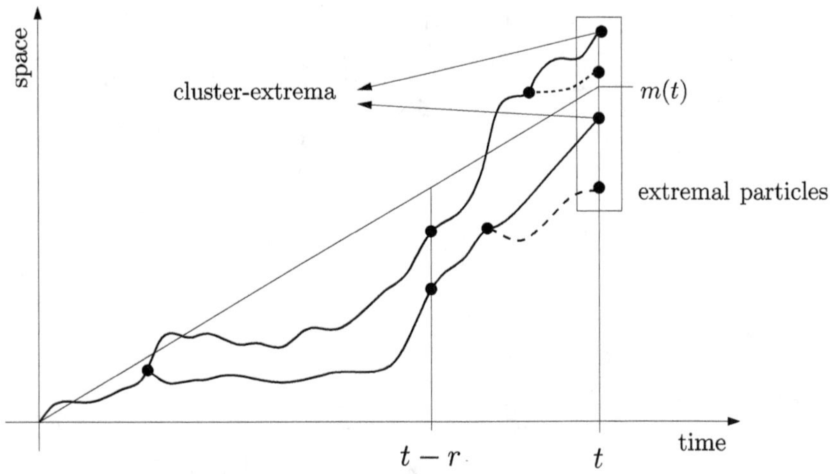

Fig. 5. Cluster-extrema (Figure 2 in [3]).

For any $s \geq 0$,

$$\mathcal{E} \stackrel{\text{law}}{=} \sum_{i\in\mathbb{N}, k\leq n^{(i)}(s)} \delta_{e_i + x_k^{(i)}(s) - \sqrt{2}s}, \tag{2.18}$$

where $\{x_k^{(i)}(s) : k \leq n^{(i)}(s), s \geq 0\}_{i\in\mathbb{N}}$ are i.i.d. BBMs. Indeed, use that for $t \to \infty$,

$$m(t) = m(t-s) + \sqrt{2}s + o(1).$$

Then use the recursive structure and the strong Markov property of the process once more to rewrite

$$\begin{aligned}\mathcal{E}_t \equiv \sum_{i\leq n(t)} \delta_{x_i(t)-m(t)} &= \sum_{i\leq n(t-s)} \sum_{k\leq n^{(i)}(s)} \delta_{x_i(t-s)+x_k^{(i)}(s)-m(t)} \\ &= \sum_{i\leq n(t-s)} \sum_{k\leq n^{(i)}(s)} \delta_{x_i(t-s)-m(t-s)+x_k^{(i)}(s)-\sqrt{2}s+o(1)}\end{aligned}$$

and take $t \to \infty$.

Remark 2.24: Recall from (2.9) that $M(s) = \max_{k\leq n(s)} x_k(s)$ is close to $m(s) = \sqrt{2}s - \frac{3}{2\sqrt{2}}\log(s) < \sqrt{2}s$ with high probability. Hence, $\max_{j\leq n(s)} \left(x_j(s) - \sqrt{2}s\right) \to -\infty$ almost surely as $s \to \infty$. Let $m^{(i)}(s) = \max_{k\leq n^{(i)}(s)} x_k^{(i)}(s)$. Then (2.18) suggests that the extremal particles are

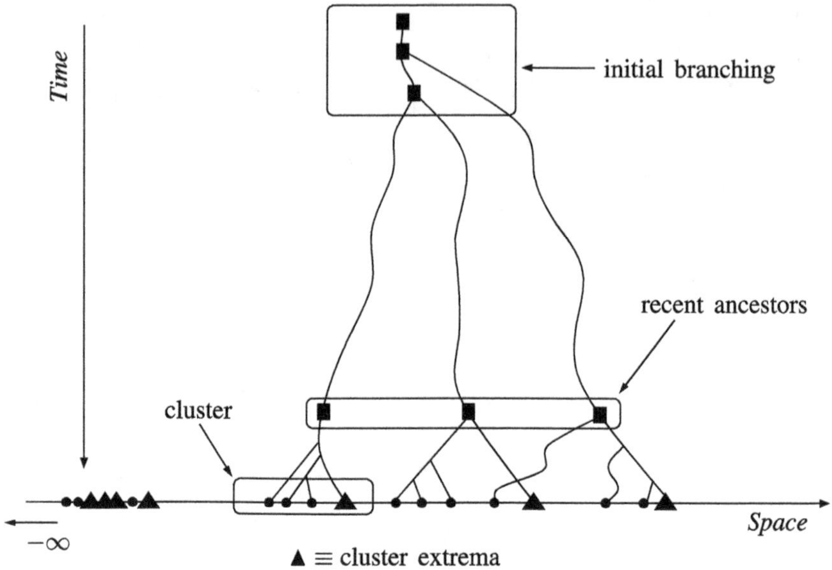

Fig. 6. Onset of the extremal process (Figure 1 of [4]).

particles with "uncharacteristically" large displacements. For more on this phenomenon, see the discussion following (3.9) in [4].

2.5. *Idea of the proof of Theorem 2.20*

Recall Theorem 2.15. Based on this result, we give a short overview of the main ideas of the proof of Theorem 2.20. A closer look at the proof of Theorem 2.15 follows in the subsequent subsection.

The fact that we obtain the same limit $\mathcal{E}^0$ for $\frac{r}{t} < q < 1 - \frac{r}{t}$ with r big enough follows from (2.15). It remains to show (2.17). To be more precise, it suffices to show that, conditionally on Z,

$$\lim_{r\to\infty}\lim_{t\to\infty} \mathcal{E}_t^{(r/t)} = PPP\big(CZ\sqrt{2}e^{-\sqrt{2}x}dx\big).$$

This is done via convergence of Laplace functionals.

Recall that if $X \sim PPP(\lambda)$, then $\mathbb{E}[e^{-\int \phi(x)X(dx)}] = e^{\int (e^{-\phi(x)}-1)\lambda(dx)}$ (cf. [12], Proposition 6.8 of the Appendix) and that the Laplace functional of a point process uniquely characterizes its distribution. For $\phi \in \mathcal{C}_c^+$ we claim that

$$\lim_{r\to\infty}\lim_{t\to\infty}\mathbb{E}\left[e^{-\int\phi(x)\mathcal{E}_t^{(r/t)}(dx)}\right]=\mathbb{E}\left[e^{-CZ\int\left(1-e^{-\phi(x)}\right)\sqrt{2}e^{-\sqrt{2}x}dx}\right]. \quad (2.19)$$

Conditional on the evolution of the BBM up to time r, we obtain

$$\mathcal{E}_t^{(r/t)} \overset{\text{law}}{=} \left\{x_j(r)+M^{(j)}(t-r)-m(t)\right\}_{j=1,\dots,n(r)},$$

where $\{x_k^{(j)}(t)\}_{k\leq n^{(j)}(t)}$, $j \in \mathbb{N}$ are i.i.d. BBM with $M^{(j)}(t) \equiv \max_{k\leq n^{(j)}(t)} x_k^{(j)}(t)$. Recall Figure 4 to see that we condition on the value of the BBM at the end of the period of free evolution and consider the evolution of the progenies of each of these particles independently from there onwards. By adding $M^{(j)}(t-r)$, we indeed select the maximal particle within the respective progenies.

Now,

$$m(t)=\sqrt{2}r+m(t-r)+o(1) \text{ for } t\to\infty.$$

By (2.9), $\mathbb{P}(M(t)-m(t)\leq x)=u(t,m(t)+x)\to w(x)$ uniformly in x as $t\to\infty$. As a result,

$$\begin{aligned}&\lim_{t\to\infty}\mathbb{E}\left[e^{-\int\phi(x)\mathcal{E}_t^{(r/t)}(dx)}\right]\\&=\lim_{t\to\infty}\mathbb{E}\left[\prod_{j=1}^{n(r)}\mathbb{E}\left[e^{-\phi\left(x_j(r)-\sqrt{2}r+M^{(j)}(t-r)-m(t-r)+o(1)\right)}\right]\right]\\&=\mathbb{E}\left[\prod_{j=1}^{n(r)}\mathbb{E}\left[e^{-\phi\left(x_j(r)-\sqrt{2}r+\bar{M}\right)}\right]\right],\end{aligned}$$

where $\overline{M}$ has w as distribution function.

Recall Remark 2.24. We can rewrite the above, using $\log(x)\sim-(1-x)$ for x close to 1, $0<x<1$, to

$$\begin{aligned}&\lim_{t\to\infty}\mathbb{E}\left[e^{-\int\phi(x)\mathcal{E}_t^{(r/t)}(dx)}\right]=\mathbb{E}\left[e^{\sum_{j=1}^{n(r)}\log\left(\mathbb{E}\left[e^{-\phi\left(x_j(r)-\sqrt{2}r+\bar{M}\right)}\right]\right)}\right]\\&\sim\mathbb{E}\left[e^{-\sum_{j=1}^{n(r)}\mathbb{E}\left[1-e^{-\phi\left(x_j(r)-\sqrt{2}r+\bar{M}\right)}\right]}\right].\end{aligned}$$

Recall the asymptotics for $\overline{M}$, that is, $1-w(x)\sim Cxe^{-\sqrt{2}x}$ for $x\to\infty$ from (2.14). Note that $w'(x)\sim C\sqrt{2}xe^{-\sqrt{2}x}$ for x big enough and recall that $Z(t) = \sum_{k=1}^{n(t)} \big(\sqrt{2}t - x_k(t)\big)e^{-\sqrt{2}(\sqrt{2}t - x_k(t))} \to Z$ for $t \to \infty$ from (2.11)–(2.12). Then the above is

$$\sim \mathbb{E}\left[e^{-\int \left(1-e^{-\phi(x)}\right) \sum_{j=1}^{n(r)} C\sqrt{2}\left(-x_j(r)+\sqrt{2}r+x\right) e^{-\sqrt{2}(-x_j(r)+\sqrt{2}r+x)} dx}\right]$$

$$\overset{r\to\infty}{\to} \mathbb{E}\left[e^{-\int C\sqrt{2}\left(1-e^{-\phi(x)}\right) Z e^{-\sqrt{2}x} dx}\right],$$

which indeed coincides with the right hand side of (2.19).

2.6. *Idea of the proof of Theorem 2.15*

The proof relies in essential parts on the localization of paths of extremal particles. For instance, look ahead at Figure 7. The path of an extremal particle up until time t can be interpreted as the path of a BM conditional on being close to $m(t)$ at time t. By Remark 2.10, in law, this is equal to considering a Brownian bridge starting at 0 and ending at $m(t)$. By Part (c) of this remark, one can approximate it by the line connecting $(0,0)$ to $(t, m(t))$ plus a Brownian bridge connecting 0 to 0. A control on the local deviation of the latter around 0 (respectively the line) is given for instance in Part (e). Of course, a more refined knowledge of the properties of Brownian bridges is necessary to make all details work but the above already gives a flavor of the ideas behind the proofs.

2.6.1. *Localization of paths of extremal particles*

In a first step, it is shown that the paths of extremal particles between times r_u and $t - r_u$ lie with high probability below the upper curve in Figure 7. To be more precise, for $\gamma > 0$, set

$$f_{t,\gamma}(s) = \begin{cases} s^\gamma, & 0 \le s \le \frac{t}{2}, \\ (t-s)^\gamma, & \frac{t}{2} \le s \le t. \end{cases}$$

Then the *upper envelope at time* t is defined as

$$U_{t,\gamma}(s) \equiv \frac{s}{t} m(t) + f_{t,\gamma}(s).$$

Theorem 2.25: *(Upper Envelope, Theorem 2.2 of [2]). Let* $0 < \gamma < 1/2$. *Let also* $y \in \mathbb{R}$ *and* $\epsilon > 0$ *be given. There exists* $r_u = r_u(\gamma, y, \epsilon)$ *such that for* $r \ge r_u$ *and for any* $t > 3r$,

$$\mathbb{P}\big(\exists k \le n(t) : x_k(s) > y + U_{t,\gamma}(s) \text{ for some } s \in [r, t-r]\big) < \epsilon.$$

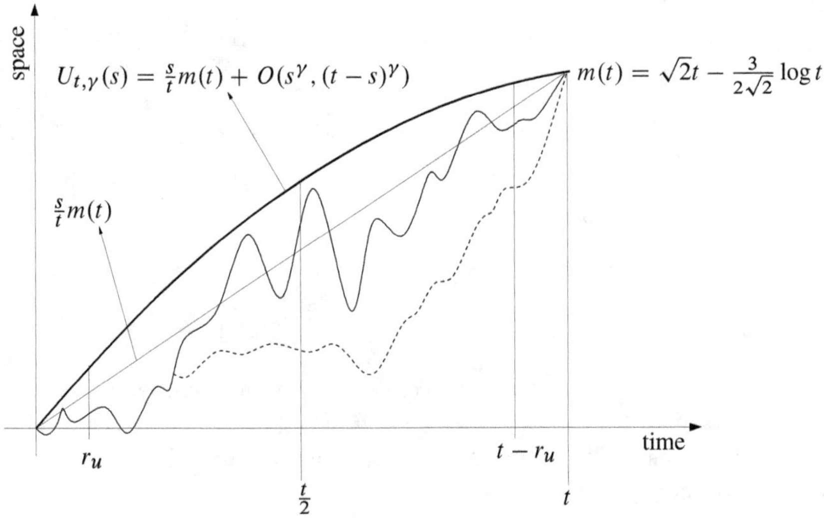

Fig. 7. The upper envelope (Figure 2.1 of [2]).

Idea of proof. Consider first the path only at integer times $\lfloor s \rfloor$ and use the uniform (in $\lfloor s \rfloor$) estimate [2], (2.7) (taken from [13], Proposition 3) on the deviation of the maximum above the centering term $m(\lfloor s \rfloor)$ at a fixed value $\lfloor s \rfloor$ of the discretization. For general s, use that if the deviation above $m(s)$ is high, then it should also be high at time $\lfloor s \rfloor$. □

Remark 2.26: (Why is this useful?) Following [2], below (2.8), we motivate the need for such path properties. For $\mathbb{E}[n(t)] = e^t$ i.i.d. BMs, $m(t) = \sqrt{2}t - \frac{1 \cdot \log(t)}{2\sqrt{2}}$ (cf. Remark 2.11) and we indeed obtain, on average, a constant number of extremal particles close to the tip,

$$\mathbb{E}[\#\{k \le n(t) : B_k(t) > m(t)\}] \sim \frac{e^t}{\sqrt{2\pi t}} e^{-\frac{m(t)^2}{2t}}$$

$$= \frac{e^t}{\sqrt{2\pi t}} e^{-\frac{2t^2 - \frac{2\sqrt{2}t\log(t)}{2\sqrt{2}} + \frac{\log(t)^2}{8}}{2t}} = \frac{1}{\sqrt{2\pi t}} e^{\frac{\log(t)}{2} - \frac{\log(t)^2}{16t}} = \frac{1}{\sqrt{2\pi}} e^{-\frac{\log(t)^2}{16t}} = O(1).$$

Note that the first asymptotics indeed follows from Lemma 2.4.

By contrast, for BBM, $m(t) = \sqrt{2}t - \frac{3 \cdot \log(t)}{2\sqrt{2}}$. For e^t i.i.d. BMs we now get

$$\mathbb{E}[\#\{k \le n(t) : B_k(t) > m(t)\}] = \frac{1}{\sqrt{2\pi t}} e^{3\frac{\log(t)}{2} - \frac{9\log(t)^2}{16t}} = O(t). \tag{2.20}$$

Thus, an approximation by i.i.d. BMs is a priori not available, unless we add further restrictions on their paths. Here, we only give a first crude estimate.

For a BM B_t, starting in 0,

$$\begin{aligned}&\mathbb{P}(\exists r \leq s \leq t-r : B_s > U_{t,\gamma}(s) \mid B_t = m(t)) \\ &= \mathbb{P}(\exists r \leq s \leq t-r : \mathfrak{z}^t_{0,m(t)}(s) > U_{t,\gamma}(s)) \\ &= \mathbb{P}(\exists r \leq s \leq t-r : \mathfrak{z}^t_{0,0}(s) > f_{t,\gamma}(s)).\end{aligned} \tag{2.21}$$

Lemma 3.4 of [2] with $Z_1 = Z_2 = t^\gamma$, $\gamma > 0$ and $r_1 = r_2 = r$ constant yields

$$\mathbb{P}(\exists r \leq s \leq t-r : \mathfrak{z}^t_{0,0}(s) > f_{t,\gamma}(s)) \leq \frac{2}{t-2r}\left(t^\gamma + \sqrt{r}\right)^2 \lesssim t^{2\gamma-1}.$$

With a bit of more work (for instance, discretization of paths) and knowledge of Brownian bridges, an upper bound of order $1/t$ can be achieved in (2.21) for r, γ as in Theorem 2.25. In combination with (2.20) this yields the right order.

The upper envelope can be replaced by a lower *"entropic envelope"* $E_{t,\alpha}$, still serving to bound the upper paths from above with high probability (cf. Figure 8). Recall Remark 2.10(c),(e). As a heuristic, suppose that at some point a path hits the line $s \mapsto \frac{s}{t}m(t)$, say at time s_0, then the remaining part of the path is distributed like a Brownian bridge connecting the hitting point $\left(s_0, \frac{s_0}{t}m(t)\right)$ with the endpoint $(t, m(t))$. The deviation of the path away from the connecting line is of order up to $(t-s_0)^{1/2}$. The upper envelope only allows for a deviation of $(t-s_0)^\gamma$, $0 < \gamma < 1/2$. As a result, a path that avoids hitting the upper envelope has to stay well below the connecting line. For more on the effect of "entropic repulsion", see the paragraph preceding Theorem 2.3 in [2].

For $\alpha > 0$, let

$$E_{t,\alpha}(s) = \frac{s}{t}m(t) - f_{t,\alpha}(s).$$

Then the following holds.

Theorem 2.27: *(Entropic Repulsion, Theorem 2.3 of [2]). Let $D \subset \mathbb{R}$ be a compact set and $0 < \alpha < 1/2$. Set $\bar{D} = \sup\{x \in D\}$. For any $\epsilon > 0$ there exists $r_e = r_e(\alpha, D, \epsilon)$ such that for $r \geq r_e$ and $t > 3r$,*

$$\mathbb{P}(\exists k \leq n(t) : x_k(t) \in m(t)+D, \textit{ but } \exists s \in [r, t-r] : x_k(s) \geq \bar{D}+E_{t,\alpha}(s)) < \epsilon.$$

Have a look ahead at Figure 9, where $0 < \alpha < \frac{1}{2} < \beta < 1$. Together with a *lower envelope* (cf. [2], Theorem 2.5), one gets:

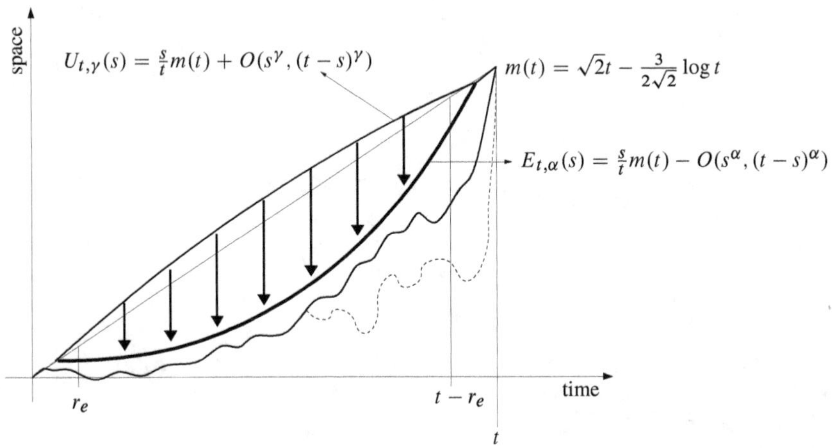

Fig. 8. Entropic repulsion (Figure 2.2 of [2]).

Theorem 2.28: *(The "Tube", Corollary 2.6 of [2]). Let $D \subset \mathbb{R}$ be a compact set. Let $0 < \alpha < \frac{1}{2} < \beta < 1$. For any $\epsilon > 0$ there exists $r_1 = r_1(\alpha, \beta, D, \epsilon)$ such that for $r \geq r_1$ and $t > 3r$,*

$$\mathbb{P}\big(\forall k \leq n(t) : x_k(t) \in m(t) + D, \\ \overline{D} + E_{t,\beta}(s) \leq x_k(s) \leq \overline{D} + E_{t,\alpha}(s)\ \forall s \in [r, t-r]\big) \geq 1 - \epsilon.$$

In what follows, we give a motivation along the lines of [2], (4.3)–(4.4) and Remark 2.4 of how to use these pathwise results to obtain statements on the genealogy.

Observe first that the expected number of pairs of particles of BBM whose (respective) path $(x(s))_{0 \leq s \leq t}$ satisfies some conditions for $s \in [r, t-r]$, say $\Sigma_t^{[r,t-r]}$, is

$$\mathbb{E}\Big[\#\Big\{(k,\ell) : k \neq \ell, x_k(\cdot), x_\ell(\cdot) \in \Sigma_t^{[r,t-r]}\Big\}\Big] \\ = C(e^t)^2 \int_0^t ds\, e^{-s} \int_{-\infty}^{\infty} dy\, p_s(y) \\ \times \mathbb{P}\Big(x(\cdot) \in \Sigma_t^{[r,t-r]} \mid x(s) = y\Big)\, \mathbb{P}\Big(x(\cdot) \in \Sigma_t^{[r \vee s, t-r]} \mid x(s) = y\Big).$$

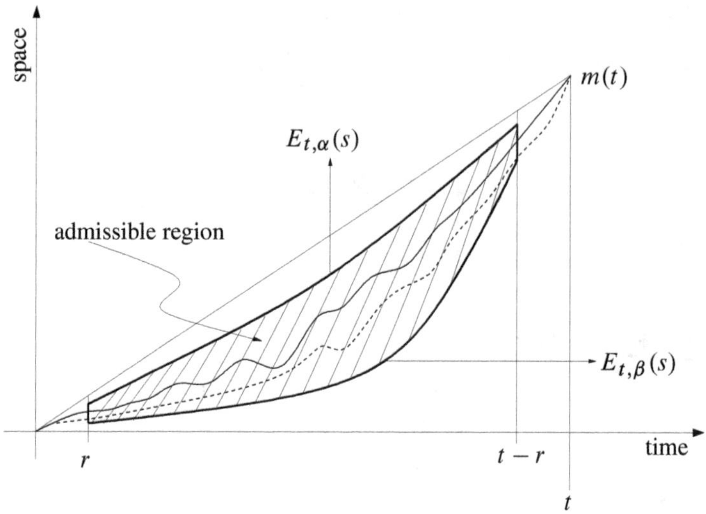

Fig. 9. The "tube" (Figure 2.3 of [2]).

The terms involved model the following.

$C(e^t)^2$: how many pairs at time t on average;

$ds\, e^{-s}$: condition on splitting at time s and

$dy\, p_s(y)$: condition on splitting at position y.

$[r \vee s, t]$: If the first particle satisfies the condition on $[0, t]$, then the second one automatically satisfies it on $[0, s]$. From s onwards, the second particle evolves independent of the first one.

If we include a condition on genetic distance as in Theorem 2.15, we get

$$\mathbb{E}\Big[\#\Big\{(k,\ell) : k \neq \ell, x_k(\cdot), x_\ell(\cdot) \in \Sigma_t^{[r,t-r]}, d(x_k(t), x_\ell(t)) \in (r, t-r)\Big\}\Big]$$
$$= Ce^t \int_r^{t-r} ds\, e^{t-s} \int_{-\infty}^{\infty} dy\, p_s(y) \mathbb{P}\Big(x(\cdot) \in \Sigma_t^{[r,t-r]} \mid x(s) = y\Big)$$
$$\times \mathbb{P}\Big(x(\cdot) \in \Sigma_t^{[s,t-r]} \mid x(s) = y\Big)\,.$$

Now proceed along the following steps.

(i) Let k, ℓ be such that $d(x_k(t), x_\ell(t)) = s \in [r, t-r]$. We show that the probability for both particles to be extremal (that is, $\overline{x_k}(t), \overline{x_\ell}(t) \in D$,

D an arbitrarily fixed compact set) is small. Consider the case

$$s \le t/2 \text{ and to simplify calculations } s = O(t)$$

in what follows. Assume that

$$x_k(t), x_\ell(t) \in m(t) + D$$

and that, due to entropic repulsion,

$$\max\{x_k(s), x_\ell(s)\} < \bar{D} + E_{t,\alpha}(s) = \bar{D} + \sqrt{2}s - \frac{s}{t}\frac{3}{2\sqrt{2}}\log(t) - s^\alpha$$
$$\sim \sqrt{2}s - s^\alpha$$

for some fixed $0 < \alpha < 1/2$ and for all $s \in [r, t-r]$.

(ii) For particles k and ℓ to reach $m(t) + D$, the MRCA of k and l (at time s) must itself produce a BBM that after a time-interval of length $t - s$ has *height at least*

$$\big(m(t) - \min(D)\big) - \big(\sqrt{2}s - s^\alpha\big) \sim \sqrt{2}(t-s) + s^\alpha.$$

(iii) *Number of possible choices for ancestor, alive at time s,* ***at height*** *$\sqrt{2}s - s^\alpha$:* At time s, there are on average e^s particles. The chance of one particle (that is, a BM) to reach $\sqrt{2}s - s^\alpha$ is of order

$$\frac{1}{\sqrt{2\pi s}}e^{-\frac{2s^2 - 2\sqrt{2}ss^\alpha + s^{2\alpha}}{2s}} = \frac{1}{\sqrt{2\pi s}}e^{-s}e^{\sqrt{2}s^\alpha}e^{-\frac{s^{2\alpha-1}}{2}} \lesssim \frac{1}{\sqrt{2\pi s}}e^{-s}e^{\sqrt{2}s^\alpha}.$$

Note that this is an over-estimate as we have on average more particles at this height than if we consider BBM. In the product (with e^s), at time s we have on average at most of order $e^{\sqrt{2}s^\alpha}$ choices.

(iv) *Chance for the progeny of an ancestor, alive at time s, to include* ***at least one*** *particle at height $\sqrt{2}(t-s) + s^\alpha$ (cf. (ii)), alive at time t:* Starting with a single particle, the probability that BBM makes it this high / far in the time-interval $t-s$ is (use that r big, $t-s \ge r$, $s = O(t)$ and $\mathbb{P}(M(t) - m(t) > x) \sim 1 - w(x) \sim Cxe^{-\sqrt{2}x}$ for $t, x \to \infty$)

$$\begin{aligned}
&\mathbb{P}\big(M(t-s) \ge \sqrt{2}(t-s) + s^\alpha\big) \\
&= \mathbb{P}\left(M(t-s) - m(t-s) \ge \frac{3}{2\sqrt{2}}\log(t-s) + s^\alpha\right) \\
&\sim 1 - w\left(\frac{3}{2\sqrt{2}}\log(t-s) + s^\alpha\right) \lesssim C(s^\alpha + \delta)e^{-\sqrt{2}(s^\alpha + \delta)} \lesssim e^{-\sqrt{2}s^\alpha}
\end{aligned}$$

for $\delta > 0$.

(v) *Chance for ancestor, alive and splitting at time s, to have* **two** *children at height $\sqrt{2}(t-s)+s^{\alpha}$, alive at time t:* This corresponds to two independent BBM starting with one particle each, to make it to height $\sqrt{2}(t-s)+s^{\alpha}$ within a time-interval of length $t-s$. By Part (iv) the probability is at most of order $\left(e^{-\sqrt{2}s^{\alpha}}\right)^2$.

(vi) *Conclusion:* Overall chance for **two** particles k, ℓ as in (i) with an ancestor **at height** $\sqrt{2}s - s^{\alpha}$ (cf. (iii)) is of order $e^{\sqrt{2}\left(s^{\alpha}-2s^{\alpha}\right)}$, so negligible.

(vii) Finally, to remove the height-condition on the ancestor in (vi), use the lower envelope (respectively the "tube"), to see that

$$\min\{x_k(s), x_\ell(s)\} \geq \bar{D} + E_{t,\beta}(s) \sim \sqrt{2}s - s^{\beta}$$

for some fixed $1/2 < \beta < 1$ and for all $s \in [r, t-r]$. Redo the above argument increment-wise, noting that the lower the ancestor is located, the higher its children have to reach (cf. (ii)).

2.7. *The extremal process of BBM*

For those readers who are interested in the complete description of the limit of the extremal process of BBM, $\mathcal{E}$, the answer can be found in [1] and [4], published simultaneously. Both articles give a description of the (weak with respect to $\phi \in \mathcal{C}_c^+$) limit $(t \to \infty)$ of the *extremal process*

$$\mathcal{E}_t = \sum_{k \leq n(t)} \delta_{x_k(t)-m(t)}.$$

See Gouéré [17] for a (French) review that presents both approaches. Bovier [12] discusses results of [4] in detail. The proofs in [4] rely on the consideration of the respective Laplace functionals (recall Example 2.8). The proofs in [1] use path localization and path decomposition techniques.

We finish this article by remarking on some changes in notation in [1] and their consequences, to facilitate the reading of the latter article for the reader who is used to the notation of this article.

Remark 2.29: (Change in Scaling). A different scaling is used in [1] compared to [4]. Instead of rightmost particles, leftmost are considered.

(1) Particles now follow a BM *with drift* 2 *and variance* 2, that is, replace B_t by $\sqrt{2}(B_t - \sqrt{2}t)$.
(2) (The exponential clocks for splitting-events still ring at rate 1 and a particle splits in two.)
(3) Instead of $M(t) = \max_{k \leq n(t)} x_k(t)$ they consider $\min_{k \leq n(t)} x_k(t)$.

Recall $m(t)$ from (2.7) and the derivative martingale $Z(t)$ from (2.11).

Remark 2.30: (Consequences of Scaling).

(1) $m(t)$ becomes $m'(t) = +\frac{3}{2}\log(t)$.
(2) $Z(t)$ becomes $\frac{1}{\sqrt{2}}\sum_{k=1}^{n(t)} x_k(t)e^{-x_k(t)}$. In [1], the authors use $Z'(t) = \sum_{k=1}^{n(t)} x_k(t)e^{-x_k(t)}$ instead and thus CZ becomes $(C/\sqrt{2})Z' = C'Z'$.
(3) $\mathbb{E}[Z'(t)] = 0$ for all $t \geq 0$.

3. Genealogy of BBM under absorption

3.1. *Population models with selection*

In this subsection, we consider branching Brownian motion (BBM) with selection. Loosely speaking, selection means killing of a set of particles. The selection mechanism(s), which we are going to define, can be interpreted as attempts to model Darwinian evolution. The location of a particle on the real line represents its "fitness" value, i.e., a measure of how well it is adapted to its environment. Due to mutations, after birth, fitness of an individual evolves randomly according to a Brownian motion. Selection progressively eliminates individuals whose fitness becomes too low. The two basic selection mechanisms described below are widely studied in the literature. Together with the biological interpretation, it is important to recognize that these mechanisms can also be viewed as in the global framework of propagation of fronts under the effect of noise.

Definition 3.1: (BBM with absorption). According to this model, selection is viewed as a moving wall with a constant speed μ. A BBM with a single particle at $x > 0$ is started and as soon as a particle hits the space time line $y = \mu t$, it is killed immediately. In 1978, Kesten introduced this model (see [22]) and showed that there exists a critical value $\mu_c = \sqrt{2}$ such that if $\mu \geq \mu_c$, then the process dies out almost surely, and if $\mu < \mu_c$, the process survives with positive probability.

Definition 3.2: (BBM under selection (N-BBM)). This particle system was inspired from a discrete model that was proposed by Brunet, Derrida and their co-authors (see [15], [16]). The N-BBM is described as follows: Particles evolve according to branching Brownian motions. At times of an independent Poisson process, each particle gives birth to two offsprings and then dies. When the number of particles exceeds N, the least fit particle gets killed, i.e., at each birth event, the left-most particle gets killed such that only the N right-most particles remain. Brunet, Derrida and their co-authors actually introduced a discrete counterpart of this model, branching random walk under selection (N-BRW), where they considered N particles undergoing branching random walks and at each birth event, only the N right-most particles remain.

Regarding the N-BBM model, the physicists studied the discrete model in extraordinary detail and made some striking predictions.

(i) They observed that the linear speed of this system differs from the speed of the right-most particle in BBM (without selection) by a quantity of the order of $(\log N)^2$.
(ii) Using numerical simulations for an approximate model, they suggested that the genealogy of the particle system on a time scale of $(\log N)^3$ can be described by the Bolthausen-Sznitman coalescent.

The first conjecture, regarding the $(\log N)^2$ correction of the linear speed for the N-BRW (branching random walk) model, was settled by Bérard and Gouéré [5] under some assumptions on the displacement distribution. Regarding the second conjecture, we first mention that this prediction is in sharp contrast with population models *without* selection. For these models, e.g., the Moran model or the Wright-Fisher model, the genealogical tree is given by the Kingman's coalescent which allows only pairwise merging of the ancestral lines. On the other hand, the Bolthausen-Sznitman coalescent allows multiple merging, i.e., many ancestral lines can merge at the same time. Heuristically, under the selection mechanism, if a particle develops a very high value of fitness, its offsprings have a better chance to survive for a longer time period and then the descendants of this particle quickly constitute a positive proportion of the population. Hence when we look back at the ancestral lines in the proper time scale, this leads to multiple merging of the ancestral lines. But because of the dependencies introduced due to selection, up to this point, no one has been able to make the above heuristics completely rigorous. In another setting where the particles enjoy more independence compared to the N-BBM particle system, Berestycki, Berestycki and Schweinsberg [7] were able to prove that the Bolthausen-Sznitman coalescent is the limiting coalescent process. They approximate (we will explain this approximation later) the N-BBM by a BBM with a linear space time absorption barrier and prove that under appropriate time scaling, properly scaled population size converges to a continuous state branching process and the genealogy of the system converges to the Bolthausen-Sznitman coalescent, respectively. In this case, the approximating process, i.e., the BBM with absorption is more tractable than the N-BBM, since the particles enjoy more independence in the latter case. Recently for the N-BBM process, Maillard [29] was able to prove that appropriately scaled total number of particles observed in $(\log N)^3$ time scale converges to a Lévy process. Though this is a considerable advancement towards proving the second conjecture, the convergence of the genealogy of the population is still open.

In the rest of this section, we will describe the result of Berestycki, Berestycki and Schweinsberg [7] in detail and discuss the idea of their proof briefly. We should mention here that their argument is fairly technical and involves a lot of carefully designed estimates. In this expository article, it will not be possible to describe their argument in detail. We will try to provide an outline of the basic steps. The other goal of this expository article is to provide a comprehensive account of some of the background materials used in [7]. We first give a brief introduction to coalescent process and continuous state branching process.

3.2. *Continuous state branching process (CSBP) and coalescent process*

3.2.1. *Continuous state branching process*

Consider a continuous time Galton Watson process $(Z_t^m : t \geq 0)$, where Z_t^m denotes the size of the population at time t starting with m individuals at time zero. Clearly, it is a continuous time discrete state space Markov process with absorbing state 0. Note that

$$Z_t^{m_1+m_2} \stackrel{d}{=} Z_t^{m_1} + Z_t^{m_2} \text{ for all } t \geq 0,$$

where Z^{m_1} and Z^{m_2} are independent. This is referred as the fundamental *branching property*. Using this property, the notion of branching process can be extended to obtain a continuous state space Markov process.

Definition 3.3: A continuous state branching process (CSBP) is a Markov process with RCLL paths taking values in $[0, \infty]$, whose transition kernels $(P_t)_{t\geq 0}$ satisfy the following branching property

$$\mathbb{P}_t(x, \cdot) * \mathbb{P}_t(y, \cdot) = \mathbb{P}_t(x + y, \cdot) \text{ for all } t, x, y \geq 0. \tag{3.1}$$

In other words, if Z^x and Z^y are two independent copies of CSBP Z started respectively at x and y, then $Z^x + Z^y$ has the same law of Z started at

$x+y$. Continuous state branching processes were introduced by Jirina [21]. Lamperti [25] observed a one-to-one correspondence between continuous state branching processes and Lévy processes with no negative jumps and showed that a CSBP can be obtained through a time change of the corresponding Lévy process. Continuous state branching processes can also be observed as the limit of Galton Watson processes (see [26]).

Assuming that the process can not make an instantaneous jump to infinity, a continuous state branching process can be characterized by a function $\psi : [0,\infty) \mapsto \mathbb{R}$ of the form

$$\psi(u) = \alpha u + \beta u^2 + \int_0^\infty (e^{-ux} - 1 + ux\mathbf{1}_{\{x\leq 1\}})\nu(dx), \tag{3.2}$$

where $\alpha \in \mathbb{R}$, $\beta \geq 0$ and ν is a measure on $\mathbb{R}^+$ satisfying $\int_0^\infty (1 \wedge x^2)\nu(dx) < \infty$. The function ψ is called the branching mechanism. If $(Z(t), t \geq 0)$ is a CSBP with branching mechanism ψ, then for all $\lambda \geq 0$ we have

$$\mathbb{E}[e^{-\lambda Z(t)} | Z(0) = a] = e^{-au_t(\lambda)}, \tag{3.3}$$

where the function $t \mapsto u_t(\lambda)$ is a solution to the differential equation

$$\frac{\partial u_t(\lambda)}{\partial t} = -\psi(u_t(\lambda)) \text{ with initial condition } u_0(\lambda) = \lambda.$$

We will be interested in a continuous state branching process $(Z(t), t \geq 0)$ with branching mechanism $\psi(u) = au + bu \log u$, where $a \in \mathbb{R}$ and $b > 0$. For such process there exists a $c \in \mathbb{R}$ such that

$$\psi(u) = -cu + b\int_0^\infty (e^{-ux} - 1 + ux\mathbf{1}_{\{x\leq 1\}})x^{-2}dx, \tag{3.4}$$

in other words, its Lévy measure has a density proportional to x^{-2}. Next we introduce coalescent process briefly.

3.2.2. *Coalescent process*

A partition π of $\mathbb{N}$ is an equivalence relation on $\mathbb{N}$. The blocks of the partition are the equivalence classes of this relation. In other words, if the elements i and j are in the same block of π then they are equivalent w.r.t. partition π. The space of partitions of $\mathbb{N}$ is denoted by $\mathcal{P}$. For each $n \in \mathbb{N}$, $\mathcal{P}_n$ denotes the space of partitions of the set $[n] = \{1, 2, \ldots, n\}$. For $\pi \in \Pi$ and for $n \in \mathbb{N}$, the restriction of π to $[n]$ is denoted by π^n. For $\pi_1, \pi_2 \in \Pi$, take

$$d(\pi_1, \pi_2) = 1/(\sup\{n \in \mathbb{N} : \pi_1^n = \pi_2^n\}).$$

The resulting space $(\mathcal{P}, d(\cdot))$ becomes complete and separable and thereby allows us to study convergence in distribution for random variables taking values in $\mathcal{P}$ or $\mathcal{P}_n$.

A coalescent process is a Markov process defined over $\mathcal{P}$ or $\mathcal{P}_n$ such that the blocks merge over time. Merging of blocks corresponds to the merging of the (corresponding) ancestral lines when the ancestral lines are traced back in time. We start with a basic coalescent model, the Kingman's coalescent, which is a fundamental model for population genetics. This model was introduced in [23]. The Kingman's coalescent on $\mathcal{P}_n$ is a continuous time strong Markov process which starts with n singleton blocks and evolves through binary coalescence such that every pair of blocks merge at rate 1 irrespective of their sizes and independent of all other blocks.

Allowing only pairwise merging is too restrictive. In particular, in the presence of natural selection, descendants of an individual with a high fitness value quickly occupy a positive fraction of the population. This motivates the study of general coalescent processes with multiple merging mechanisms. Pitman [31] showed that for each finite measure Λ on $[0,1]$, there exists a coalescent process with multiple merging mechanism as follows.

Definition 3.4: For each finite measure Λ on $[0,1]$, there exists a Markov process defined over $\mathcal{P}$, such that the restriction of the process to each $\mathcal{P}_n, n \in \mathbb{N}$ is a Markov chain with the following transition rates: when the partition has b blocks, each k-tuple of blocks is merging to form a single block at rate

$$\lambda_{b,k} = \int_0^1 x^{k-2}(1-x)^{b-k}\Lambda(dx) \text{ for all } 2 \le k \le b. \tag{3.5}$$

This coalescent process is called the Λ-coalescent.

Consider an array of non-negative numbers $(\lambda_{b,k})_{2\le k\le b}$ satisfying (3.5), then these numbers satisfy in particular the following recursion relation:

$$\lambda_{b,k} = \lambda_{b+1,k} + \lambda_{b+1,k+1}. \tag{3.6}$$

The interpretation is that, given that k blocks among the first b blocks merge together, the $(b+1)$-st block could either participate in the merger or not. Pitman [31] further showed that for any array of non-negative numbers $(\lambda_{b,k})_{2\le k\le b}$ satisfying (3.5), there exists a finite measure Λ on the interval $[0,1]$ such that (3.5) holds.

Definition 3.5: For the choice Λ as the uniform measure over $[0,1]$, the corresponding Λ-coalescent is known as the Bolthausen-Sznitman coalescent.

The Bolthausen-Sznitman coalescent was derived by Bolthausen and Sznitman from Ruelle's probability cascades, in the context of the Sherrington-Kirkpatrick spin glass model in mathematical physics [11] and later Pitman [31] showed that it is the Λ-coalescent with the choice Λ as the uniform measure over $[0, 1]$. This coalescent process is believed to be the universal scaling limit for models which can be described by "random travelling waves". We refer the interested reader to look at Berestycki's lecture notes [6] for an excellent introduction on the coalescent processes.

Bertoin and Le Gall [8] showed how to define the notion of genealogy of a continuous state branching process. We will explain this in more detail in Subsection 3.5.1. They found that the genealogy of any CSBP with branching mechanism $\psi(u) = u \log u$ (known as Neveu's CSBP) is given by the Bolthausen-Sznitman coalescent [8]. In [10], these results were extended for any CSBP with branching mechanism $\psi(u) = au + bu \log u$.

3.3. *Branching Brownian motion under absorption*

3.3.1. *Description of the model*

In this section we describe the model of Berestycki, Berestycki and Schweinsberg [7] and state their result precisely. In order to study the asymptotic properties, consider a sequence of collection of BBM's with drifts in the near critical regime. For any positive integer N, we start with a family of branching Brownian motions $(X_N(t), t \geq 0)$ with drift

$$\mu = \mu_N = \sqrt{2 - \frac{2\pi^2}{(\log N + 3 \log \log N)^2}}, \tag{3.7}$$

to the left and particles are killed when they reach the origin. The starting configurations of the family of branching Brownian motions are assumed to satisfy some technical conditions, which will be described later because, at this moment, they may not look natural. These assumptions essentially ensure that the system starts with order N particles. Later we will see that under these assumptions, this system roughly behaves like a system of N particles in the time scale of interest and in that sense this process approximates N-BBM.

Fix $t > 0$. Select a random sample of n individuals from this population at time $(\log N)^3 t$ and label them uniformly at random by the integers

$1, \ldots, n$. For $0 \le s \le 2\pi t$, let $\Pi_N(s)$ denote the partition of $\{1, \ldots, n\}$ such that the integers i and j belong to the same block if and only if i-th and j-th particles are descended from the same ancestor at time $(\log N)^3(t - s/2\pi)$. The following theorem is the main result of [7].

Theorem 3.6: *Under the initial assumptions, as $N \to \infty$, the sequence of processes $(\Pi_N(s), 0 \le s \le 2\pi t)$ converges in the sense of finite dimensional distributions to the Bolthausen-Sznitman coalescent process $(\Pi(s),\ 0 \le s \le 2\pi t)$.*

We reiterate that in case of population models without selection, the genealogical tree converges to the Kingman's coalescent process which allows only *pairwise merging* of the ancestral lines. The above result shows that the behaviour of the genealogy changes completely with the introduction of selection and the resulting limiting coalescent process allows *multiple merging.*

3.3.2. *The basic idea of the proof*

We first describe the initial assumptions under which Theorem 3.6 holds. In order to do that, we need to introduce some notation. Let $(X_N(t), t \ge 0)$ denote the N-th family of BBM's with drifts. Let $M_N(t)$ denote the total number of particles at time t. The spatial locations of these $M_N(t)$ particles at time t are denoted by $X_{1,N}(t) \ge X_{2,N}(t) \ge \cdots \ge X_{M_N(t),N}(t)$. Set

$$L = L_N := \frac{1}{\sqrt{2}}(\log N + 3 \log\log N).$$

We further need to define three more processes:

$$Z_N(t) = \sum_{i=1}^{M_N(t)} e^{\mu X_{i,N}(t)} \sin\Big(\frac{\pi X_{i,N}(t)}{L}\Big) \mathbf{1}_{\{X_{i,N(t)} \le L\}} \tag{3.8}$$

$$Y_N(t) = \sum_{i=1}^{M_N(t)} e^{\mu X_{i,N}(t)} \tag{3.9}$$

$$V_N(t) = \frac{1}{N(\log N)^2} Z_N((\log N)^3 t). \tag{3.10}$$

It is useful to observe that only particles to the left of L contribute to $Z_N(t)$. Later it will turn out that $Z_N(t)$ gives a good measure of the 'size' of the process in the sense that it approximates the size of the population well for a short time after t.

We are ready to state the assumptions now.

(A_1) As $N \to \infty$, the distribution of $V_N(0)$ converges to ν where ν is probability distribution on $[0, \infty)$.
(A_2) As $N \to \infty$, the random variable $Y_N(0)/N(\log N)^3$ converges to zero in probability as $N \to \infty$.

We mentioned already that these initial assumptions are to ensure that the system stays roughly of the order of N. From here onward, throughout this article we assume that both the assumptions (A_1) and (A_2) continue to hold.

In order to explain these initial assumptions and the choice of the drift parameter, we present some predictions made by the physicists, Brunet, Derrida and their co-authors ([15], [16]) regarding the behavior of the N-BRW particle system.

(1) Most of the times, the particles are in a meta-stable state. In this state, the cloud of particles (also called the front) has a diameter of $\log N + O(1)$. The empirical density of particles seen from the left-most is proportional to $e^{-x} \sin(\pi x/\log N)$, and the system moves at a linear speed $v_{\text{cutoff}} = 1 - \pi^2/(2(\log N)^2)$. Hence most of the particles are at $O(1)$ distance from the left-most particle.
(2) Occasionally particles move too far to the right which causes perturbation from the meta-stable state. To explain this phenomena in more detail, we fix a point at the barycenter in the cloud and say that a particle reaches a point y if it moves to distance y from that fixed point. It is predicted that a particle moving up to the $\log N + x$ (from the fixed point) after a relaxation time of order $\log^2 N$, causes a shift of the front given by

$$\Delta = \log(1 + Ce^x/(\log N)^3) \text{ where } C \text{ is a positive constant.}$$

Hence, according to the predicted behaviour, in order to have a 'detectable' effect on the position of the front, a particle has to reach a point $\log N + 3 \log \log N$.
(3) Assuming that such an event, where a particle reaches the point $\log N + x$, happens at a rate proportional to e^{-x}, the time taken by a particle to come close to $\log N + 3 \log \log N$ (and thus causing shifts of the front) is of the order of $(\log N)^3 N$.

These heuristics suggest that for any particle system, which well approximates the N-BBM, as long as the particles do not reach $\log N + 3 \log \log N$ (from the center), the size of the population does not change much and those who are able to hit $\log N + 3 \log\log N$ require different analysis. Next we describe some results on BBM within a strip with killing boundaries which will be used in the sequel.

Fix $L > 0$ and $x \in (0, L)$. Consider a Brownian motion starting from x at time 0 and assume that the process is killed as soon as it hits 0 or L. Then the density of the process at time t is (see [27])

$$\nu_t(x,y) = \frac{2}{L} \sum_{n=1}^{\infty} e^{-\pi^2 n^2 t/(2L^2)} \sin\left(\frac{n\pi x}{L}\right) \sin\left(\frac{n\pi y}{L}\right). \tag{3.11}$$

Using (3.11), one can obtain the density of particles at time t for a branching Brownian motion, starting from x, with birth rate 1 and drift $\mu(> 0)$ to the left in which particles are killed as soon as they hit 0 or L. An application of many-to-one lemma (Lemma 2.4) together with a Girsanov factor of $e^{\mu(x-y)-\mu^2 t/2}$ due to the drift give that the density at $y \in (0, L)$ at time t is given by,

$$q_t(x,y) = e^{(1-\mu^2/2)t+\mu(x-y)} \frac{2}{L} \sum_{n=1}^{\infty} e^{-\pi^2 n^2 t/(2L^2)} \sin\left(\frac{n\pi x}{L}\right) \sin\left(\frac{n\pi y}{L}\right), \tag{3.12}$$

in the sense that for $B \subset (0, L)$, the expected number of particles in B at time t is $\int_B q_t(x, y)dy$. Using the fact that $|\sin ny| \leq n \sin y$ for any $n \in \mathbb{N}$ and $y \in [0, \pi]$, it follows that after a large time t, the first term in (3.12) dominates and this approximates $q_t(x, y)$ as

$$p_t(x,y) = \frac{2}{L} e^{(1-\mu^2/2-\pi^2/(2L^2))t} e^{\mu x} \sin\left(\frac{\pi x}{L}\right) e^{-\mu y} \sin\left(\frac{\pi y}{L}\right). \tag{3.13}$$

There are a some important observations to make from (3.13). Firstly, in (3.13) the parameter t appears only in the first exponential factor. On the other hand, according to the earlier predictions for any process, which approximates the N-BBM well, a shift in front will be detected if and only if some particle reaches $\log N + 3 \log\log N$, and the contribution of these particles to the population are different from the rest of the particles and should be analyzed separately. Taking $L = L_N = \frac{1}{\sqrt{2}}(\log N + 3 \log\log N)$ and $\mu = \mu_N$ as in (3.7), we have $(1 - \mu^2/2 - \pi^2/(2L^2)) = 0$, and (3.13) suggests that after sufficiently large time, the total number of particles that never reach L does not fluctuate much with time and stays roughly constant over time. This explains the choice of the parameters $\mu = \mu_N$ and $L = L_N$.

Secondly, the formula is proportional to $e^{\mu x} \sin(\pi x/L)$, which summing over the positions of all particles gives $Z_N(t)$. Hence $Z_N(t)$ provides a useful measure of the "size" of the process. It follows that Assumption (A_1) ensures that the number of particles in the system is of order N. On the other hand, Assumption (A_2), i.e., the condition on $Y_N(0)$ ensures that no single particle is likely to have descendants that constitute a large fraction of the population after a short time.

Basically, the evolving particle system can be divided into two parts. The number of particles which stay inside $(0, L)$ behaves like a deterministic process. Through first and second moment calculations it can be shown that under the initial assumptions, the variance of the number of particles is a small constant times N^2. Thus, an application of weak law of large numbers shows that the number of particles which stay inside behave deterministically. The other part of the process, which is in some sense 'the real stochastic part', contains the contribution of those highly favored particles. Now starting with a single particle at L, it can be shown that the typical right most descendant of this particle will reach a level that exceeds L only by a constant. Hence, the contribution of descendants of this particle to the population at time t can be estimated by using (3.13) with L and $L+\alpha$ in places of x and L respectively, where $\alpha > 0$. This motivates introducing another absorption barrier $L_A = \frac{1}{\sqrt{2}}(\log N + 3\log\log N - A)$, where A is a large constant. The authors of [7] extensively used this additional absorption barrier for explicit calculations. In order to understand the contribution of the 'highly favoured' particles, we need to understand at what rate particles reach the level L_A and need to estimate the contribution of a typical particle after reaching L_A as well. It can be shown that, under the initial assumptions, the time after which a particles reaches L_A is of the order $(\log N)^3$ and this explains the $(\log N)^3$ time scaling. Through a second moment calculation, it could be further established that the expected number of particles that reach L_A within a time-interval of length $\theta(\log N)^3$ (for small $\theta > 0$), conditional on the event that at least one particle reaches L_A, is bounded by a constant.

The authors of [7] further showed that if a particle starts at L_A, and y is a large constant, then the number of descendants of the particle that reach $L_A - y$ is approximately $y^{-1}e^{\sqrt{2}y}W$, where W is a random variable with tail

$$\mathbb{P}(W > x) \sim 1/x \text{ as } x \to \infty. \tag{3.14}$$

This understanding about the contribution of the particles, which hit L_A, to the population makes it possible to study the limit of the process $(\frac{1}{2\pi N}M_N((\log N)^3t), t > 0)$, where $M_N((\log N)^3t)$ denotes population size at time $(\log N)^3 t$, and proves the following:

Theorem 3.7: *Assuming (A_1) and (A_2) hold, as $N \to \infty$, the finite dimensional distributions of the process $(\frac{1}{2\pi N} M_N((\log N)^3 t), t > 0)$ converge to the finite dimensional distributions of the CSBP with branching mechanism $\psi(u) = au + 2\pi^2 u \log u$ started with distribution ν (specified as in (A_1)) at time zero.*

About the above theorem, the following observations can be made.

Remark 3.8:

(i) Because of technical reasons, this convergence holds only for $t > 0$.
(ii) This result proves that the size of the population remains of the order of N at $(\log N)^3$ time scaling.
(iii) Because of fluctuations, it is not possible to have process convergence while working with the Skorohod's J_1 topology. But it might be of interest to see whether it is possible to achieve process convergence with respect to some other topology (e.g., Skorohod's M_1 topology).

This theorem is the backbone of proving Theorem 3.6. For a continuous state branching process, Bertoin and Le Gall [8] were the first to define the notion of genealogy precisely and they further proved that the genealogy of Neveu's CSBP is given by the Bolthausen-Sznitman coalescent. In a later work [10], it was shown that the genealogy of any CSBP with branching mechanism $\psi(u) = au+bu\log u$ is given by the Bolthausen-Sznitman coalescent. The proof of Theorem 3.6 uses the above result and "flow of bridges", a tool developed by Bertoin and Le Gall in [9] to study convergence of coalescent processes. It boils down to showing that the flow of bridges representing the N-th population genealogy converge (in the sense of finite dimensional distributions) to the flow of bridges for the limiting CSBP whose coalescent process is given by the Bolthausen-Sznitman coalescent. Hence, conceptually (3.14) is the reason behind having the Bolthausen-Sznitman coalescent as the limiting coalescent process. As the tail probability is given by $1/x$ for large x, the Lévy measure of the limiting CSBP has a density proportional to x^{-2} and therefore the corresponding coalescent process is given by the Bolthausen-Sznitman coalescent. In the next two sections, we explain these ideas in detail. First we describe the basic steps for proving Theorem 3.7.

3.4. *Idea of the proof of Theorem 3.7*

The above discussion explains that the proof of Theorem 3.7 can be divided into the following parts:

(i) Estimating the number of particles that do not reach L_A and stay inside $(0, L_A)$ throughout.
(ii) Estimating the number of particles that hit L_A.
(iii) Understanding the contribution of these 'favorable' particles to the population.

In the remaining, we try to explain each of these parts in more detail.

3.4.1. *Particles that do not reach L_A*

Fix $K > 0$ and $x \in (0, K)$. Consider a BBM starting at x with unit birth rate and drift $\mu(> 0)$ to the left and the particles are killed as soon as they reach 0 or K. The results discussed here will be used later for the choice $K = L_A$. Recall that the quantity $p_t(x, y)$ defined as in (3.13) approximates the density of a particle at $y \in (0, K)$ after a large time. Using this density expression, one can calculate first and second moments for several other quantities related to BBM on a strip. These estimates are used to obtain first and second moments for the number of particles. Our construction suggests that these are actually those particles whose ancestors stay inside $(0, L_A)$ throughout.

Consider a branching Brownian motion in a strip with a drift μ to the left, in which each particle gives birth at rate 1 and is killed upon reaching 0 or K. Let $M(t)$ denote the total number of particles at time t and the positions at time t are denoted by $X_1(t) \leq X_2(t) \leq \cdots \leq X_{M(t)}(t)$. Consider the following quantities defined below:

$$
\begin{aligned}
V(t) &= \sum_{i=1}^{M(t)} X_i(t) e^{\mu X_i(t) + (\mu^2/2 - 1)t}, \\
Y(t) &= \sum_{i=1}^{M(t)} e^{\mu X_i(t)}, \qquad\qquad (3.15) \\
Z(t) &= \sum_{i=1}^{M(t)} e^{\mu X_i(t)} \sin\left(\frac{\pi X_i(t)}{K}\right).
\end{aligned}
$$

Under the above set up, from Lemma 2 of [18], it follows that the process $(V_t, t \geq 0)$ is a martingale with respect to the canonical filtration. Using

(3.13) together with (3.12), one can show that

$$\mathbb{E}(M(t)) = \frac{2}{K}e^{(1-\mu^2/2-\pi^2/(2K^2))t}(1+D_1)Z(0)\int_0^K e^{-\mu y}\sin(\frac{\pi y}{K})dy,$$
$$\mathbb{E}(Y(t)) = \frac{4}{\pi}e^{(1-\mu^2/2-\pi^2/(2K^2))t}(1+D_2)Z(0),$$
$$\mathbb{E}[Z(t)] = \frac{4}{\pi}e^{(1-\mu^2/2-\pi^2/(2K^2))t}Z(0),$$

where D_1 and D_2 are constants such that

$$|D_1| \vee |D_2| \leq \frac{\sum_{n=2}^{\infty} n^2 e^{-\pi^2 n^2 t/2K^2}}{e^{-\pi^2 t/2K^2}}. \tag{3.16}$$

For details of these calculations, we refer the reader to Lemma 7 of [7]. These estimates help us to calculate the first moment for the number of particles staying inside $(0, L_A)$. But an estimate of second moment is also required. The next lemma will be used to calculate the second moments.

Lemma 3.9: *(cf. Lemma 8 in [7]). Under the above set up, let $f : (0,K) \mapsto [0,\infty)$ be a measurable function. Then*

$$\mathbb{E}[(\sum_{i=1}^{M(t)} f(X_i(t)))^2] = \int_0^K f(y)^2 q_t(x,y)dy$$
$$+ 2\int_0^t\int_0^K q_s(x,z)(\int_0^K f(y)q_{t-s}(z,y)dy)^2 dzds,$$

where $q_s(x,z)$ is defined as in (3.12).

The earlier calculations enable us to find an upper bound for the second moment of $Z(t)$.

Lemma 3.10: *(cf. Lemma 9 of [7]). Suppose that $1-\mu^2-\pi^2/(2K^2) < 0$ and there exist positive constants C_1, C_2 such that $C_1K^2 \leq 1 \leq C_2/(1-\mu^2/2)$. Then there exists some positive constant C depending on μ, C_1 and C_2 but not on x and K such that*

$$\mathbb{E}[Z(t)^2] \leq Ce^{\mu x}e^{\mu K}(1/K^2 + 1/K^4).$$

Idea of proof. The earlier lemma with the choice $f(y) = e^{\mu y}\sin(\pi y/K)$ gives

$$\mathbb{E}(Z(t))^2 = \int_0^K e^{2\mu y}\sin(\pi y/K)^2 q_t(x,y)dy$$
$$+ 2\int_0^t\int_0^K q_s(x,z)(\int_0^K e^{\mu y}\sin(\pi y/K)q_{t-s}(z,y)dy)^2 dzds. \tag{3.17}$$

In order to prove the lemma, proper upper bounds are required to be used for both the terms. For the first term in (3.17), using $1-\mu^2-\pi^2/(2K^2)\le 0$ for $x,y\in(0,K)$, we have that

$$q_t(x,y)\le \frac{C}{K}e^{\mu(x-y)}\sin(\pi x/K)\sin(\pi y/K) \tag{3.18}$$

for some positive constant C.

The trigonometric fact that $\sin(\pi y/K)^2\le 1$ together with (3.18) and $\int_0^K e^{\mu y}\sin(\pi y/K)dy\le Ce^{\mu K}/K$, give the following bound for the first term:

$$\int_0^K e^{2\mu y}\sin(\pi y/K)^2 q_t(x,y)dy\le \frac{Ce^{\mu x}e^{\mu K}}{K^2}. \tag{3.19}$$

To deal with the second term, the Green's function for the Brownian motion in a strip will be used. Towards that we need to introduce some notation.

In the sequel $(B_t, t>0)$ will be used to denote one-dimensional Brownian motion without drift. Define the Green's function $G(x,y)$ such that if B_t starts from $x\in(0,K)$ at time zero and if $\tau=\inf\{t\ge 0: B_t\notin(0,K)\}$, then for all bounded measurable functions g, we have

$$\mathbb{E}[\int_0^\tau g(B_t)dt]=\int_0^K G(x,y)g(y)dy.$$

The Green's function $G(x,y)$ is given by

$$G(x,y)=\begin{cases}2x(K-y)/K & \text{if } y\ge x\\ 2y(K-y)/K & \text{if } y\le x.\end{cases}$$

Let $\nu_t(x,y)$ denote the density at time t at of a Brownian particle started at $x\in(0,K)$ and killed when it reaches 0 or K. Note that $\int_0^\infty \nu_s(x,y)ds=G(x,y)$. The relation with the Green's function follows from the observation that for $s\le t$, we have

$$q_s(x,y)=e^{\mu(x-y)+(1-\mu^2/2)s}\nu_s(x,y).$$

The rest of the proof follows using calculus and carefully designed upper bounds. For details we refer the reader to Lemma 9 of [7]. □

Let $(\mathcal{F}_t: t\ge 0)$ be the natural filtration of $(X_N(t), t\ge 0)$. Though this filtration depends on N, but in order to keep the notation simple, we suppress this dependence in the notation. Fix $u,s>0$ and consider the time-interval $[u,u+s]$. Using the earlier results, given $\mathcal{F}_u$, conditional first and second moments of the number of particles whose ancestors stay inside $(0,L_A)$ throughout can be obtained. In order to do that, fix small $\theta>0$

and define $t_k = (u + \theta k s)(\log N)^3$ where $0 \leq k \leq \theta^{-1}$. In other words, the time-interval $[u(\log N)^3, (u+s)(\log N)^3]$ is divided into 'small' intervals of width $\theta(\log N)^3$.

In the remaining part of this subsection, 'good' event $G_{N,\theta^{-1}}(\epsilon)$, where $\epsilon > 0$ is a constant, will be of main interest to us, in the sense that as $N \to \infty$, this event occurs with high probability. In this article we chose not to give the precise definition of $G_{N,\theta^{-1}}(\epsilon)$ as it is fairly technical and refer the interested reader to Section 3.1 of [7]. Note that, so far we have introduced three constants A, θ and ϵ and typically we assume that A is large and both ϵ and θ are small positive constants. We are not giving the exact constraints that these constants need to satisfy and refer to Equations (32), (33), (34), (35) in [7].

For all large N, the event $G_{N,\theta^{-1}}(\epsilon)$ basically ensures that all the Brownian particles at time t_j for $j = 0, 1, \ldots, \theta^{-1}$ are to the left of L_A. Without giving further details we comment that under the assumptions (A_1) and (A_2), for all large N such an event occurs with high probability (rigorously proved in Proposition 23 of [7]). Note that for any $0 \leq k \leq \theta^{-1}$ the event $G_{N,k}$ is defined analogously and this event basically ensures that all the Brownian particles at time t_j for $j = 0, 1, \ldots, k$ are to the left of L_A. With this set up, studying the evolution of particles whose ancestors stay inside $(0, L_A)$ at all time points, is equivalent studying branching Brownian motion moving with drift $\mu = \mu_N$ and particles are killed upon reaching 0 or L_A.

Fix $t \in [t_{k-1}, t_k]$. For $1 \leq i \leq M_N(t)$, the 'i'-th particle is in $S(t)$ if ancestor of the particle $X_{i,N}(t)$ at time s is inside $(0, L_A)$ for all $s \in [t_{k-1}, t]$. For $t \in [t_{k-1}, t_k]$, the random variables are defined as

$$\begin{aligned} Z_{N,1}(t) &= \sum_{i=1}^{M_N(t)} e^{\mu X_{i,N}(t)} \sin\Big(\frac{\pi X_{i,N}(t)}{L}\Big)\mathbf{1}_{\{i \in S(t)\}} \text{ and} \\ Z'_{N,1}(t) &= \sum_{i=1}^{M_N(t)} e^{\mu X_{i,N}(t)} \sin\Big(\frac{\pi X_{i,N}(t)}{L_A}\Big)\mathbf{1}_{\{i \in S(t)\}}. \end{aligned} \tag{3.20}$$

Observe that the process $(Z'_{N,1}(t), t \geq t_{k-1})$ uses BBM with drift μ to the left and killing barriers at 0 and L_A. The idea is to use $Z'_{N,1}(t)$ to study $Z_{N,1}(t)$ since $Z'_{N,1}(t)$ approximates $Z_{N,1}(t)$ well (see Lemma 10 of [7]). Because of the Markov property of BBM, using (3.17) we can determine the conditional mean of $Z'_{N,1}(t)$ given $\mathcal{F}_{t_{k-1}}$. Since $Z'_{N,1}(t)$ approximates $Z_{N,1}(t)$ well, this estimates the conditional mean of $Z_{N,1}(t)$ given $\mathcal{F}_{t_{k-1}}$ as well and gives

$$\mathbb{E}[Z_{N,1}(t_k)|\mathcal{F}_{t_{k-1}}] = Z_N(t_{k-1})(1 - 2\pi^2 A\theta s + O(A^2\theta^2)) + o(N(\log N)^2). \tag{3.21}$$

The next lemma gives the conditional variance of $Z'_{N,1}(t_k)$ given $\mathcal{F}_{t_{k-1}}$.

Lemma 3.11: *(cf. Lemma 12 of [7]). On the event $G_{N,k-1}$ we have*

$$Var(Z'_{N,1}(t_k)|\mathcal{F}_{t_{k-1}}) \leq C\theta N(\log N)^2 e^{-A}(Z_N(t_{k-1})) + o(N(\log N)^2).$$

For a detailed proof of this lemma, we refer interested reader to Lemma 12 of [7].

After obtaining the variance, one can apply the conditional form of Chebyshev's inequality to obtain

$$\mathbb{P}(|Z_{N,1}(t_k) - Z_{N,1}(t_{k-1})| > 4e^{-A/4}N(\log N)^2|\mathcal{F}_{t_{k-1}}) \leq C\theta e^{-A/2}\epsilon^{-1/2}(1+o(1)).$$

The next step is to use Lemma 3.11 to determine the conditional variance for the number of particles at time t_k whose ancestor at time t is in $(0, L_A)$ for all $t \in [t_{k-1}, t_k]$, given $\mathcal{F}_{t_{k-1}}$. Assuming $A > 0$, define

$$M'_N(t_k) = \sum_{i=1}^{M_N(t_k)} \mathbf{1}_{\{i \in S(t_k)\}}.$$

Proposition 3.12: *(cf. Proposition 14 of [7]). On $G_{N,k-1}$, there exists constant C such that*

$$Var(M'_N(t_k)|\mathcal{F}_{t_{k-1}}) \leq C\theta\epsilon^{-1/2}N^2(1 + o(1)).$$

This proposition is quite technical and we chose not to discuss the proof here. It is important to observe that, with the above proposition, an application of weak law of large numbers gives that the number of particles who never hit L_A or more precisely the number of particles whose ancestors stay inside $(0, L_A)$ throughout, essentially behaves like a deterministic process.

3.4.2. *Particles that hit L_A*

Now we need to estimate the number of particles hitting L_A. For $k \in \mathbb{N}$, let R_k denote the number of particles that are killed by hitting L_A between times t_{k-1} and t_k. We comment that the particles which hit L_A before time t_{k-1}, can contribute to R_k as well. We need to calculate the conditional mean and second moment of R_k given $\mathcal{F}_{t_{k-1}}$. To do that we consider BBM starting with a single particle $x \in (0, L_A)$ with drift μ to the left and the particles are killed upon reaching 0 or L_A. Let R denote the number of particles that hit L_A between time t and $t+\kappa$ where $\kappa \in (0,1)$.

Then the mean of R is given as (Lemma 15 of [7])

$$\mathbb{E}(R) = 2\pi e^A \kappa e^{(1-\mu^2/2-\pi^2/2L_A^2)} e^{\mu x} \sin(\frac{\pi x}{L_A}) \frac{(1+D)(1+o(1))(1+o(\kappa))}{N(\log N)^5}, \tag{3.22}$$

where $|D|$ is bounded by the right hand side of (3.16).

Here we give brief heuristics for the proof of (3.22). We first try to obtain an upper bound for $\mathbb{E}(R)$. Suppose that there is a particle at y at time t. Since we are considering drift to the left, it always reduces the probability that a Brownian particle reaches L_A. Since on average e^κ many descendants are produced within the time period $[t, t+\kappa]$, an application of many-to-one lemma (Lemma 2.4) gives the following as an upper bound for the number of descendants of the Brownian particle at y at time t, that reach L_A:

$$e^\kappa \mathbb{P}(\max\{B_t : t \in [0,\kappa]\} \geq L_A - y) = 2e^\kappa \mathbb{P}(B_\kappa \geq L_A - y).$$

The equality follows from the reflection principle. Hence we have

$$\mathbb{E}(R) \leq (1+D) \int_0^{L_A} p_t(x,y) 2e^\kappa \mathbb{P}(B_\kappa \geq L_A - y) dy,$$

where $|D|$ is bounded by right hand side of (3.16).

For a lower bound we can ignore the branching and then the probability that a Brownian particle starting from y at time t hits L_A before time $t+\kappa$ and before hitting 0 gives a lower bound. For $y \geq \mu\kappa$, this gives

$$\mathbb{P}\big(\max\{B_t : t \in [0,\kappa]\} \geq L_A - y + \mu\kappa\big) - \mathbb{P}\big(\min\{B_t : t \in [0,\kappa]\} \leq -y + \mu\kappa\big).$$

This gives a lower bound for $\mathbb{E}(R)$ as

$$\begin{aligned}\mathbb{E}(R) \geq (1+D) \int_{L_A/2}^{L_A} & p_t(x,y)\big(\mathbb{P}\big(\max\{B_t : t \in [0,\kappa]\} \geq L_A - y + \mu\kappa\big) \\ & - \mathbb{P}\big(\min\{B_t : t \in [0,\kappa]\} \leq -y + \mu\kappa\big)\big) dy.\end{aligned}$$

Using these bounds, after some calculations one can show that the right-hand side of (3.22) gives an upper as well as lower bound for $\mathbb{E}(R)$. For more details we refer the interested reader to Lemma 15 of [7].

Using (3.22), the conditional mean and second moment of R_k given $\mathcal{F}_{t_{k-1}}$ can be determined. On the event $G_{N,k-1}$, the conditional mean of R_k given $\mathcal{F}_{t_{k-1}}$ is given by (Proposition 16 of [7]),

$$\mathbb{E}(R_k | \mathcal{F}_{t_{k-1}}) = 2\pi e^A \frac{Z_N(t_{k-1})\theta s}{N(\log N)^2}(1 + O(|A|\theta)) + o(1). \tag{3.23}$$

On the other hand, on the event $G_{N,k-1}$, the conditional second moment of R_k given $\mathcal{F}_{t_{k-1}}$ is given by (Proposition 18 of [7]),

$$\mathbb{E}(R_k^2|\mathcal{F}_{t_{k-1}}) \leq \frac{C\theta e^A Z_N(t_{k-1})}{N(\log N)^2} + o(1). \tag{3.24}$$

Remark 3.13:

(i) Together with Markov's inequality, (3.23) shows that when θ is small, during most intervals $[t_{k-1}, t_k]$ particles do not able to reach L_A.

(ii) Further using the second moment method, (3.24) gives that

$$\mathbb{E}[R_k|R_k > 0, \mathcal{F}_{t_{k-1}}] \leq C(1 + o(1)).$$

In other words, conditional on the event that at least one particle is able to reach L_A, the expected number of particles that reach L_A is bounded by a constant which does not depend on θ or A.

(3.23) and (3.24) give us estimates for the number of particles that hit L_A during the time-interval $[t_{k-1}, t_k]$ given $\mathcal{F}_{t_{k-1}}$. Next we try to understand the contribution of each of these particles, who able to hit L_A, to the population. Before that, it should be mentioned that using the above estimates it can be shown that (Proposition 23 of [7])

$$\lim_{\epsilon\to 0} \sup_{\theta} \big(\limsup_{N\to\infty}(1 - \mathbb{P}(G_{N,\theta^{-1}}(\epsilon)))\big) = 0 \tag{3.25}$$

(with ϵ as in (3.28) to follow). Because of the above result, in order to study asymptotic results in N, it is essentially enough to restrict to the 'good' event $G_{N,k}$.

3.4.3. *Convergence to the CSBP*

After estimating the number of particles hitting L_A, it is required to have an understanding about how many descendants of these particles contribute to the process at later times. The contribution of such particles to the population in the distant future can be approximated by the number of descendants that reach $L_A - y$ where y is some large constant. Henceforth, the next step is to try to estimate the number of descendants that reach $L_A - y$ for large y. In order to capture this contribution, divide $Z_N(t_k)$ into two random variables. Recall that R_k particles hit L_A between times t_{k-1} and t_k and the time points at which these particles reach L_A are denoted by $u_1 < u_2 < \cdots < u_{R_k}$. Next we write $Z_N(t_k)$ as sum of two random variables in the following way:

$$Z_N(t_k) = Z_{N,1}(t_k) + Z_{N,2}(t_k),$$

where,

$$Z_{N,2}(t_k) = \sum_{i=1}^{M_N(t_k)} e^{\mu X_{i,N}(t_k)} \sin\Big(\frac{\pi X_{i,N}(t_k)}{L}\Big) \mathbf{1}_{\{i \notin S(t_k)\}} \mathbf{1}_{\{X_{i,N}(t_k) \le L\}}. \tag{3.26}$$

In other words, $Z_{N,2}(t_k)$ denotes the contribution of those particles at time t_k that are descended from the R_k particles that reach L_A at one of these time points $u_1, u_2, \dots, u_{R_k}$.

The fluctuations of $Z_N(t_k)$ will be largely due to that of $Z_{N,2}(t_k)$, since an application of weak law of large numbers shows that the process $Z_{N,1}(t)$ roughly behaves like a deterministic process. To study the contributions of $Z_{N,2}(t_k)$ on the event $G_{N,k-1}$, two cases will be of our interest mainly:

(i) when the contribution of these particles are 'small' (bounded by $\epsilon N(\log N)^2$);
(ii) when the contribution of these particles leads to a 'big' jump of the process (by at least an amount of $rN(\log N)^2$).

In particular the interest is to study the following probability

$$\mathbb{P}((Z_N(t_k) - Z_N(t_{k-1})) > rN(\log N)^2 | \mathcal{F}_{t_{k-1}}) \text{ and} \tag{3.27}$$

to bound the following expectation

$$\mathbb{E}\big[(Z_N(t_k) - Z_N(t_{k-1}))\mathbf{1}_{\{Z_N(t_k) - Z_N(t_{k-1})\} \le \epsilon N(\log N)^2\}} | \mathcal{F}_{t_{k-1}}\big]. \tag{3.28}$$

To study the above, consider a particle that reaches L_A at time $u_j \in (t_{k-1}, t_k]$ and let V_{j_k} denote the number of descendants of this particle that reach $(L_A - y) + (t - u_j)(\sqrt{2} - \mu)$ for the first time, i.e., for all $u \in [u_j, t)$, the ancestors of these descendants stay in the interval $(L_A - y + (u - u_j)(\sqrt{2} - \mu), \infty)$. This is equivalent to studying the number of descendants that would reach $(L_A - y) + (t - u_j)(\sqrt{2} - \mu)$ at time t for some t if particles are killed on reaching this level. In order to understand the distribution of V_{j_k}, we need to introduce another random variable.

Consider a BBM with drift $\sqrt{2}$ to the left starting with a single particle at 0. Let Z_y be the number of particles that are killed on reaching $-y$. Since we are working with a critical branching Brownian motion, the random variable Z_y is finite almost surely. Further, from the strong Markov property of BBM, it follows that Z_{w+y} has the same distribution as the sum of Z_w independent random variables, each having the same distribution as Z_y. This implies that the process $(Z_s, s \ge 0)$ is a continuous-time branching process. It is important to observe that the random variable V_{j_k} has the same distribution as Z_y for large y, where the factor $(t-u_j)(\sqrt{2}-\mu)$ is due to adjustment for different drifts. Observe that the BBM corresponding to Z_y has drift $-\sqrt{2}$ instead of $-\mu$. The next proposition, which appears on page 238 of [30], introduces another random variable that will be very important for rest of this article.

Proposition 3.14: *(cf. Proposition 24 of [7]). There exists a random variable W such that*

$$\lim_{y\to\infty} ye^{-\sqrt{2}y}Z_y = W.$$

Further for all $u \in \mathbb{R}$, we have $\mathbb{E}[e^{-e^{-\sqrt{2}u}W}] = \psi(u)$, where $\psi : \mathbb{R} \mapsto [0,1]$ satisfies Kolmogorov's equation

$$\frac{1}{2}\psi'' - \sqrt{2}\psi' = \psi(1-\psi).$$

Because of the relation with the random variable V_{j_k}, the tail decay of the random variable W is important to us. The authors of [7] showed that (Proposition 27 of [7])

$$\mathbb{P}(W > x) \sim \frac{B}{x} \tag{3.29}$$

for some constant $B > 0$ as $x \to \infty$.

As mentioned earlier, (3.29) is the main reason for having the Bolthausen-Sznitman coalescent as the limiting genealogy process. The principal strategy is to study the Laplace transform $\mathbb{E}[e^{-\lambda W}]$ for small λ and then apply a Tauberian theorem.

Next we use the tail decay estimate for W, to study (3.27) and (3.28). Observe that for fixed k, conditional on $\mathcal{F}_{t_{k-1}}$ and conditional on $R_k = r$, the random variables $W_{1,k}, W_{2,k}, \ldots, W_{r,k}$ such that these random variables are i.i.d. copies of W.

Intuitively, the contribution of $Z_{N,2}(t_k)$ from the j-th particle to hit L_A can be approximated by $\pi\sqrt{2}e^{-A}N(\log N)^2 W_{j,k}$. The authors of [7] carefully bounded the probabilities of several unlikely events and proved that this is indeed the case. More precisely, for a 'long jump' contribution, i.e., in order to have $Z_N(t_k) - Z_N(t_{k-1}) > rN(\log N)^2$ we must have (with high probability) $W_{j,k} > r/(\pi\sqrt{2}e^{-A})$ for some $1 \le j \le R_k$. From (3.29) it follows that the probability $\mathbb{P}(W_{j,k} > r/(\pi\sqrt{2}e^{-A}))$ is proportional to $1/r$. Consequently, for the scaled process $(Z_N((\log N)^3 t)/(N(\log N)^2), t \ge 0)$, the rate of jumps of size at least r is proportional to $1/r$, and this is consistent with the CSBP with branching mechanism $\psi(u) = au + 2\pi^2 u \log u$, since its Lévy measure has a density proportional to x^{-2} (see (3.4)).

With this heuristics and through proper estimations of (3.27) and (3.28), it can be shown that fluctuation of the scaled process $(Z_N((\log N)^3 t)/(N(\log N)^2), t \geq 0)$ matches with the infinitesimal generator of the CSBP with branching mechanism

$$\psi(u) = au + 2\pi^2 u \log(u).$$

In other words as $N \to \infty$, finite dimensional distributions of the scaled process $(Z_N((\log N)^3 t)/(N(\log N)^2), t \geq 0)$ converges to that of a CSBP with branching mechanism $\psi(u) = au + 2\pi^2 u \log u$ where $a \in \mathbb{R}$ some constant and started with distribution ν (as in Assumption A_1) at time zero (see Proposition 1 of [7]). Theorem 3.7 basically follows from the above, since the process $Z_N(\cdot)$ well approximates the size of the population after a short time. More precisely, it can be shown that for any fixed $t > 0$, as $N \to \infty$,

$$|\frac{1}{2\pi N} M_N((\log N)^3 t) - Z_N((\log N)^3 t)/(N(\log N)^2)| \to 0,$$

in probability and as the process $(Z_N((\log N)^3 t)/(N(\log N)^2), t \geq 0)$ converges to a CSBP, this completes the proof of Theorem 3.7.

3.5. *Idea of the proof of Theorem 3.6*

In the final section, we discuss the proof of Theorem 3.6 in detail. Theorem 3.7 shows that the scaled population in $(\log N)^3$ time scaling converges to a continuous state branching process with branching mechanism $\psi(u) = au + 2\pi^2 u \log u$. It is known that the genealogy process for such a CSBP is given by the Bolthausen-Sznitmann coalescent process (see [10]).

3.5.1. *Notion of genealogy for a CSBP*

We first discuss how to define the notion of genealogy for a continuous state branching process. Note that, for a CSBP, recovering the genealogical information is non-trivial, as there is no graphical tree representation. Bertoin and Le Gall [8] showed how to obtain genealogical information for such a process. In order to explain the heuristics, we start with the usual Galton Watson branching process and try to recover the genealogy information in a mathematical way. Suppose for all $t \geq 0$ the population at time t, denoted by Z_t, is labeled in a certain way which respects ancestry information.

Fix $0 < s < t$ and for each $1 \leq i \leq Z_s$, consider the sub-populations at time t generated from $\{1, \ldots, i-1\}$ and $\{1, \ldots, i\}$ at time s respectively.

From this construction it is evident how to identify the individuals from Z_t which are descended from the 'i'-th individual from Z_s. Observe that this way of representing the genealogical information only uses the fact that for all $1 \leq i \leq Z_s$, the Galton-Watson branching processes starting from the set $\{1, \ldots, i\}$ can be constructed on the *same* probability space. We are going to use the same intuition for a CSBP.

In order to highlight the role of the initial value, we write $Z_t = Z(t, a)$ to denote that the CSBP $(Z_t : t \geq 0)$ starts from $Z_0 = a \in \mathbb{R}^+$. By the fundamental branching property, we have

$$Z(\cdot, a) + Z'(\cdot, b) \stackrel{d}{=} Z(\cdot, a + b),$$

where $Z'(\cdot, b)$ is independent of $Z(\cdot, a)$ and has the same distribution as $Z(\cdot, b)$. This implies that there exists a probability space such that both the CSBP's, $Z(\cdot, a)$ and $Z(\cdot, a + b)$, are defined over it with $Z(\cdot, a+b) - Z(\cdot, a)$ is independent of $Z(\cdot, a)$ and has the same law as $Z(\cdot, b)$. Now fix $0 \leq a_1 < a_2 < a_3 < \infty$ and consider three processes defined over some probability space such that:

(i) $Z'(\cdot, a_2 - a_1)$ is independent of $Z(\cdot, a_1)$ and has the same distribution as $Z(\cdot, a_2 - a_1)$;
(ii) $Z''(\cdot, a_3 - a_2)$ is independent of both $Z(\cdot, a_1)$ and $Z'(\cdot, a_2 - a_1)$ and has the same distribution as $Z(\cdot, a_3 - a_2)$.

Using these three processes it is possible to construct copies of $Z(\cdot, a_i) : 1 \leq i \leq 3$ on the same probability space such that for $1 \leq j < i \leq 3$, the process $Z(\cdot, a_i) - Z(\cdot, a_j)$ has the law $Z(\cdot, a_i - a_j)$ and independent of $Z(\cdot, a_l) : 1 \leq l \leq j$.

For any $k \in \mathbb{N}$, this construction can be carried over for k points $0 \leq a_1 < a_2 < \cdots < a_k < \infty$ and this construction does not depend on the order in which we pick these k points. By Kolmogorov's theorem, we can construct a process $(Z(t, a) : t \geq 0$ and $a \geq 0)$ such that $Z(\cdot, 0) = 0$ and for every $a, b \geq 0$, the process $Z(\cdot, a + b) - Z(\cdot, a)$ has law $Z(\cdot, b)$ and is independent of the family $\{Z(\cdot, c) : 0 \leq c \leq a\}$. Hence it follows that for any $t > 0$ the process $Z(t, \cdot)$ has *independent and stationary increments.* Further the process $Z(t, \cdot)$ is non-decreasing a.s. In other words $Z(t, \cdot)$ is a 'subordinator' (defined below).

Definition 3.15: A subordinator is a non-decreasing Lévy process taking values in $\mathbb{R}$. Here we will work with modification which has RCLL paths almost surely.

The next proposition (Proposition 1 of [8]) gives a deeper connection between these two processes. Assume the semigroup of the CSBP $(Z(t,a) : t \geq 0)$ can be studied through the Laplace transform characterized as below,

$$\mathbb{E}(e^{-\lambda Z_t}|Z_0 = a) = \exp(-au_t(\lambda)).$$

Recall that, if the branching mechanism of the CSBP is given by $\psi(u)$, then the function $u_t(\lambda)$ satisfies

$$\frac{\partial u_t(\lambda)}{\partial t} = -\psi(u_t(\lambda)), \quad u_0(\lambda) = \lambda.$$

Proposition 3.16: *(cf. Proposition 1 in [8]) On some probability space, there exists a process*

$$(S^{(s,t)}(a), 0 \leq s \leq t \text{ and } a \geq 0) \text{ such that:}$$

(i) For every $0 \leq s \leq t$, $(S^{(s,t)} = S^{(s,t)}(a), a \geq 0)$ is a subordinator with Laplace exponent $u_{t-s}(\cdot)$.
(ii) For every integer $p \geq 2$ and $0 \leq t_1 \leq \cdots \leq t_p$, the subordinators $S^{(t_1,t_2)}, \ldots, S^{(t_{p-1},t_p)}$ are independent and

$$S^{(t_1,t_p)}(a) = S^{(t_{p-1},t_p)} \circ \cdots \circ S^{(t_1,t_2)}(a); \text{ for all } a \geq 0 \text{ a.s..}$$

Finally, the processes $(S^{(0,t)}(a), t \geq 0$ and $a \geq 0)$ and $(Z(t,a) : t \geq 0$ and $a \geq 0)$ have the same finite dimensional marginals.

The above proposition enables us to recover the genealogical information.

Definition 3.17: For every $b, c \geq 0$ and $0 \leq s < t$, we say that the individual c in the population at time t has the individual b in the population at time s as ancestor if b is a jump time of $S^{(s,t)}$ and

$$S^{(s,t)}(-b) < c < S^{(s,t)}(b).$$

Observe that the set of individuals in the population at time t having an ancestor at time s is a random open subset of $[0, \infty)$ with canonical decomposition $\bigcup(S^{(s,t)}(b-), S^{(s,t)}(b))$ where the union is taken over the jump times. Its complement can be identified as the closed range of $S^{(s,t)}$. Under the assumption that $S^{(s,t)}$ has zero drift, the closed range has zero Lebesgue measure a.s. This implies that the individuals in the population at time t having no ancestor at time s are of Lebesgue measure zero a.s.. Clearly an individual in the population at time t can have at most one ancestor at time s.

Fix $0 \le r < s < t$. For an individual d at time t with ancestor c at time s, where c has ancestor b at time zero, we have

$$S^{(s,t)}(c-) < d < S^{(s,t)}(c) \text{ and } S^{(s,t)}(b-) < c < S^{(s,t)}(b).$$

Since $S^{(r,t)} = S^{(s,t)} \circ S^{(r,s)}$, by monotonicity we have

$$S^{(r,t)}(b-) < d < S^{(r,t)}(b),$$

i.e., the individual d at time t has ancestor b at time r. This shows that the genealogies obtained through this construction is *consistent.*

Bertoin and Le Gall [8] further proved that the genealogy for the CSBP with branching mechanism $\psi(u) = u\log(u)$, studied by Neveu, is given by the Bolthausen-Sznitman coalescent process. Later, it was proved that the genealogy for a CSBP with branching mechanism $\psi(u) = au + bu\log(u)$, is also given by the Bolthausen-Sznitmann coalescent process (see [10]).

3.5.2. *Convergence of coalescent processes and 'flow of bridges'*

Next we are going to discuss how to study convergence of genealogies using 'flow' of bridges, a technique introduced by Bertoin and Le Gall [9]. Following [9], we give the following definitions.

Definition 3.18: A bridge $(B(t) : t \in [0,1])$ is a non-decreasing random process with values in $[0,1]$ with exchangeable increments and right-continuous paths such that $B(0) = 0$ and $B(1) = 1$.

A bridge gives a partition of $\mathbb{N}$ in the following way:

Definition 3.19: Let $\{V_n : n \in \mathbb{N}\}$ be an i.i.d. collection of $U(0,1)$ random variables. For a bridge $(B(t) : t \in [0,1])$, the corresponding random partition of $\mathbb{N}$ is denoted by $\Pi(B)$ and is given by the equivalence relation,

$$i \sim_{\Pi(B)} j \Leftrightarrow B^{-1}(V_i) = B^{-1}(V_j) \text{ for } i,j \in \mathbb{N}.$$

Here B^{-1} denotes the usual right inverse given by

$$B^{-1}(r) = \inf\{s \in [0,1] : B(s) > r\} \text{ for } r \in [0,1) \text{ and } B^{-1}(1) = 1.$$

Observe that since B has exchangeable increments, $\Pi(B)$ is an exchangeable partition. Next we consider a collection of bridges $(B_{s,t}, 0 \le s \le t)$ defined on the same probability space.

Definition 3.20: We call $(B_{s,t}(x), 0 \le s \le t, 0 \le x \le 1)$ a flow of bridges if it is a collection of bridges that additionally satisfies the following conditions:

- For every $s < t < u$, we have $B_{s,u} = B_{t,u} \circ B_{s,t}$.
- The law of $B_{s,t}$ only depends on $t - s$.
- If $s_1 < s_2 < \cdots < s_n$, then the bridges $B_{s_1,s_2}, \ldots, B_{s_{n-1},s_n}$ are independent.
- $B_{0,0} = \text{Id}$ (identity mapping from $[0,1]$ to itself) and $B_{0,t} \to \text{Id}$ as $t \to 0$ in probability, in the sense of Skorohod's topology.

Note that we are using a different convention for the time parameters than in [9]. The bridge $B_{s,t}$ defined here would be called $B_{-s,-t}$ in [9]. It is important to observe that bridges are considered to have exchangeable increments in [9].

Bertoin and Le Gall [9] proved that there exists a one-to-one correspondence between 'flow' of bridges and so-called exchangeable coalescent processes. The following continuity lemma taken from [9], explains the usefulness of studying the 'flow' of bridges in the context of convergence of genealogies. Let $D([0,1], \mathbb{R}_+)$ be the space of RCLL functions with non-decreasing paths defined over $[0,1]$ such that $f(0) = 0$ and $f(1) = 1$ for all $f \in D([0,1], \mathbb{R}_+)$. In what follows, $\Rightarrow$ is used to denote convergence in distribution.

Lemma 3.21: *Consider a sequence of bridges $\{B_n : n \in \mathbb{N}\}$. Then the following are equivalent:*

(i) The exchangeable partition $\Pi(B_n)$ converges in distribution to $\Pi(B_\infty)$ as $n \to \infty$.
(ii) $B_n \Rightarrow B_\infty$ in $D([0,1], \mathbb{R}_+)$ endowed with the Skorohod topology.

More generally, if we start from a sequence $\{(B_n) : n \in \mathbb{N}\}$ of bridges and assume that the exchangeable partition $\Pi(B_n)$ converges in distribution, then B_n converges in distribution to B_∞ and the limit has to be of the form $\Pi(B_\infty)$. With this, we end this subsection and comment that, in order to prove Theorem 3.6, we need to construct a sequence of flows of approximating bridges representing the genealogy for the N-th BBM family and show that the sequence converges to an appropriate flows of bridges representing the Bolthausen-Sznitman coalescent.

3.5.3. *Proof of Theorem 3.6*

In this section, we describe the construction of flows of bridges required to prove Theorem 3.6. We first need to come up with flows of discrete bridges $(B^N_{s,t}(x), 0 \le x \le 1, 0 \le s \le t)$ such that for any $0 = t_0 < t_1 < \cdots < t_k = t$, we have the equality of distributions

$$(\Pi(B^N_{t_{k-1},t_k}),\ldots,\Pi(B^N_{t_0,t_k})) \stackrel{d}{=} (\Pi_N(2\pi(t-t_{k-1})),\ldots,\Pi_N(2\pi(t-t_0))),$$

where $\Pi_N(2\pi(t-t_i))$ is the coalescent process obtained by looking at the genealogy of $n, n \in \mathbb{N}$ randomly picked individuals from the population at time $(\log N)^3 t$ over the time period $[(\log N)^3 t_i, (\log N)^3 t]$ at $(\log N)^3$ time scale. Because of Lemma 3.21, showing that, as $N \to \infty$,

$$(B^N_{t_{k-1},t_k},\ldots,B^N_{t_0,t_k}) \Rightarrow (B_{t_{k-1},t},\ldots,B_{t_1,t})$$

where $(B_{s,t}(x) : 0 \le x \le 1, 0 \le s \le t)$ represents the corresponding flow of bridges for the limiting CSBP as in Theorem 3.7, completes the proof. Strictly speaking, in order to complete the proof we need an additional lemma which strengthens Lemma 3.21 as the approximating bridges are no longer required to have exchangeable increments. We describe the construction of flows of bridges next.

We first construct the flow of bridges for the limiting CSBP $(Z_t, t \ge 0)$. Proposition 3.16 ensures that we can consider a two-parameter process $(Z(t,a) : t \ge 0, a \ge 0)$ such that t is the time parameter and a is the initial population size, and associate a flow of subordinators $(S^{(s,t)}(a), 0 \le s \le t, a \ge 0)$ where $S^{(s,t)}(a)$ can be understood as the descendants in the population at time t of the first a individuals in the population at time s.

For $z > 0$, suppose that we start with $Z(0) = z$. For each $s \le t$ the renormalized process is defined as

$$B_{s,t}(x) = S^{(s,t)}(xS^{(0,s)}(z))/S^{(0,t)}(z).$$

It is not difficult to check that the family $(B_{s,t}(x), 0 \le s \le t, 0 \le x \le 1)$ gives a flow of bridges.

We now construct the flow of approximating bridges. Fix $K \in \mathbb{N}$ and the time points $0 = t_0 < t_1 < \cdots < t_K$. For $0 \le i < j \le K$, we define a process $(B^N_{t_i,t_j}(s), s \in [0,1])$. Next we label the branching Brownian motion particles in such a way that respects ancestry. To do that, we start with $(v_{i,j}, i \ge 0, 0 \le j \le K)$, a collection of i.i.d. uniform $(0,1)$ random variables independent of the BBM X_N. The particles at time 0 are labeled as $u_{i,0} = v_{i,0}$ for $i \le M_N(0)$. For $j \ge 1$ and $i \le M_N((\log N)^3 t_j)$, $u_{i,j}$ is a vector of length $j+1$ which are defined inductively by setting

$$u_{i,j} = (u_{p(i),j-1}, v_{i,j}),$$

where $u_{p(i),j-1}$ is the label of the parent particle at time $(\log N)^3 t_{j-1}$. Then the particles are ordered using lexicographic ordering. For $0 \le j \le K$

and $i \le M_N((\log N)^3 t_j)$, let $x_{i,j}$ denote the spatial location of the i-th individual w.r.t. this ordering at time $(\log N)^3 t_j$. The assigned weight to this individual given by:

$$w(i,j) = \begin{cases} \frac{1}{Z_N((\log N)^3 t_j)} e^{\mu x_{i,j}} \sin(\pi x_{i,j}/L) \mathbf{1}_{\{x_{i,j} \le L\}} & \text{if } 0 \le j \le K-1 \\ \frac{1}{M_N((\log N)^3 t_K)} & \text{if } j = K. \end{cases} \tag{3.30}$$

Observe that the particles are weighted proportional to their contribution to $Z_N(\cdot)$ (which gives a good approximation of the population size) except at time $t_K(\log N)^3$. At time $t_K(\log N)^3$ all particles are weighted equally. For $0 \le j \le k \le K$, let $A_i(j,k)$ denote the set of descendants at time $(\log N)^3 t_k$ of the i-th individual at time $(\log N)^3 t_j$. For $0 \le y \le 1$ and $0 \le j \le K$ define

$$L_j(y) = \max\{l \in \mathbb{N} : \sum_{i=1}^{l} w(i,j) \le y\},$$

with the condition that maximum of an empty set is zero.

Now we are ready to construct our discrete bridges as follows: For $0 \le y \le 1$ and $0 \le j \le k \le K$, let

$$B^N_{t_j,t_k}(y) = \sum_{i=1}^{L_j(y)} \sum_{m \in A_i(j,k)} w(m,k).$$

It is not hard to check that, for these approximating bridges and for $0 \le i < j < k \le K$, we have

(i) $B^N_{t_i,t_k} = B^N_{t_j,t_k} \circ B^N_{t_i,t_j}$ and
(ii) $(B^N_{t_i,t_k})^{-1} = (B^N_{t_i,t_j})^{-1} \circ (B^N_{t_j,t_k})^{-1}$.

It is important to observe that these processes are *not* exactly bridges in the sense of Definition 3.18, as they do *not* have exchangeable increments.

Fix $m \ge 1$ and let $0 = u_0 < u_1 < \cdots < u_m = 1$. In order to obtain convergence of flow of bridges, the first steps is to show that for each fixed i with $1 \le i \le K$

$$(B^N_{0,t_i}(u_j))_{j=1}^m \Rightarrow (B_{0,t_i}(u_j))_{j=1}^m. \tag{3.31}$$

In order to prove the above convergence, the increments $(B^N_{0,t_i}(u_j) - B^N_{0,t_i}(u_{j-1}))_{j=1}^m$ are required to be studied, which corresponds to the subset of the population at time 0 associated with the quantiles $[u_{j-1}, u_j)$ when the population is weighted according to (3.30). In other words this gives a partition of the initial random subgroups and we need to show that, if the population at time zero is divided into m random subgroups, then under some reasonable assumptions Theorem 3.7 must hold for each of these subgroups.

Suppose the initial population is divided into m random subgroups $S_1, \ldots, S_m$. For $1 \leq i \leq m$, let $Z_{i,N}(\cdot)$ denote the contribution of S_i to the sum $Z_N(\cdot)$ and let $M_{i,N}(\cdot)$ denote the size of the sub-population descended from S_i. Then together with the assumptions A_1 and A_2, a necessary condition is that the initial joint distribution of $(Z_{i,N}(0)/N(\log N)^2)_{i=1}^m$ converges to some probability measure ρ on $[0,\infty)^m$. Lemma 50 of [7], proved that the necessary condition is sufficient, and as $N \to \infty$, the finite dimensional distributions of the m-dimensional vector valued process

$$\{(\frac{Z_{i,N}((\log N)^3 t)}{N(\log N)^2})_{i=1}^m, t > 0\}$$

jointly converges to the finite dimensional distributions of $\{(Z_i(t))_{i=1}^m, t > 0\}$, where $(Z_i(0))_{i=1}^m$ has distribution ρ, and conditional on $(Z_i(0))_{i=1}^m$, each Z_i evolves independently as a CSBP with branching mechanism $\psi(u) = au + 2\pi^2 u \log u$. Observing the fact that conditional on the initial configuration, the m components evolve independently, the proof follows from Theorem 3.7. Using the above convergence, (3.31) can be established.

Next we need to establish joint convergence of bridges in terms of finite dimensional distributions. Towards this, for any $k \in \mathbb{N}$, consider k time points $0 = t_0 < t_1 < \cdots < t_k$. For the time-interval $[t_i(\log N)^3, t_{i+1}(\log N)^3]$ and for any $0 = u_0 < u_1 < \cdots < u_m = 1$, the strong Markov property together with (3.31) give the joint convergence

$$(B^N_{t_i,t_{i+1}}(u_j) : 1 \leq j \leq m)_{0 \leq i \leq k-1} \Rightarrow (B_{t_i,t_{i+1}}(u_j) : 1 \leq j \leq m)_{0 \leq i \leq k-1}. \tag{3.32}$$

Together with a tightness condition (Lemma 55 of [7]), Equation (3.32) establishes convergence of the flow of approximating bridges to the flow of bridges corresponding to the limiting CSB. Now it is important to observe that Lemma 3.21 is not directly applicable here to give Theorem 3.6 as the approximating bridges do not have exchangeable increments. Lemma 57 of [7] proves a stronger version of Lemma 3.21 in the sense that the approximating bridges need not have exchangeable increments, and thereby completes the proof of Theorem 3.6.

Acknowledgments

Sandra Kliem would like to thank M. Roberts and L.-P. Arguin, A. Bovier and N. Kistler for allowing the use of their pictures in this article. Her work was partially supported by the DFG through the SPP Priority Programme 1590 and by the Deutsche Forschungsgemeinschaft (DFG, German Research Foundation) - Projectnumber 39309207. Both of the authors would like to thank the IMS for supporting their visit to the IMS in Singapore. Finally, the authors would like to thank an anonymous referee for the careful reading and the valuable comments which helped to improve this exposition.

References

1. E. Aïdékon, J. Berestycki, É. Brunet and Z. Shi, "Branching Brownian motion seen from its tip", *Probab. Theory Related Fields*, **157**(1-2):405–451, 2013.
2. L.-P. Arguin, A. Bovier and N. Kistler, "Genealogy of extremal particles of branching Brownian motion", *Comm. Pure Appl. Math.*, **64**(12):1647–1676, 2011.
3. L.-P. Arguin, A. Bovier and N. Kistler, "Poissonian statistics in the extremal process of branching Brownian motion", *Ann. Appl. Probab.*, **22**(4):1693–1711, 2012.
4. L.-P. Arguin, A. Bovier and N. Kistler, "The extremal process of branching Brownian motion", *Probab. Theory Related Fields*, **157**(3-4):535–574, 2013.
5. J. Bérard and J. B. Gouéré, "Brunet-Derrida behavior of branching-selection particle systems on the line", *Comm. Math. Phys.*, **298**:323–342, 2010.
6. N. Berestycki, "Recent progress in coalescent theory", *Ensios Matematicos*, **16**:1–193, 2009.
7. J. Berestycki, N. Berestycki and J. Schweinsberg, "The genealogy of branching Brownian motion with absorption", *Ann. Probab.*, **41**:527–618, 2013.
8. J. Bertoin and J. F. Le Gall, "The Bolthausen-Sznitman coalescent and the genealogy of continuous-state branching processes", *Probab. Theory Related Fields*, **117**:249–266, 2000.
9. J. Bertoin and J. F. Le Gall, "Stochastic flows associated to coalescent processes", *Probab. Theory Related Fields*, **126**:261–288, 2003.
10. M. Birkner, J. Blath, M. Capaldo, A. Etheridge, M. Möhle, J. Schweinsberg and A. Wakolbinger, "Alpha-stable branching and beta-coalescents", *Electron. J. Probab.*, **10**:303–325, 2005.
11. E. Bolthausen and A.-S. Sznitman, "On Ruelle's probability cascades and an abstract cavity method", *Comm. Math. Phys.*, **197**:247–276, 1998.
12. A. Bovier, "From spin glasses to branching Brownian motion—and back?", In *Random walks, random fields, and disordered systems*, volume 2144 of *Lecture Notes in Math.*, pages 1–64. Springer, Cham, 2015.
13. M. Bramson, "Maximal displacement of branching Brownian motion", *Comm. Pure Appl. Math.*, **31**(5):531–581, 1978.

14. M. Bramson, "Convergence of solutions of the Kolmogorov equation to travelling waves", *Mem. Amer. Math. Soc.*, **44**(285), 1983.
15. E. Brunet, B. Derrida, A. H. Mueller and S. Munier, "Noisy traveling waves: Effect of selection on genealogies", *Europhys. Lett.*, **76**:1–7, 2006.
16. E. Brunet, B. Derrida, A. H. Mueller and S. Munier, "Effect of selection on ancestry: An exactly soluble case and its phenomenological generalization", *Phys. Rev. E*, **76**:1–7, 2006.
17. J.-B. Gouéré, "Le mouvement brownien branchant vu depuis sa particule la plus à gauche (d'après Arguin-Bovier-Kistler et Aïdékon-Berestycki-Brunet-Shi)", *Astérisque*, (361),271–298, 2014.
18. J. W. Harris and S. C. Harris, "Survival probabilities for branching Brownian motion with absorption", *Electron. Commun. Probab.*, **12**:81–92, 2007.
19. S. C. Harris and M. I. Roberts, "The many-to-few lemma and multiple spines", *Ann. Inst. H. Poincaré Probab. Statist.*, **53**(1):226–242, 2017.
20. N. L. Johnson, S. Kotz and N. Balakrishnan, "Continuous univariate distributions. Vol. 2", second ed., *Wiley Series in Probability and Mathematical Statistics: Applied Probability and Statistics*, John Wiley & Sons, Inc., New York, 1995.
21. M. Jirina, "Stochastic branching processes with continuous state space", *Czechoslovak Math. J.*, **8(83)**:292–313, 1958.
22. H. Kesten, "Branching Brownian motion with absorption", *Stochastic Process. Appl.*, **7**:9–47, 1978.
23. J. F. C. Kingman, "The coalescent", *Stochastic Process. Appl.*, **13**:235–248, 1982.
24. S. P. Lalley and T. Sellke, "A conditional limit theorem for the frontier of a branching Brownian motion", *Ann. Probab.*, **15**(3):1052–1061, 1987.
25. J. Lamperti, "Continuous state branching processes", *Bull. Amer. Math. Soc.*, **73**:382–386, 1967.
26. J. Lamperti, "The limit of a sequence of branching processes", *Z. Wahrsch. Verw. Gebiete*, **7**:271–288, 1967.
27. G. F. Lawler, "Introduction to Stochastic Processes", 2nd ed., Chapman & Hall/CRC, Boca Raton, FL.
28. M. R. Leadbetter, G. Lindgren and H. Rootzén, "Extremes and related properties of random sequences and processes", *Springer Series in Statistics*, Springer, New York-Berlin, 1983.
29. P. Maillard, "Speed and fluctuations of N-particle branching Brownian motion with spatial selection", *Probab. Theory Related Fields*, **166**: 1061–1173, 2016.
30. J. Neveu, "Multiplicative martingales for spatial branching processes", In *Seminar on Stochastic Processes, 1987 (Princeton, NJ, 1987), Progr. Probab. Statist.*, **15**: 223–242. Birkhauser, Boston, MA.
31. J. Pitman, "Coalescents with multiple collisions", *Ann. Probab.*, **27**:1870–1902, 1999.
32. S. I. Resnick, "Extreme values, regular variation, and point processes", *Applied Probability. A Series of the Applied Probability Trust*, volume 4, Springer, New York, 1987.

33. M. I. Roberts, "A simple path to asymptotics for the frontier of a branching Brownian motion", *Ann. Probab.*, **41**(5):3518–3541, 2013.
34. S. Sawyer, "Branching diffusion processes in population genetics", *Adv. in Appl. Probab.*, **8**:659–689, 1976.

Markovian Dynamics of Exchangeable Arrays

Jiří Černý
Department of Mathematics and Computer Science
University of Basel
Spiegelgasse 1
4051 Basel, Switzerland
jiri.cerny@unibas.ch

Anton Klimovsky
Fakultät für Mathematik
Universität Duisburg-Essen
Thea-Leymann-Str. 9
D-45117 Essen, Germany
anton.klymovskiy@uni-due.de

We study Markov processes with values in the space of general two-dimensional arrays whose distribution is exchangeable. The results of this paper are inspired by the theory of exchangeable dynamical random graphs developed by H. Crane in [4, 5].

1. Introduction

The goal of this paper is to survey some of the recent results on the Markovian dynamics of exchangeable random graphs due to Harry Crane [4, 5] and generalise them to the context of dynamics of exchangeable random arrays whose entries do not necessarily take values in a finite set.

The paper extends the results presented by the authors in the learning session "Genealogies of particles on dynamic random networks" during the Programme "Genealogies of Interacting Particle Systems" of Institute for Mathematical Sciences in August 2017. The learning session concentrated on various aspects of the dynamics of random graphs and, in particular, of particle systems on such graphs. While the original theory due to H. Crane cannot be applied directly in this context, the results of this paper could be relevant, e.g., for exchangeable Markovian dynamics of particle systems on weighted exchangeable dynamical random graphs.

Our results closely follow [4, 5]. However, as we cannot use the fact that the entries of the array take values in a finite space, some of the proofs require non-trivial modifications, which, in our opinion, sometimes make them cleaner. Many of the results of [4, 5] depend very strongly on the finiteness of the entry space and cannot be proved easily in the general context. Those are omitted here.

Acknowledgements. The work was done partially while the authors were visiting the Institute for Mathematical Sciences, National University of Singapore in 2017. The visit was supported by the Institute. AK's visit was also partially supported by the German Academic Exchange Service (DAAD).

2. Exchangeable random arrays

We will consider arrays with values in an arbitrary Polish space S. This space will be endowed with its Borel σ-field $\mathcal{B}(S)$ and a compatible metric d_S, which we assume to be bounded by 1. We write $\mathcal{P}(S)$ for the set of all probability measures on $(S, \mathcal{B}(S))$ endowed with the topology of weak convergence, which is a Polish space as well.

A random S-valued array is a collection $Y = (Y_{ij})_{ij \in \mathbb{N}}$ of S-valued random variables on some probability space $(\Omega, \mathcal{A}, P)$. Otherwise said, Y is $\mathbb{S} := S^{\mathbb{N}^2}$-valued random variable. We endow $\mathbb{S}$ with the product topology and the compatible metric $d_{\mathbb{S}}(y, y') = \sum_{i,j \in \mathbb{N}} 2^{-i-j} d_S(y_{ij}, y'_{ij})$.

For an arbitrary set $A \subset \mathbb{N}$, we define $Y|_A = (Y_{ij})_{i,j \in A}$ to be the restriction of Y to the index set A. In particular, with $[n] := \{1, \dots, n\}$, $Y|_{[n]}$ is its restriction to the first n coordinates, $Y|_{[n]} \in \mathbb{S}_n := S^{n^2}$.

Similarly, for every probability distribution ν on $\mathbb{S}$ (or on $\mathbb{S}_m$, $m \geq n$), we denote by $\nu|_{[n]}$ its image under the canonical restriction from $\mathbb{S}$ (or $\mathbb{S}_m$) to $\mathbb{S}_n$. It is a known fact that a sequence of probability measures $(\mu^k)_{k \geq 1}$ on $\mathbb{S}$ converges weakly to $\mu \in \mathcal{P}(\mathbb{S})$, iff all restrictions $\mu^k|_{[n]} \in \mathcal{P}(\mathbb{S}_n)$, converge weakly in $\mathcal{P}(\mathbb{S}_n)$, or equivalently $\mu^k(f) \to \mu(f)$, for every bounded continuous cylinder function f on $\mathbb{S}$.

Let Σ be the set of all permutations of integers, that is the set of all bijections of $\mathbb{N}$ which fix all but finitely many values; Σ_n denotes the set of all permutations of $[n]$. For an array Y and $\boldsymbol{\pi} = (\pi_1, \pi_2) \in \Sigma^2$, we define a new array $Y^{\boldsymbol{\pi}}$ by $Y^{\boldsymbol{\pi}}_{ij} = Y_{\pi_1(i)\pi_2(j)}$. For $\pi \in \Sigma$, we also define Y^{π} by

$Y^{\pi}_{ij} = Y_{\pi(i)\pi(j)}$. An array Y is called *exchangeable* if

$$Y \overset{\text{law}}{=} Y^{\boldsymbol{\pi}}, \qquad \text{for every } \boldsymbol{\pi} \in \Sigma^2. \tag{2.1}$$

An array Y is called *weakly exchangeable*[a] if it is symmetric (i.e., $Y_{ij} = Y_{ji}$) and

$$Y \overset{\text{law}}{=} Y^{\pi}, \qquad \text{for every } \pi \in \Sigma. \tag{2.2}$$

The key result of the theory of random arrays is their characterisation due to Aldous [1] and Hoover [7] which can be viewed as a "two-dimensional version" of de Finetti's theorem.

Theorem 2.1: (a) *If $(Y_{ij})_{i,j\in\mathbb{N}}$ is an S-valued exchangeable array, then there exists a measurable function $f\colon [0,1]^4 \to S$ such that $Y \overset{\text{law}}{=} Y^{\star}$, where*

$$Y^{\star}_{ij} = f(U, U_i, U'_j, U_{ij}), \tag{2.3}$$

and U, $(U_i)_{i\in\mathbb{N}}$, $(U'_i)_{i\in\mathbb{N}}$, and $(U_{ij})_{i,j\in\mathbb{N}}$ are independent collections of i.i.d. Uniform($[0,1]$) random variables.

(b) *If $(Y_{ij})_{i,j\in\mathbb{N}}$ is an S-valued weakly exchangeable array, then the analogous statement holds with a function $f\colon [0,1]^4 \to S$ satisfying $f(\cdot, x, y, \cdot) = f(\cdot, y, x, \cdot)$, and with*

$$Y^{\star}_{ij} = Y^{\star}_{ji} = f(U, U_i, U_j, U_{ij}), \qquad i \geq j. \tag{2.4}$$

The representing function f of the Aldous-Hoover theorem is not uniquely determined. E.g., in the case (a), if two functions f and f' satisfy $f'(a,b,c,d) = f(T_1(a), T_2(b), T_3(c), T_4(d))$ for some measure preserving transformations $T_1, \ldots, T_4$ of $[0,1]$, then the corresponding exchangeable arrays have the same distribution.

A (weakly) exchangeable array is called *dissociated* if

$$(Y_{ij} : i,j \leq n) \text{ is independent of } (Y_{ij} : i,j > n), \text{ for each } n. \tag{2.5}$$

It is obvious that if the function f in the representation of Theorem 2.1 does not depend on the first coordinate, then Y is dissociated. Converse statement hold as well, see Corollary 14.13 in [2].

Dissociated arrays play a similar role as i.i.d. sequences do in the theory of exchangeable sequences: Every (weakly) exchangeable array is a mixture of (weakly) exchangeable dissociated arrays. To state this more formally, we need more definitions.

---

[a]The terminology is slightly misleading: due to the symmetry requirement, the weak exchangeability is not weaker than the exchangeability.

A set $A \in \mathcal{B}(\mathbb{S})$ is called *exchangeable* if $A = A^\pi$ for every $\pi \in \Sigma^2$, where $A^\pi = \{y^\pi : y \in A\}$ and $y^\pi_{ij} = y_{\pi_1(i)\pi_2(j)}$. The collection $\mathcal{E}_S \subset \mathcal{B}(\mathbb{S})$ of all exchangeable sets is called the *exchangeable σ-field.* For an exchangeable array Y, we define $\mathcal{E}_Y = \{Y^{-1}(A) : A \in \mathcal{E}_S\}$. We use $\mathcal{D}_S \subset \mathcal{P}(\mathbb{S})$ to denote the set of all distributions of dissociated exchangeable arrays, which is a closed subset of $\mathcal{P}(\mathbb{S})$. We write $\tilde{\mathcal{D}}_S$ for the set of all distributions of dissociated weakly exchangeable arrays.

The following proposition follows from Proposition 14.8 and Theorem 12.10 of [2].

Proposition 2.2: (a) *A (weakly) exchangeable array Y is dissociated iff $P(A) \in \{0,1\}$ for every $A \in \mathcal{E}_Y$, that is its exchangeable σ-field is P-trivial.*

(b) *Let Y be a (weakly) exchangeable array and α its regular conditional distribution given $\mathcal{E}_Y$. Then, $\alpha(\omega) \in \mathcal{D}_S$ (resp. $\alpha(\omega) \in \tilde{\mathcal{D}}_S$) for P-a.e. ω. Moreover, the distribution μ_Y of Y can be written as*

$$\mu_Y(\cdot) = \int_{\mathcal{D}_S} \nu(\cdot)\Lambda_Y(\mathrm{d}\nu) \tag{2.6}$$

for a uniquely determined probability measure Λ_Y on $\mathcal{D}_S$ (resp. $\tilde{\mathcal{D}}_S$).

An important feature of exchangeable arrays is that regular conditional distribution α of Y given $\mathcal{E}_Y$ can, a.s., be recovered from a realisation of Y by the following procedure. For $m \geq n$, and $y \in \mathbb{S}$, let $t^{y,n}_m \in \mathcal{P}(\mathbb{S}_n)$ be defined by

$$t^{y,n}_m = \frac{1}{((m)_n)^2} \sum_{\psi_1,\psi_2} \delta_{(y_{\psi_1(i),\psi_2(j)})_{i,j\in[n]}}, \tag{2.7}$$

where the sum runs over all injections $\psi_1, \psi_2 \colon [n] \to [m]$ and $(m)_n = m(m-1)\dots(m-n+1)$. $t^{y,n}_m$ can be viewed as the empirical distribution of $n \times n$ sub-arrays in the array $y|_{[m]}$. We further define

$$t^{y,n} = \lim_{m\to\infty} t^{y,n}_m \tag{2.8}$$

whenever this limit exists in the weak sense, and set $|y| = (t^{y,n})_{n\geq 1}$ whenever all $t^{y,n}$, $n \geq 1$, exist.

It follows from the construction that the probability measures $t^{y,n}_m$, $n = 1,\dots,m$, are consistent in the sense that $t^{y,n}_m|_{[n-1]} = t^{y,n-1}_m$ for every $2 \leq n \leq m$. This consistence transfers to the limit, that is

$$t^{y,n}|_{[n-1]} = t^{y,n-1}, \qquad \text{for every } n \geq 2. \tag{2.9}$$

Therefore, in view of Kolmogorov's extension theorem, $|y|$, when it exists, can be viewed as an element of $\mathcal{P}(\mathbb{S})$.

In the weakly exchangeable case, we set $\tilde{t}^{y,n}_m$ by

$$\tilde{t}^{y,n}_m = \frac{1}{(m)_n} \sum_{\psi} \delta_{(y_{\psi(i),\psi(j)})_{i,j\in[n]}}, \tag{2.10}$$

where the sum runs over all injections π from $[n]$ to $[m]$. We then define $\tilde{t}^{y,n}$ and $|y| = (\tilde{t}^{y,n})_{n\geq 1}$ analogously as in the exchangeable case.

The next proposition establishes the connection between $|Y|$ and its regular conditional distribution α.

Proposition 2.3: (a) *Let Y be a (weakly) exchangeable array and α its regular conditional distribution given $\mathcal{E}_Y$. Then, for P-a.e. ω, $|Y(\omega)|$ exists and equals to $\alpha(\omega)$. In particular, $|Y(\omega)| \in \mathcal{D}_S$ (resp. $|Y(\omega)| \in \tilde{\mathcal{D}}_S$), P-a.s.*

(b) *If Y is dissociated, then $|Y|$ exists a.s. and coincides with the distribution of Y.*

Proof: We only sketch the argument. We assume first that the law of Y is dissociated and denote it by $\alpha \in \mathcal{D}_S$. In this case, we should show that, P-a.s., for all $n \geq 1$, $\lim_{m\to\infty} t^{Y,n}_m = \alpha|_{[n]}$ weakly. Let $f\colon \mathbb{S}_n \to \mathbb{R}$ be a bounded continuous function. Then,

$$t^{Y,n}_m(f) = \frac{1}{((m)_n)^2} \sum_{\psi_1,\psi_2} f\big((Y_{\psi_1(i),\psi_2(j)})_{i,j\in[n]}\big). \tag{2.11}$$

Using the fact that Y is dissociated and thus Y_{ij} is independent of $Y_{i'j'}$ when $i \neq i'$ and $j \neq j'$, it is then straightforward to extend the usual law-of-large-numbers type arguments to show that $\lim_{m\to\infty} t^{Y,n}_m(f) = \alpha|_{[n]}(f)$, P-a.s. To show that this convergence holds P-a.s. jointly for all such f, one can then adapt the proof of Varadarajan's Theorem (see, e.g., Theorem 11.4.1 in [6]), which completes the proof in the dissociated case.

In the general case, denoting by α the regular conditional distribution of Y given $\mathcal{E}_Y$, using that $\alpha \in \mathcal{D}_S$ by Proposition 2.2, and the claim in the dissociated case, we obtain

$$P[|Y| = \alpha] = E\big[P[|Y| = \alpha|\mathcal{E}_Y]\big] = 1, \tag{2.12}$$

which completes the proof. □

Remark 2.4: For the rest of the paper, it will be suitable to extend the definition of $|y|$ to all possible $y \in \mathbb{S}$. For those $y \in \mathbb{S}$ for which some of the limits $t^{y,n}$ do not exist, we define $|y| = \partial$, where $\partial \notin \mathcal{P}(\mathbb{S})$ is an arbitrary symbol. In addition, for y such that $|y|$ exists but is not in $\mathcal{D}_S$, we set $|y| = \partial$ as well. By Proposition 2.3, we can then view $|y|$ as a map from $\mathbb{S}$ to $\mathcal{D}^\star_S := \mathcal{D}_S \cup \{\partial\}$.

As can be seen from the previous results, the differences between exchangeable and weakly exchangeable arrays are mostly a matter of notation. That is why, from now on, we mostly focus on the exchangeable case; the corresponding statements for the weakly exchangeable case can be derived easily.

2.1. *Relation to exchangeable graphs and graph limits*

The above construction is a straightforward generalisation of the graph limit construction from the theory of dense random graphs, which we recall briefly.

A (vertex) *exchangeable random graph* is a random graph G with countably many vertices labelled by $\mathbb{N}$ whose distribution is invariant under permutations of the labels. By considering the adjacency matrix $(G_{ij})_{i,j\in\mathbb{N}}$ of this graph, it can be viewed as a $\{0,1\}$-valued weakly exchangeable array whose diagonal entries are 0.

Graph limits were introduced by Lovász and Szegedy [8] while studying sequences of dense graphs. They encode the limiting density of finite subgraphs in an infinite graph. Formally, let $\mathcal{G}_n$ be the set of all graphs with n vertices labelled by $[n]$. For $m \geq n$ and $F \in \mathcal{G}_n$ and $G \in \mathcal{G}_m$, let $\operatorname{ind}(F,G)$ be the number of injections $\psi\colon [n] \to [m]$ such that $G_{\psi(i)\psi(j)} = F_{ij}$ for all $i,j \in [n]$. Then, for every infinite graph G with vertices labelled by $\mathbb{N}$, one can define the "density of F in G"

$$t(F,G) = \lim_{m\to\infty} \frac{\operatorname{ind}(F, G|_{[m]})}{(m)_n}, \qquad F \in \mathcal{G}_n. \tag{2.13}$$

It can be checked easily that $t(\cdot, G)$, restricted to $\mathcal{G}_n$, if it exists, is a probability measure on $\mathcal{G}_n$. This probability measure, in fact, coincides with the measure $t^{G,n}$ that was introduced in (2.10), when graphs are identified when their adjacency matrices.

By construction, every $t(\cdot, G)$ is invariant under action of Σ,

$$t(F^\pi, G) = t(F,G), \quad \text{for every } F \in \mathcal{G}_n,\ \pi \in \Sigma_n. \tag{2.14}$$

Similarly, the following consistency relation, corresponding to (2.9) above, holds:

$$t(F,G) = \sum_{\bar F \in \mathcal{G}_m : \bar F|_{[n]} = F} t(\bar F, G). \tag{2.15}$$

That means that $(t(F,G))_{F\in\cup_n\mathcal{G}_n}$, if it exists for every $F \in \cup_n \mathcal{G}_n$, can be viewed (again in the sense of Kolmogorov's extension theorem) as a distribution of a random graph, which must be exchangeable due to (2.14). This distribution corresponds to $|y|$ of the previous section.

3. Dynamics of exchangeable arrays

We now turn to the main goal of this paper, the investigation of processes $X = (X(t))_{t\in T}$ taking values in the space $\mathbb{S}$ of two-dimensional S-valued arrays. Here, T denotes the set of times which can be both discrete, $T = \mathbb{N}_0$, or continuous $T = [0, \infty)$.

In the continuous-time case, we assume that the sample paths of X are càdlàg. Since we endowed $\mathbb{S}$ with the product topology, this is the case iff every restriction $X|_{[n]}$ has càdlàg paths in $\mathbb{S}_n$, or equivalently, $t \mapsto X_{ij}(t)$ is càdlàg for every $i, j \in \mathbb{N}$. We write, $D(S)$ for the space of all càdlàg functions from T to S, endowed with the usual Skorokhod topology. The previous reasoning implies that $D(\mathbb{S}) = (D(S))^{\mathbb{N}^2}$.

By convention, every function on T is càdlàg in the discrete-time case. This allows us to use the adjective 'càdlàg' without specifying which case we consider.

A $\mathbb{S}$-valued process X is called *exchangeable*, if

$$X^{\pi} := (X^{\pi}(t))_{t\in T} \overset{\text{law}}{=} X, \qquad \text{for every } \pi \in \Sigma^2. \tag{3.1}$$

Equivalently, viewing X as an array of functions $(t \mapsto X_{ij}(t))_{i,j\in\mathbb{N}}$, it is often useful to regard X as an exchangeable $D(S)$-valued array. Corresponding to this point of view, we define an exchangeable σ-field, $\mathcal{E}_X$, associated to the whole process,

$$\mathcal{E}_X = \{X^{-1}(A) : A \in \mathcal{E}_{D(S)}\}, \tag{3.2}$$

where $\mathcal{E}_{D(S)}$ is defined as $\mathcal{E}_{\mathbb{S}}$ with $D(S)$ playing the role of S.

The process X is a *Markov process* when the Markov property holds, that is the past $(X(s))_{s\le t}$ and the future $(X(s))_{s\ge t}$ are conditionally independent given the present $X(t)$ for all $t \in T$. The following proposition gives criteria implying the exchangeability of a Markov process. Its straightforward proof is left to the reader.

Proposition 3.1: *Let X be an $\mathbb{S}$-valued Markov process with transition probability kernel*

$$p_{s,t}(x, A) := P[X(t) \in A \mid X(s) = x], \qquad s < t \in T, A \in \mathcal{B}(\mathbb{S}). \tag{3.3}$$

Then, X is exchangeable if

(a) *its initial state* $X(0)$ *is an* S*-valued exchangeable array, that is*

$$X(0)^{\pi} \stackrel{\text{law}}{=} X(0); \tag{3.4}$$

(b) *its transition kernels are invariant under action of* Σ^2*, that is for every* $\pi \in \Sigma^2$*,* $s < t \in T$*,* $x \in \mathbb{S}$*, and* $A \in \mathcal{B}(\mathbb{S})$

$$p_{s,t}(x^{\pi}, A^{\pi}) = p_{s,t}(x, A). \tag{3.5}$$

For convenience, we mostly omit "$\mathbb{S}$-valued" from the terminology and say, e.g., "exchangeable Markov process" instead of "$\mathbb{S}$-valued exchangeable Markov process".

We now study how exchangeable Markov processes interact with the "projection" operation $\mathbb{S} \ni y \mapsto |y| \in \mathcal{D}_S^{\star}$, cf. Remark 2.4. Our first result implies that the projection of $X(t)$ is in $\mathcal{D}_S$, a.s., simultaneously for all $t \in T$, that is one can, a.s., project the process X on the space $\mathcal{D}_S$ of (the distributions of) dissociated exchangeable arrays, cf. Proposition 2.3. Remark that Markov property is not assumed.

Theorem 3.2: *Let* X *be an exchangeable process with càdlàg sample paths. Then,* P*-a.s.,* $|X(t)| \in \mathcal{D}_S$ *for all* $t \in T$*.*

Proof: In the discrete-time case, it suffices to observe that $X(t)$ is an exchangeable S-valued array for every $t \in \mathbb{N}_0$. By Proposition 2.3, $|X(t)| \in \mathcal{D}_S$, P-a.s., and the claim follows, since $\mathbb{N}_0$ is countable.

In the continuous-time case, we view X as a $D(S)$-valued exchangeable array, cf. the remark below (3.1), and assume that this array is dissociated first. Using Proposition 2.3 with $D(S)$ in place of S, recalling that $|Y|$ there denotes the sequence of limits $(t^{Y,n})_{n\in\mathbb{N}}$, we see that for every $n \in \mathbb{N}$, the sequence $t_m^{X,n}$ of probability measures on $D(\mathbb{S}_n)$ converges weakly as $m \to \infty$ P-a.s. to some $t^{X,n} \in \mathcal{P}(D(\mathbb{S}_n))$. Moreover, since we assume that X is dissociated, $t^{X,n}$ is a.s. deterministic and coincides with the distribution of $X|_{[n]}$, by Proposition 2.3(b).

Let J_n^X be the (deterministic) set of times defined by

$$J_n^X = \{t \in T : t^{X,n}(\{x \in D(\mathbb{S}_n) : t \text{ is a jump point of } x\}) > 0\}. \tag{3.6}$$

By the general theory of probability measures on Skorokhod spaces, see Chapter 15 in [3], J_n^X is at most countable. Therefore, using the same argument as in the discrete case, P-a.s., $|X(t)| \in \mathcal{D}_S$ for all $t \in \cup_n J_n^X$.

For $t \in T \backslash \cup_n J_n^X$, the coordinate projections $\phi_t \colon D(\mathbb{S}_n) \ni x \mapsto x(t) \in \mathbb{S}_n$ are $t^{X,n}$-a.s. continuous. By Theorem 5.1 of [3], the weak convergence of

$t_m^{X,n}$ then implies the existence of the weak limit $t^{X(t),n} := \phi_t \circ t^{X,n} = \lim_{m\to\infty} \phi_t \circ t_m^{X,n} = \lim_{m\to\infty} t_m^{X(t),n}$. The limit measures $t^{X(t),n} \in \mathcal{P}(\mathbb{S}_n)$ are consistent and dissociated, as $t_m^{X,n}$ are, and thus determine a probability measure $|X(t)| \in \mathcal{D}_S$, P-a.s., simultaneously for all $t \in T \setminus \cup_n J_n^X$.

The last two paragraphs together imply that for a dissociated X, $P[|X(t)| \in \mathcal{D}_S \text{ for all } t \in T] = 1$.

A general exchangeable càdlàg process X can be written as a mixture of dissociated processes by conditioning on $\mathcal{E}_X$, by Proposition 2.2. Therefore,

$$\begin{aligned} &P[|X(t)| \in \mathcal{D}_S \text{ for all } t \in T] \\ &\quad = \int_\Omega P[|X(t)| \in \mathcal{D}_S \text{ for all } t \in T \mid \mathcal{E}_X](\omega) P(\mathrm{d}\omega). \end{aligned} \tag{3.7}$$

Under $P[\cdot \mid \mathcal{E}_X]$, the law of X is dissociated, and thus the integrand equals 1, a.s., by the previous paragraph. This completes the proof. □

From Proposition 2.3, we know that $|X(t)|$ is a regular conditional distribution of $X(t)$ given its own exchangeable σ-field $\mathcal{E}_{X(t)}$. In general, however, $\mathcal{E}_{X(t)}$ does not need to agree with $\mathcal{E}_X$. We now show that $|X(t)|$ is also a regular conditional distribution of $X(t)$ given $\mathcal{E}_X$.

Lemma 3.3: (a) *For every $t \in T$,*

$$\mathcal{E}_{X(t)} \subset \mathcal{E}_X.$$

(b) *Let α^X be the regular conditional distribution of X given $\mathcal{E}_X$. Then, P-a.s.,*

$$\alpha^X(\omega, X(t) \in \cdot) = |X(t)|(\omega, \cdot). \tag{3.8}$$

Or, equivalently, denoting by ϕ_t the projection $D(\mathbb{S}) \ni x \mapsto x(t) \in \mathbb{S}$,

$$\phi_t \circ \alpha^X = |X(t)|. \tag{3.9}$$

Proof: (a) Let $B \in \mathcal{E}_S$. Then $\phi_t^{-1}(B) \in \mathcal{E}_{D(\mathbb{S})}$, and thus $X^{-1}(\phi_t^{-1}(B)) \in \mathcal{E}_X$. In addition,

$$\begin{aligned} X^{-1}(\phi_t^{-1}(B)) &= \{\omega \in \Omega : X(\omega) \in \phi_t^{-1}(B)\} \\ &= \{\omega \in \Omega : (\phi_t \circ X)(\omega) \in B\} \\ &= \{\omega \in \Omega : X(t)(\omega) \in B\} = X(t)^{-1}(B). \end{aligned} \tag{3.10}$$

Since, by definition, $\mathcal{E}_{X(t)} = \{X(t)^{-1}(B) : B \in \mathcal{E}_S\}$, it follows that $\mathcal{E}_{X(t)} \subset \mathcal{E}_X$, as claimed.

(b) Heuristically, the proof uses the fact that $|X(t)|$ is a dissociated distribution, and dissociated distributions are extremal in the set of all exchangeable distributions.

By properties of regular conditional distributions, for every $C \in \mathcal{E}_X$, and every bounded measurable $f\colon \mathbb{S} \to \mathbb{R}$,

$$E[\mathbf{1}_C f(X(t))] = \int_\Omega \mathbf{1}_C(\omega)\alpha^X(\omega, f \circ \phi_t)P(\mathrm{d}\omega). \tag{3.11}$$

By conditioning on $\mathcal{E}_{X(t)}$, we obtain

$$(3.11) = \int_\Omega P(\mathrm{d}\omega') \int_\Omega P(\mathrm{d}\omega \mid \mathcal{E}_{X(t)})(\omega')\mathbf{1}_C(\omega)\alpha^X(\omega, f \circ \phi_t). \tag{3.12}$$

Choosing $C \in \mathcal{E}_{X(t)} \subset \mathcal{E}_X$ and using that $\mathbf{1}_C(\omega) = \mathbf{1}_C(\omega')$, $P(\cdot \mid \mathcal{E}_{X(t)})(\omega')$-a.s., in this case, we get

$$(3.11) = \int_\Omega P(\mathrm{d}\omega')\mathbf{1}_C(\omega') \int_\Omega P(\mathrm{d}\omega \mid \mathcal{E}_{X(t)})(\omega')\alpha^X(\omega, f \circ \phi_t). \tag{3.13}$$

Observe that, as function of ω', the inner integral is $\mathcal{E}_{X(t)}$ measurable. Therefore,

$$\int_\Omega P(\mathrm{d}\omega \mid \mathcal{E}_{X(t)})(\phi_t \circ \alpha^X)(\omega) \tag{3.14}$$

is a version of regular conditional distribution of $X(t)$ given $\mathcal{E}_{X(t)}$, that is it equals $|X(t)|$, P-a.s. However, $|X(t)|$ is dissociated, and thus extremal in the set of all exchangeable probability distributions. Therefore, necessarily, $(\phi_t \circ \alpha^X)(\omega) = |X(t)|(\omega)$ must hold true for $P(\cdot \mid \mathcal{E}_{X(t)})$-a.e. ω. This then implies that $\phi_t \circ \alpha^X = |X(t)|$, P-a.s., as claimed. □

Theorem 3.4: *Let X be an exchangeable process with càdlàg sample paths. Then, the projection $|X| = (|X(t)|)_{t\in T}$ has P-a.s. càdlàg sample paths.*

Proof: Assume first that the distribution of X is dissociated, that is $\mathcal{E}_X$ is P-trivial. Then, $|X(t)|$ exists P-a.s. simultaneously for all $t \in T$, and when it exists, it coincides with the distribution of $X(t)$. Since the trajectories of X are càdlàg, $\lim_{s\downarrow t} X(s) = X(t)$ pointwise and thus in distribution, implying $|X(t)|$ is right continuous.

On the other hand, let $X^-_{ij}(t) = \lim_{s\uparrow t} X_{ij}(t)$. Then, $X^-(t)$ is exchangeable array, and, by the same arguments as in the proof of Theorem 3.2, one can show that $|X^-(t)|$ exists a.s. simultaneously for all $t \in T$. As the distribution of X is dissociated, $|X^-(t)|$ agrees a.s. with the distribution of

$X^-(t)$. Repeating the argument from the first part of the proof, we obtain $\lim_{s\uparrow t}|X(s)| = |X^-(t)|$.

For a general exchangeable process X, we write its distribution as a mixture of dissociated distributions by conditioning on $\mathcal{E}_X$,

$$P(X \in \cdot) = \int_\Omega P(\mathrm{d}\omega)P(X \in \cdot \mid \mathcal{E}_X)(\omega). \tag{3.15}$$

Under $P(\cdot \mid \mathcal{E}_X)$, the distribution of X is dissociated, and $|X(t)|$ agrees with regular conditional distribution of $X(t)$ given $\mathcal{E}_X$, by Lemma 3.3. So, by the previous reasoning, $P(t \mapsto |X(t)|$ is càdlàg $\mid \mathcal{E}_X) = 1$ a.s., and the claim follows. □

Theorem 3.5: *Let X be an exchangeable Markov process with càdlàg sample paths. Then, $|X|$ is a $\mathcal{D}^\star_{\mathbb{S}}$-valued Markov process with a.s. càdlàg sample paths.*

Proof: The càdlàg property follows from Theorem 3.4. We thus need to show that the Markov property is preserved by the map $\mathbb{S} \ni y \mapsto |y| \in \mathcal{D}^\star_S$. To this end, it is sufficient to show that for every $s < t$, and $A \subset \mathcal{D}_S$ measurable

$$p_{s,t}(x, A^{\leftarrow}) = p_{s,t}(x', A^{\leftarrow}), \qquad \text{for every } x, x' \in \mathbb{S} \text{ with } |x| = |x'|, \tag{3.16}$$

where $A^{\leftarrow} = \{x \in \mathbb{S} : |x| \in A\}$.

To prove this, observe first $A^{\leftarrow} = (A^{\leftarrow})^{\pi}$ and thus, by the exchangeability (3.3) of the transition kernel

$$p_{s,t}(x, A^{\leftarrow}) = p_{s,t}(x^{\pi}, (A^{\leftarrow})^{\pi}) = p_{s,t}(x^{\pi}, A^{\leftarrow}), \qquad \text{for all } \pi \in \Sigma^2. \tag{3.17}$$

Hence, $x \mapsto p_{s,t}(x, A^{\leftarrow})$ is $\mathcal{E}_S$ measurable. In addition, by the same arguments as in Corollary 3.10 of [2], $\mathcal{E}_S$ agrees with the σ field generated by the map $x \mapsto |x|$, which implies the claim. □

4. Jumps of discrete-time Markov processes

In this and the next section, we study in detail the structure of the jumps of time-homogeneous exchangeable Markov processes. We first consider processes in discrete time, where the situation is rather simple.

Lemma 4.1: *Let X be an exchangeable Markov process in discrete time. Then, the array $J_{ij}(t) = \mathbf{1}\{X_{ij}(t-1) \neq X_{ij}(t)\}$ encoding its jumps at time $t \geq 1$ is also exchangeable. As consequence, only the following two possibilities occur a.s.*

- *X is constant at t, that is $X_{ij}(t-1) = X_{ij}(t)$ for all $(i,j) \in \mathbb{N}^2$.*
- *There is a positive proportion of entries which jump, that is*

$$\lim_{n\to\infty} \frac{1}{n^2} \sum_{1\le i,j\le n} \mathbf{1}\{X_{ij}(t-1) \neq X_{ij}(t)\} > 0. \tag{4.1}$$

Proof: The exchangeability of $J_{ij}(t)$ follows directly from the exchangeability of X. $J(t)$ is then $\{0,1\}$-valued exchangeable array. The proportion of ones in this array equals $t^{J,1}(\{1\})$ in the notation of Section 2. Therefore, by Proposition 2.3, it exists a.s. If it is zero, then the array J_{ij} must be identically 0, a.s. Otherwise, there is a positive proportion of entries that jump. □

4.1. *Restrictions of Markov exchangeable processes are not Markov*

If one is interested not only in the occurrence of jumps, but also in their "sizes", this argument can be pushed even further, similarly as in [4]. For $t \ge 1$, consider S^2-valued array $Z_{ij} := (X_{ij}(t-1), X_{ij}(t))$, which is again exchangeable. By Proposition 2.3 (with S^2 in place of S), for every $n \in \mathbb{N}$, the limit $t^{Z,n} \in \mathcal{P}(\mathbb{S}_n^2)$ exists a.s.

The measure $t^{Z,n}$ can be used to construct a new Markov transition kernel q_n on $\mathbb{S}^n$, by disintegrating $t^{Z,n}$ with respect to its first marginal $t^{X(t-1),n}$,

$$t^{Z,n}(\mathrm{d}y_1, \mathrm{d}y_2) = t^{X(t-1),n}(\mathrm{d}y_1) q^n_{t-1,t}(y_1, \mathrm{d}y_2), \tag{4.2}$$

or, in the case when S is finite, simply by defining

$$q^n_{t-1,t}(y_1,y_2) = \frac{t^{Z,n}(\{(y_1,y_2)\})}{t^{X(t-1),n}(\{y_1\})}, \qquad y_1, y_2 \in \mathbb{S}_n, \tag{4.3}$$

(and $q^n_{t-1,t}(y_1,y_2) = \delta_{y_1,y_2}$ in the case when $t^{X(t-1),n}(\{y_1\}) = 0$). Since, by Proposition 2.3, $t^{X,n}$ agrees with the distribution of $X|_{[n]}$ given $\mathcal{E}_X$, it is tempting to interpret the kernels q_n as transition kernels of $X|_{[n]}$ (at least conditionally on $\mathcal{E}_X$), as is done in [4]: Proposition 4.8 of [4] contains, among others, the following claim (stated in the notation of the present paper):

> Let $X = (X_t)_{t\in T}$ be a time-homogeneous exchangeable Markov process, with T being finite. Conditioned on $\mathcal{E}_X$, X is dissociated, and, moreover, for every $n \in \mathbb{N}$, the restriction $X|_{[n]}$ of X to $\mathbb{S}_n$

is (conditionally) a time-inhomogeneous Markov chain with transition probabilities $q^n_{t-1,t}$.

We now provide a counterexample for a part of this claim, namely that $X|_{[n]}$ is (conditionally) Markov. We will also see that the transition kernel of $X|_{[n]}$ is not q^n.

Example 4.2: We work in the setting of exchangeable random graphs, similarly as in [4]. That is $X_{ij}(t)$ denotes the adjacency matrix of a random exchangeable graph, which can thus be viewed as $\{0,1\}$-valued weakly exchangeable array with zeros on the diagonal. We fix $T = \{0, 1, \dots, N\}$ for a large N.

To construct the process, let ξ_i, $i \in \mathbb{N}$, be i.i.d. Bernoulli$(\frac{1}{2})$ random variables. In the initial configuration $X(0)$, we draw an edge between vertices $i \neq j$ (i.e., we set $X_{ij}(0) = 1$) with probability $p_{ij}(\xi)$, where

$$p_{ij}(\xi) = \begin{cases} \frac{1}{4}, & \text{if } \xi_i = \xi_j = 0, \\ \frac{1}{2}, & \text{if } \xi_i \neq \xi_j, \\ \frac{3}{4}, & \text{if } \xi_i = \xi_j = 1. \end{cases} \tag{4.4}$$

All edges are drawn independently.

To define the dynamics, for every $x \in \mathbb{S}$, we define

$$\xi_i(x) = \mathbf{1}\Big\{ \limsup_{n\to\infty} \frac{1}{n} \sum_{j=1}^{n} x_{ij} > \tfrac{1}{2} \Big\}. \tag{4.5}$$

Given the configuration of X at time t, we construct $X(t+1)$ as follows:

- If $\xi_i(X(t)) = \xi_j(X(t)) = 0$, then X_{ij} does not change, that is $X_{ij}(t+1) = X_{ij}(t)$.
- Otherwise, X_{ij} is refreshed according to $p_{ij}(X(t))$, that is $X_{ij}(t)$ is a Bernoulli$(p_{ij}(\xi(X(t)))$ random variable, chosen independently of all other $X_{ij}(t)$'s.

It is easy to see that the process X is weakly exchangeable. And, by construction, it is obviously Markov. In addition, the law of large numbers implies that $\xi_i(X(0)) = \xi_i$ a.s., and thus $X(1)$, and inductively also $X(t)$, $t \geq 1$, have the same distribution as $X(0)$.

The exchangeable σ-field $\mathcal{E}_X$ is P-trivial in this example, since X is dissociated by construction. Hence, conditioning on $\mathcal{E}_X$ does not have any effect.

On the other hand, the functions $\xi_i(X(t))$ cannot be determined from any finite restriction $X(t)|_{[n]}$. That is, ξ's are "hidden variables" for the restriction $X|_{[n]}$, and while conditionally on ξ, $X|_{[n]}$ is Markov, it is not Markov unconditionally.

To prove this, fix $n = 2$, that is consider only the state of the edge connecting the vertices 1 and 2. Then, by an easy computation taking into account all possible values of ξ_1 and ξ_2, we obtain that $P(X_{12}(t+1) = 1 \mid X_{12}(t) = 1) = \frac{21}{32}$. On the other hand, $P(X_{12}(N) = 1 \mid X_{12}(t) = 1, \forall t < N)$ can be made arbitrarily close to one by choosing N large, because if we know that $X_{12}(t) = 1$ for all $t < N$, then very likely $\xi_1 = \xi_2 = 0$ and thus X_{12} never flips:

$$\begin{aligned} P(X_{12}(N) = 1 \mid X_{12}(t) = 1, \forall t < N) &= \frac{P(X_{12}(N) = 1, \forall t \le N)}{P(X_{12}(N) = 1, \forall t < N)} \\ &= \frac{\frac{1}{4}\cdot\frac{1}{4}\cdot 1 + \frac{1}{2}\cdot(\frac{1}{2})^N + \frac{1}{4}\cdot(\frac{3}{4})^N}{\frac{1}{4}\cdot\frac{1}{4}\cdot 1 + \frac{1}{2}\cdot(\frac{1}{2})^{N-1} + \frac{1}{4}\cdot(\frac{3}{4})^{N-1}} \xrightarrow{N\to\infty} 1. \end{aligned} \tag{4.6}$$

This implies that X_{12} is not Markov.

Remark 4.3:

(a) On the technical level, the problem of [4] lies in the fact that the relation (14) therein, which gives certain consistency for the kernels q_n, does not hold true, in general. This can hinder the Markov property of the finite restrictions as shown in Example 4.2.

(b) However, in Section 6 (see Theorem 6.3), we show that under the additional assumption that the "global" Markov process X has the Feller property (cf., Definition 6.1), all the "local" restrictions $X|_{[n]}$ are indeed Markov (and Feller). See also Remark 6.4.

5. Jumps of continuous-time Markov processes

We now study exchangeable Markov processes in continuous time. Similarly as in discrete time (see Lemma 4.1), we describe the possible jumps of this process. The structure here is richer, because the process is indexed by an uncountable set of times. So, certain events which have probability 0 in the discrete settings can occur.

Theorem 5.1: *Let X be exchangeable Markov process with càdlàg paths in continuous time, and let $J \subset (0,\infty)$ be the (random) set of times when $t \mapsto X_t$ is discontinuous. Then, a.s., J can be written as a disjoint union $J = J^1 \dot\cup J^2 \dot\cup J^3$, where*

- *J^1 is the set of times, where a positive proportion of entries of X jumps,*

$$J^1 := \Big\{t > 0 : \lim_{n\to\infty} \frac{1}{n^2} \sum_{1\le i,j\le n} \mathbf{1}\{X^-_{ij}(t) \neq X_{ij}(t)\} > 0\Big\}, \quad (5.1)$$

- *J^2 is the set of times, where a positive proportion of entries in one row or column of X jumps, $J^2 = J^{2,c} \dot\cup J^{2,r}$ with*

$$\begin{aligned} J^{2,r} = \Big\{t > 0 : \exists! i \in \mathbb{N} \text{ s.t. } X^-_{i'j}(t) = X_{i'j}(t), \forall i' \neq i, j \in \mathbb{N}, \\ \text{and } \lim_{n\to\infty} \frac{1}{n} \sum_{j=1}^{n} \mathbf{1}\{X^-_{ij}(t) \neq X_{ij}(t)\} > 0\Big\}, \end{aligned} \quad (5.2)$$

$$\begin{aligned} J^{2,c} = \Big\{t > 0 : \exists! j \in \mathbb{N} \text{ s.t. } X^-_{ij'}(t) = X_{ij'}(t), \forall j' \neq j, i \in \mathbb{N}, \\ \text{and } \lim_{n\to\infty} \frac{1}{n} \sum_{i=1}^{n} \mathbf{1}\{X^-_{ij}(t) \neq X_{ij}(t)\} > 0\Big\}, \end{aligned} \quad (5.3)$$

- *J^3 is the set of times, where a unique entry jumps,*

$$\begin{aligned} J^3 = \big\{t > 0 : \exists! i, j \in \mathbb{N} \text{ s.t. } X_{ij} \text{ is discontinuous at } t, \text{ and} \\ X_{i'j'} \text{ is continuous at } t, \forall (i',j') \neq (i,j)\big\}. \end{aligned} \quad (5.4)$$

Proof: We follow the proof of Theorem 3.6 of [4]. By conditioning on $\mathcal{E}_X$, we may assume without loss of generality that X is dissociated.[b] Similarly as in the proof of Theorem 3.2, let J^X be the set of times when X jumps with positive probability,

$$J^X := \{t > 0 : P(X \text{ is discontinuous at } t) > 0\}. \quad (5.5)$$

As we remarked previously, this set is at most countable. Hence, by considering the arrays $(\mathbf{1}\{X^-_{ij}(t) \neq X_{ij}(t)\})_{ij}$, $t \in J^X$, and using the same arguments as in the proof of Lemma 4.1, we obtain that $J^X \subset J^1$.

We now consider times $t \in (0,\infty) \setminus J^X =: C^X$. We first claim that for such t, the proportion of entries that jump must be 0, that is $J^X \supset J^1$. Indeed, since X is dissociated, by Proposition 2.3, for every $t \in (0,\infty)$,

$$P\big(X^-_{12}(t) \neq X_{12}(t)\big) = \lim_{n\to\infty} \frac{1}{n^2} \sum_{1\le i,j\le n} \mathbf{1}\big\{X^-_{ij}(t) \neq X_{ij}(t)\big\}, \quad (5.6)$$

[b]Remark that the conditioning on $\mathcal{E}_X$ effectively removes the Markov property of X, since $\mathcal{E}_X$ contains information about the whole trajectory. In particular, information about certain jumps of X is contained in $\mathcal{E}_X$.

so if the right-hand side is positive, so must be the left-hand side, implying $t \in J^X$.

We further claim that at $t \in C^X$ it is a.s. impossible that two entries that are not in the same row or in the same column jump at the same time. To see this, fix $i \neq k$ and $j \neq l$ and write J_{ij} for the set of times when X_{ij} jumps. Then,

$$P(J_{ij} \cap J_{kl} \cap C^X \neq \emptyset) = E\big[P(J_{ij} \cap J_{kl} \cap C^X \neq \emptyset \mid J_{kl})\big]. \tag{5.7}$$

The set J_{kl} is at most countable and J_{ij} and J_{kl} are independent because X is dissociated. Therefore, the conditional probability in the last formula satisfies

$$\begin{aligned} P(J_{ij} \cap J_{kl} \cap C^X \neq \emptyset \mid J_{kl}) &\leq P(J_{ij} \cap C^X \neq \emptyset \mid J_{kl}) \\ &= P(J_{ij} \cap C^X \neq \emptyset) = 0, \end{aligned} \tag{5.8}$$

where the last equality follows from the definition of C^X. This yields the claim.

It remains to be shown that if X_{ij} jumps at t, then either it is the only entry that jumps, or that there is a positive proportion of entries that jump in i-th row or j-th column. To see this, it is sufficient to observe that $X_{i\cdot} := (X_{ij})_{j \in \mathbb{N}}$ is an exchangeable $D(S)$-valued sequence. In general, $X_{i\cdot}$ is not Markov, but we do not need it to be. By conditioning on its exchangeable field $\mathcal{E}_i$ (which, in general, is not related to $\mathcal{E}_X$), $X_{i\cdot}$ is becomes an i.i.d. sequence, by de Finetti's theorem. We may then repeat the arguments of the previous paragraphs (applied to sequences instead of arrays) to show that $J^{2,r} = \dot{\bigcup}_{i \in \mathbb{N}} J_i^{2,r}$, where $J_i^{2,r}$ is the set of times when the row i jumps with a positive probability,

$$J_i^{2,r} = \{t > 0 : P(X_{i\cdot}^-(t) \neq X_{i\cdot}(t) \mid \mathcal{E}_i) > 0\}, \tag{5.9}$$

and, out of $J_i^{2,r}$ there are no simultaneous jumps of two entries X_{ij} and $X_{ij'}$ with $j \neq j'$, that is

$$P((J_{ij} \cap J_{ij'}) \setminus J_i^{2,r} \neq \emptyset \mid \mathcal{E}_i) = 0. \tag{5.10}$$

This completes the proof. □

Inspection of the previous proof allows us to deduce the following claim about the discontinuities of the projection $|X|$.

Corollary 5.2: *Let $J^{|X|}$ be the set of times when $t \mapsto |X(t)|$ is discontinuous. Then, $J^{|X|} \subset J^1$, where J^1 is as in* Theorem 5.1.

The inclusion in the previous theorem might be strict. As an example, consider the process started from $X_{ij}(0)$ being i.i.d. Bernoulli($\frac{1}{2}$), where all entries are refreshed simultaneously by an independent i.i.d. Bernoulli($\frac{1}{2}$) array at times of jump of a standard Poisson process N_t. In this case, for every $t \geq 0$, $|X(t)|$ is the distribution of the i.i.d. Bernoulli($\frac{1}{2}$) array, that is $J^{|X|} = \emptyset$. On the other hand, J^1 agrees with the set of jumps of N_t.

6. The Feller property

In the last part of this paper, we discuss the conditions under which exchangeable $\mathbb{S}$-valued Markov processes in continuous time have the Feller property.

Recall the following.

Definition 6.1: A time-homogeneous S-valued Markov process with transition kernels $p_t(\cdot, \cdot)$ is called *Feller* if

(a) For every $g \in C_b(S)$, $t \geq 0$ and $y \in S$, the map $x \mapsto \int g(y)p_t(x, \mathrm{d}y)$ is continuous.
(b) For every $x \in S$ and $g \in C_b(S)$,

$$\lim_{t \downarrow 0} \int g(y)p_t(x, \mathrm{d}y) = g(x). \tag{6.1}$$

It is easy to construct exchangeable Markov processes that are not Feller. E.g., the process considered in Example 4.2 does not satisfy (a) of the Feller property. To see this, take $g(y) = y_{12}$, $y \in \mathbb{S}$, and observe that for every $t > 0$ there is $\varepsilon_t > 0$ such that if $x_{12} = 1$, then

$$\int g(y)p_t(x, \mathrm{d}y) \begin{cases} = 1, & \text{if } \xi_1(x) = 0 \text{ and } \xi_2(x) = 0, \\ < 1 - \varepsilon_t, & \text{otherwise.} \end{cases} \tag{6.2}$$

Inspecting, the definition (4.5) of $\xi_i(x)$, it is easy to see that it is not continuous function of x, and thus $X(t)$ is not Feller.

This example indicates one possibility of how the Feller property can be violated by exchangeable Markov processes: If the transition kernel depends on "non-local exchangeable quantities", then the process is not Feller. We now show that this is essentially the only mechanism, how the Feller property can be violated.

The following definition imposes a very strong "locality" of the distribution of X.

Definition 6.2: An exchangeable Markov process X is called *consistent* if its every restriction $X|_{[n]}$ to $\mathbb{S}_n$ is Markov with respect to its own natural filtration.

Theorem 6.3: *For a time-homogeneous exchangeable Markov process X, the following are equivalent:*

(i) *X is consistent and every $X|_{[n]}$ is a Feller process on $\mathbb{S}_n$.*
(ii) *X is Feller.*

Proof: We begin with the following observation. Let $\mathcal{B}_n \subset \mathcal{B}(\mathbb{S})$ be the σ-field generated by the canonical projection from $\mathbb{S}$ to $\mathbb{S}_n$. Then, X is consistent iff its transition kernel satisfies

$$x \mapsto p_t(x, A) \quad \text{is } \mathcal{B}_n\text{-measurable for every } A \in \mathcal{B}_n \text{ and } t \geq 0, \tag{6.3}$$

or, in the case when S is finite,

$$P[X|_{[n]}(t) = y \mid X(0) = x] = P[X|_{[n]}(t) = y \mid X(0) = x'] \tag{6.4}$$

for every $t \geq 0$, $y \in \mathbb{S}_n$ and $x, x' \in \mathbb{S}$ such that $x|_{[n]} = x'_{[n]}$.

(i) $\Longrightarrow$ (ii): Part (a) of the Feller property is equivalent to $x \mapsto p_t(x, \cdot)$ is continuous when $\mathcal{P}(\mathbb{S})$ is endowed with the topology of weak convergence. As remarked in the introduction, this is equivalent to $x \mapsto \int g(y) p_t(x, \mathrm{d}y)$ is continuous for every $t \geq 0$ and every cylinder function g, that is for every $\mathcal{B}_n$-measurable g, $n \geq 1$. Since the restriction $X|_{[n]}$ is Markov by assumption, denoting by $p_t^n(\cdot, \cdot)$ its transition kernel, for $g \in \mathcal{B}_n$,

$$\int_{\mathbb{S}} g(y) p_t(x, \mathrm{d}y) = \int_{\mathbb{S}_n} g(y) p_t^n(x|_{[n]}, \mathrm{d}y). \tag{6.5}$$

Since $X|_{[n]}$ is assumed to be Feller, the right-hand side is a continuous function of $x|_{[n]}$. Since $x^k \to x$ in $\mathbb{S}$ implies $x^k|_{[n]} \to x|_{[n]}$ in $\mathbb{S}_n$, the continuity of the left-hand side follows.

(ii) $\Longrightarrow$ (i): We first show that X is consistent by showing that it satisfies (6.3). This is equivalent to

$$\int g(y) p_t(x, \mathrm{d}y) = \int g(y) p_t(x', \mathrm{d}y) \tag{6.6}$$

for all x' with $x'|_{[n]} = x|_{[n]}$ and for all bounded continuous $\mathcal{B}_n$-measurable functions g. X is assumed to be Feller, so the right-hand side is continuous in x', so it is sufficient to verify (6.6) for a dense set of x' satisfying $x'|_{[n]} = x|_{[n]}$.

To this end, we use the exchangeability (3.5). Let

$$\Sigma_{(n)} = \{\pi \in \Sigma : \pi(i) = i, \forall i \leq n\} \tag{6.7}$$

be the set of permutations of $\mathbb{N}$ that coincide with the identity on $[n]$. Let g be a bounded continuous $\mathcal{B}_n$-measurable function. Then, $g(y) = g(y^{\boldsymbol{\pi}})$ for every $y \in \mathbb{S}$ and $\boldsymbol{\pi} \in \Sigma^2_{(n)}$. Therefore, (3.5) implies that

$$\int g(y)p_t(x, \mathrm{d}y) = \int g(y)p_t(x^{\boldsymbol{\pi}}, \mathrm{d}y), \quad \text{for every } \boldsymbol{\pi} \in \Sigma^2_{(n)}. \tag{6.8}$$

Hence, to prove (6.6) it suffices show that there is $\bar{y} \in \mathbb{S}$ with $\bar{y}|_{[n]} = x|_{[n]}$ such that the set $\{\bar{y}^{\boldsymbol{\pi}} : \boldsymbol{\pi} \in \Sigma^2_{(n)}\}$ is dense in $\{x' : x'|_{[n]} = x|_{[n]}\}$.

We construct such $\bar{y}$ by picking it randomly. To this end, let $U \subset S$ be a countable dense subset of S, and let ρ be a probability measure on U such that $\rho(x) > 0$ for every $x \in U$. Let Y be a $\mathbb{S}$-valued random variable on some auxiliary probability space $(\tilde{\Omega}, \tilde{P})$ such that $Y|_{[n]} = x|_{[n]}$, $\tilde{P}$-a.s., and Y_{ij}, $i > n$ or $j > n$, are i.i.d. ρ-distributed.

Then, for $m \geq n$ and $z \in \mathbb{S}_m$ such that $z|_{[n]} = x_{[n]}$ and $z_{ij} \in U$ for $(i, j) \in [m]^2 \setminus [n]^2$, we have $\tilde{P}(Y|_{[m]} = z) > 0$. So, by the 0-1 law, there is $\tilde{P}$-a.s. $\boldsymbol{\pi} \in \Sigma^2_{(n)}$ such that $Y^{\boldsymbol{\pi}}|_{[m]} = z$. This implies that $\{Y^{\boldsymbol{\pi}} : \pi \in \Sigma^2_{(n)}\}$ is dense in $\{x' : x'|_{[n]} = x|_{[n]}\}$, P-a.s., which is more than sufficient for the existence of $\bar{y}$ of the last paragraph. This completes the proof that X is consistent.

The consistency then implies that $X|_{[n]}$ is Markov with respect to its natural filtration for every $n \geq 1$. The Feller property of $X|_{[n]}$ is then a direct consequence of the Feller property of X. □

Remark 6.4: If S is finite, then $\mathbb{S}_n$ is finite as well. Every càdlàg Markov process on a finite state space is Feller. Therefore, in this case, the consistency of X is equivalent to Feller property. This was proved in the exchangeable random graph case in [5].

References

1. D. J. Aldous. Representations for partially exchangeable arrays of random variables. *J. Multivariate Anal.*, 11(4):581–598, 1981.
2. D. J. Aldous. Exchangeability and related topics. In *École d'été de probabilités de Saint-Flour, XIII—1983*, volume 1117 of *Lecture Notes in Math.*, pages 1–198. Springer, Berlin, 1985.
3. P. Billingsley. *Convergence of probability measures.* Wiley Series in Probability and Statistics: Probability and Statistics. John Wiley & Sons Inc., New York, second edition, 1999.
4. H. Crane. Dynamic random networks and their graph limits. *Ann. Appl. Probab.*, 26(2):691–721, 2016.
5. H. Crane. Exchangeable graph-valued Feller processes. *Probab. Theory Related Fields*, 168(3-4):849–899, 2017.

6. R. M. Dudley. *Real analysis and probability*, volume 74 of *Cambridge Studies in Advanced Mathematics*. Cambridge University Press, Cambridge, 2002. Revised reprint of the 1989 original.
7. D. Hoover. Relations on probability spaces and arrays of random variables. Preprint, Institute for Advanced Study, Princeton, NJ, 1979.
8. L. Lovász and B. Szegedy. Limits of dense graph sequences. *J. Combin. Theory Ser. B*, 96(6):933–957, 2006.

Part II

Contributed Articles

PATRICIA Bridges

Steven N. Evans

Department of Statistics
University of California
Berkeley, CA 94720-3860, USA
evans@stat.berkeley.edu

Anton Wakolbinger

Institut für Mathematik
Goethe-Universität
60054 Frankfurt am Main, Germany
wakolbinger@math.uni-frankfurt.de

A *radix sort tree* arises when storing distinct infinite binary words in the leaves of a binary tree such that for any two words their common prefixes coincide with the common prefixes of the corresponding two leaves. If one deletes the out-degree 1 vertices in the radix sort tree and "closes up the gaps", then the resulting *PATRICIA tree* maintains all the information that is necessary for sorting the infinite words into lexicographic order.

We investigate the PATRICIA chains — the tree-valued Markov chains that arise when successively building the PATRICIA trees for the collection of infinite binary words $Z_1, \ldots, Z_n$, $n = 1, 2, \ldots$, where the source words $Z_1, Z_2, \ldots$ are independent and have a common diffuse distribution on $\{0,1\}^\infty$. It turns out that the PATRICIA chains share a common collection of backward transition probabilities and that these are the same as those of a chain introduced by Rémy for successively generating uniform random binary trees with larger and larger numbers of leaves. This means that the infinite bridges of any PATRICIA chain (that is, the chains obtained by conditioning a PATRICIA chain on its remote future) coincide with the infinite bridges of the Rémy chain. The infinite bridges of the Rémy chain are characterized concretely in Evans, Grübel, and Wakolbinger (2017) and we recall that characterization here while adding some details and clarifications.

1. Introduction

We first fix some notation. Denote by $\{0,1\}^\star := \bigsqcup_{k=0}^{\infty}\{0,1\}^k$ the set of finite tuples or *words* drawn from the alphabet $\{0,1\}$ (with the empty word $\emptyset$ allowed) – the symbol $\bigsqcup$ emphasizes that this is a disjoint union. Write an ℓ-tuple $v = (v_1, \ldots, v_\ell) \in \{0,1\}^\star$ more simply as $v_1 \ldots v_\ell$ and set $|v| = \ell$.

Define a partial order on $\{0,1\}^\star \cup \{0,1\}^\infty$ by declaring that $u < v$ if and only if u is a strict prefix of v; that is, $u < v$ if and only if $u = u_1 \ldots u_k$ for some $k \geq 0$, v is of the form $v_1 \ldots v_\ell \in \{0,1\}^\star$ where $k < \ell$ or $v_1 v_2 \ldots \in \{0,1\}^\infty$, and $v_1 \ldots v_k = u_1 \ldots u_k$. The empty word is the unique minimal element for this partial order.

A *rooted binary tree* (or *binary tree* for short) is a non-empty subset $\mathbf{t}$ of $\{0,1\}^\star$ with the property that if $v \in \mathbf{t}$ and $u \in \{0,1\}^\star$ is such that $u < v$, then $u \in \mathbf{t}$. The vertex $\emptyset$ (that is, the empty word) belongs to any such tree $\mathbf{t}$ and is the *root* of $\mathbf{t}$. For a finite rooted binary tree $\mathbf{t}$ the *leaves* of $\mathbf{t}$ are the maximal elements of $\mathbf{t}$ for the partial order $<$. A finite rooted binary tree is uniquely determined by its leaves: it is the smallest rooted binary tree that contains the set of leaves and it consists of the leaves and the elements $u \in \{0,1\}^\star$ such that $u < v$ for some leaf v.

Suppose now that $z_1, \ldots, z_n \in \{0,1\}^\infty$ are distinct infinite binary words. For each $i \in [n]$ we may construct a finite binary word y_i that is an initial segment of z_i such that $y_1, \ldots, y_n$ are the distinct leaves of a binary tree and $y_1, \ldots, y_n$ are the minimal length words with this property. The resulting binary tree is called the *radix sort tree* defined by the infinite words $z_1, \ldots, z_n$: a depth first search of this tree visits the leaves in an order that coincides with the lexicographic order of the corresponding infinite words. The radix sort tree stores more information than is necessary for the purpose of sorting the infinite binary words into lexicographic order. More precisely, if one deletes the out-degree 1 vertices in the radix sort tree and "closes up the gaps", then a depth first search of the resulting *PATRICIA tree* still visits the leaves in an order that coincides with the lexicographic order of the corresponding infinite words. PATRICIA is an acronym for "Practical Algorithm To Retrieve Information Coded In Alphanumeric". PATRICIA trees were invented independently in [22], [12]. Note that a PATRICIA tree is a *full binary tree*: each non-leaf vertex of the tree has two children; that is, if the finite binary word $v = v_1 \ldots v_m$ is a vertex of the tree that is not a leaf, then both of the words $v_1 \ldots v_m 0$ and $v_1 \ldots v_m 1$ are also vertices of the tree.

Suppose now that $Z_1, Z_2, \ldots$ is an infinite i.i.d. sequence of random elements of $\{0,1\}^\infty$ with common diffuse distribution ν. Let ${}^\nu\bar{R}_n$, $n \in \mathbb{N}$, be the PATRICIA tree constructed from $Z_1, \ldots, Z_n$. We show in Proposition 3.7 that $({}^\nu\bar{R}_n)_{n\in\mathbb{N}}$ is a Markov chain which we call a PATRICIA chain. Features of PATRICIA trees for random inputs were first studied in [15] and this topic has since been the subject of quite a large literature (see, for example, [17], [19], [29], [20], [18], [28], [10], [6], [14], [5], [3], [23], [2], [4], [27], [13]).

Our aim in this paper is to characterize the infinite bridges of the PATRICIA chains; that is, for each ν we wish to characterize the family of Markov chains that have the same backward transition probabilities as $({}^\nu\bar{R}_n)_{n\in\mathbb{N}}$. As we observe in Proposition 3.7, the backward transition probabilities of $({}^\nu\bar{R}_n)_{n\in\mathbb{N}}$ are the same for all ν and this common backward transition mechanism may be described as follows: pick a leaf uniformly at random, delete it and its sibling, and close up the gap in the tree if there is one. It follows that all the PATRICIA chains have the same infinite bridges.

Before we proceed, we need to recall the Rémy chain of [24] (see, also, [1], [21], [16]). The state space of this chain is also the set of full binary trees. Writing $\aleph$ for the full binary tree with two leaves, the Rémy chain evolves forwards in time as follows:

- Pick a vertex v uniformly at random.
- Cut off the subtree rooted at v and set it aside.
- Attach a copy of the tree $\aleph$ to the end of the edge that previously led to v.
- Re-attach the subtree that was rooted at v in uniformly at random to one of the two leaves in the copy of $\aleph$.

The Rémy chain started with the trivial tree consisting of a single vertex has the important feature that after n steps it is uniformly distributed over the set of full binary trees with n leaves. The backward transition probabilities of the Rémy chain were identified in [7] and they coincide with those of the PATRICIA chains. It follows that the common family of infinite bridges for the PATRICIA chains is the same as that of the Rémy chain and the latter was determined in [7].

An outline of the remainder of the paper is as follows.

In Section 2 we introduce some notation and recall more formally the construction of a radix sort tree and a PATRICIA tree from a set of infinite binary words. In Section 3 we begin considering the stochastic processes built by applying the radix sort tree and PATRICIA tree constructions to sequences of independent identically distributed infinite binary words with common diffuse distribution ν. We observe that these processes are Markov, show that all the radix sort chains have the same family of backward transition probabilities, and demonstrate that the same is true of the PATRICIA chains. We also investigate the issue that different diffuse probabilities ν can give rise to the same PATRICIA chain.

Section 4 initiates the study of infinite bridges *per se*. Since both in the case of the radix sort chains and of the PATRICIA chains the backward transition probabilities are the same for all ν, we concentrate on the case where ν is fair coin-tossing measure. We also recall that the distribution of any infinite bridge for either the radix sort chain or the PATRICIA chain will be a mixture of extremal infinite bridge distributions and that an infinite bridge is extremal if and only if it has an almost surely trivial tail σ-field, so it suffices for the determination of the distributions of general infinite bridges to characterize the extremal ones. We recall from [8] that the extremal infinite bridges for the radix sort chain are nothing other than the radix sort chains with general diffuse input distribution ν. By the observation that all the PATRICIA chains have the same backward transition probabilities, it follows that a PATRICIA chain with general diffuse input distribution ν is an infinite bridge for our "reference" PATRICIA chain with fair coin-tossing input distribution. Because of the characterization of the extremal infinite bridges for the radix sort chain that we have just recalled and the fact that if Φ is the "delete the out-degree 1 vertices" map that turns a binary tree into a full binary tree and $(R_n^\infty)_{n\in\mathbb{N}}$ is an infinite bridge for the radix sort chain, then $(\Phi(R_n^\infty))_{n\in\mathbb{N}}$ is an infinite bridge for the PATRICIA chain (see Proposition 4.4), one might expect that such infinite PATRICIA bridges exhaust the collection of extremal infinite PATRICIA bridges. We present an example to show that this is not the case. Rather, to describe the totality of the extremal infinite PATRICIA bridges we need to recall from [7] the description of the extremal infinite bridges for the Rémy chain because, as we have already noted, these two families coincide.

This retelling takes up the remainder of the paper with the exception of the brief interpolation Section 5 where we present some examples of infinite PATRICIA bridges and give an indication of how differently such chains can behave by examining the asymptotic growth of the successive trees they produce.

A crucial element of the development in [7] was the introduction of the notion of a *didendritic system* with a given label set. This class of objects is a generalization of the class of finite leaf-labeled full binary trees that includes objects with infinitely many "leaves". Section 6 contains a review of some facts about didendritic systems. In particular, we show that the class of didendritic systems with finite label sets is in a natural bijective correspondence with the class of finite leaf-labeled full binary trees. This fact was claimed in [7] with a slightly different definition of didendritic system but we show by an example that under the axioms in [7] this assertion is false. We stress that this gap does not illegitimate the further development in [7] which does not depend on the details of the definition of a didendritic system but only on the two facts that a finite didendritic system is effectively a finite leaf-labeled binary tree and that the class of didentritic systems is closed under projective limits in a natural way – both of which are true for our new definition. We also develop in an explicit manner an alternative description of the class of finite didendritic systems that was somewhat implicit in [7]. This alternative description, which is obtained in Proposition 6.7, is crucial for the later exposition.

We recall in Section 7 that there is a bijective correspondence between infinite PATRICIA (equivalently, Rémy) bridges and random didentric systems with label set $\mathbb{N}$ that are exchangeable in a natural sense. Moreover, the extremal infinite PATRICIA bridges correspond to the exchangeable random didendritic systems with label set $\mathbb{N}$ that are ergodic, where ergodicity is equivalent to the property that the random didendritic systems induced on disjoint subsets of $\mathbb{N}$ are independent. In order to characterize the class consisting of all PATRICIA bridges it therefore suffices to determine the class of ergodic exchangeable random didendritic systems. We recall the concrete representation of the latter from [7] in Section 8. The ingredients of this representation are a rooted complete separable $\mathbb{R}$-tree, a diffuse probability measure on the $\mathbb{R}$-tree, and a possibly random mechanism for giving a left-versus-right ordering to the tree built by sampling countably many points from the $\mathbb{R}$-tree according to the given probability measure.

Finally, we note that an alternative concrete representation of extremal infinite Rémy bridges has recently been given in [11] as part of a program that extends the study of Markov chains with Rémy-like transition probabilities to classes of discrete structures other than binary trees; for example, [11] also considers the infinite bridges investigated in [9] that are similar to Rémy or PATRICIA infinite bridges but vertices of the successive trees may have more than two offspring and there is no left-to-right ordering of offspring.

2. Radix sort trees and PATRICIA trees

For a finite rooted binary tree $\mathbf{t}$, we write $\mathbf{L}(\mathbf{t})$ for the set of leaves of $\mathbf{t}$. For $y_1, \ldots, y_m \in \{0,1\}^\star$, write

$$\mathbf{T}(y_1, \ldots, y_m) := \bigcup_{j=1}^{m} \{u \in \{0,1\}^\star : u \leq y_j\}$$

for the smallest finite rooted binary tree containing $y_1, \ldots, y_m \in \{0,1\}^\star$; the leaves of this tree form a subset of $\{y_1, \ldots, y_m\}$ and this subset is proper if and only if $y_i < y_j$ for some pair $1 \leq i \neq j \leq m$.

A collection $z_1, \ldots, z_n$ of distinct elements of $\{0,1\}^\infty$ determines a finite rooted binary tree in the following manner. For $1 \leq i \leq n$ let y_i be the minimal length prefix of z_i that differs from all prefixes of z_j, $j \neq i$. The words $y_1, \ldots, y_n$ are distinct and, moreover, they are incomparable for the partial order $<$. The *radix sort tree determined by the input* $z_1, \ldots, z_n$ is the finite rooted binary tree $\mathbf{R}(z_1, \ldots, z_n)$ with leaves $y_1, \ldots, y_n$; that is,

$$\mathbf{R}(z_1, \ldots, z_n) := \mathbf{T}(y_1, \ldots, y_n).$$

Remark 2.1: Observe that

$$\mathbf{R}(z_1, \ldots, z_n) = \mathbf{R}(z_{\sigma(1)}, \ldots, z_{\sigma(n)})$$

for any permutation σ of $[n]$.

Notation 2.2: Denote by $\mathbb{S}$ the class of binary trees that can arise as radix sort trees. A binary tree $\mathbf{s}$ belongs to $\mathbb{S}$ if and only if $\mathbf{s} = \{\emptyset\}$ or $\mathbf{s}$ has at least two leaves and for any leaf $u_1 \ldots u_p \in \mathbf{s}$ the word $u_1 \ldots u_{p-1}\bar{u}_p$ is also a vertex of $\mathbf{s}$, where we set $\bar{0} := 1$ and $\bar{1} := 0$. Write $\mathbb{S}_n$, $n \in \mathbb{N}$, for the elements of $\mathbb{S}$ with n leaves (in particular, $\mathbb{S}_1$ contains only the trivial tree with the single vertex $\{\emptyset\}$).

One of the reasons for constructing the radix sort tree for a set of inputs $z_1, \ldots, z_n$ is that if $y_1, \ldots, y_n$ are the leaves of the tree (where the indexing is such that $y_i < z_i$ for $1 \leq i \leq n$), then the lexicographic ordering of $z_1, \ldots, z_n$ is the same as that of $y_1, \ldots, y_n$; that is, σ is the unique permutation of $[n]$ such that $z_{\sigma(i)}$, $i \in [n]$, are increasing in the lexicographic order if and only if σ is the unique permutation of $[n]$ such that $y_{\sigma(i)}$, $i \in [n]$, are increasing in the lexicographic order. We now describe another procedure for associating a finite set of inputs with the leaves of a finite binary tree that shares this feature but typically uses a finite binary tree with fewer vertices.

Notation 2.3: Denote by $\bar{\mathbb{S}}$ the set of finite binary tree for which all vertices have out-degree 2 or 0. That is, $\bar{\mathbb{S}}$ consists of finite binary trees $\mathbf{t}$ such that if $v \in \mathbf{t}$, then either $v \in \mathbf{L}(\mathbf{t})$ or both $v0 \in \mathbf{t}$ and $v1 \in \mathbf{t}$. We call the elements of $\bar{\mathbb{S}}$ *full finite binary trees.* Write $\bar{\mathbb{S}}_n$, $n \in \mathbb{N}$, for the elements of $\bar{\mathbb{S}}$ with n leaves (in particular, $\bar{\mathbb{S}}_1$ contains only the trivial tree with the single vertex $\{\emptyset\}$). Note that $\#\bar{\mathbb{S}}_n = C_{n-1}$, where $C_m := \frac{1}{m+1}\binom{2m}{m}$ is the mth *Catalan number* (see, for example, Bijective Exercise 7 of [26]).

Definition 2.4: The *PATRICIA contraction* is a map $\Phi : \mathbb{S} \to \bar{\mathbb{S}}$. It maps $\mathbb{S}_n$ onto $\bar{\mathbb{S}}_n$, $n \in \mathbb{N}$. It is defined recursively as follows. For $n = 1$, put $\Phi(\{\emptyset\}) := \emptyset$. Now consider $\mathbf{s} \in \mathbb{S}_n$ for $n \geq 2$. There is a unique maximal m and $u_1 \dots u_m \in \{0,1\}^m$ such that $u_1 \dots u_m < y$ for every leaf $y \in \mathbf{L}(\mathbf{s})$. We have a decomposition

$$\mathbf{s} = \{\{\emptyset\}, u_1, u_1u_2, \dots, u_1u_2\dots u_m\} \sqcup u_1 \dots u_m 0\mathbf{s}^{(0)} \sqcup u_1 \dots u_m 1\mathbf{s}^{(1)}$$

for two binary trees $\mathbf{s}^{(0)}, \mathbf{s}^{(1)} \in \mathbb{S}$ that both have fewer leaves than $\mathbf{s}$. Put

$$\Phi(\mathbf{s}) := \{\emptyset\} \sqcup 0\Phi(\mathbf{s}^{(0)}) \sqcup 1\Phi(\mathbf{s}^{(1)}).$$

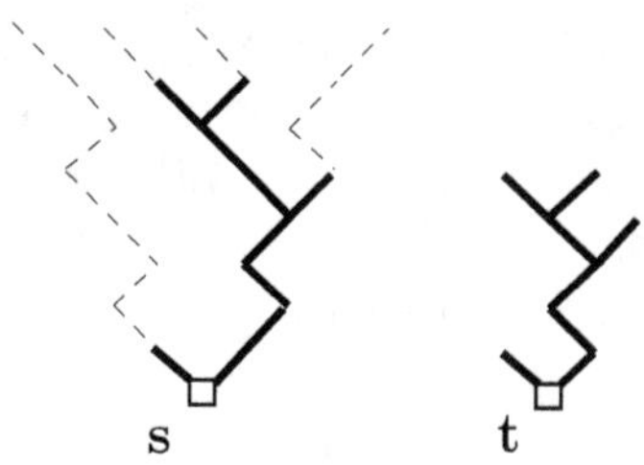

Fig. 1. Initial segments of 4 elements $z_1, z_2, z_3, z_4 \in \{0,1\}^\infty$, the radix sort tree $\mathbf{s} = \mathbf{R}(z_1, z_2, z_3, z_4)$ determined by z_1, z_2, z_3, z_4 and its PATRICIA contraction $\mathbf{t} = \Phi(\mathbf{s})$

Remark 2.5: The PATRICIA contraction of $\mathbf{s} \in \mathbb{S}$ is the unique $\mathbf{t} \in \bar{\mathbb{S}}$ with the following properties.

- Writing $\mathbf{S}(\mathbf{s})$ (respectively, $\mathbf{S}(\mathbf{t})$) for the vertices of $\mathbf{s}$ (respectively, $\mathbf{t}$) with out-degree 2, there is a bijective correspondence between $\mathbf{S}(\mathbf{s})$ and $\mathbf{S}(\mathbf{t})$ and a bijective correspondence between $\mathbf{L}(\mathbf{s})$ and $\mathbf{L}(\mathbf{t})$.
- Suppose that $w, x \in \mathbf{s}$ and $y, z \in \mathbf{t}$ are such that w corresponds to y and x corresponds to z. Then $w = u_1 \dots u_p$ and $x = u_1 \dots u_p u_{p+1} \dots u_q$ for some $u_1, \dots, u_q \in \{0,1\}$ with

$u_{p+1} = 0$ (resp. $u_{p+1} = 1$) if and only if $y = v_1 \dots v_m$ and $z = v_1 \dots v_m v_{m+1} \dots v_n$ for some $v_1, \dots, v_n \in \{0,1\}$ with $v_{m+1} = 0$ (resp. $v_{m+1} = 1$).

3. Radix sort chains and PATRICIA chains

Definition 3.1: Given a diffuse probability measure ν on $\{0,1\}^\infty$, let $(Z_n)_{n\in\mathbb{N}}$ be a sequence of independent identically distributed $\{0,1\}^\infty$-valued random variables with common distribution ν. The corresponding *radix sort chain* $({}^\nu R_n)_{n\in\mathbb{N}}$ is defined by ${}^\nu R_n := \mathbf{R}(Z_1, \dots, Z_n)$, $n \in \mathbb{N}$, and the corresponding *PATRICIA chain* $({}^\nu \bar{R}_n)_{n\in\mathbb{N}}$ is defined by ${}^\nu \bar{R}_n = \Phi({}^\nu R_n)$, $n \in \mathbb{N}$. Setting $\gamma := \pi^\infty$, where $\pi(\{0\}) = \pi(\{1\}) = \frac{1}{2}$, we will write $R_n := {}^\gamma R_n$ and $\bar{R}_n := {}^\gamma \bar{R}_n$.

The processes $({}^\nu R_n)_{n\in\mathbb{N}}$ and $({}^\nu \bar{R}_n)_{n\in\mathbb{N}}$ are indeed Markovian. This observation does not seem to have appeared explicitly in the literature except in [8] for the former process. We will state this formally along with a description of the respective backward transition probabilities, but we first need some notation.

Notation 3.2: Consider $\mathbf{t} \in \mathbb{S}_{n+1}$ and let $v = v_1 \dots v_m$ be a leaf of $\mathbf{t}$.

Suppose first that $v_1 \dots v_{m-1}\bar{v}_m$ is not a leaf of $\mathbf{t}$. Let $\kappa(\mathbf{t}, v) \in \mathbb{S}_n$ be the tree $\mathbf{t} \setminus \{v\}$. That is, $\kappa(\mathbf{t}, v)$ is the tree with the same leaf set as $\mathbf{t}$ except that v has been removed.

On the other hand, suppose that $v_1 \dots v_{m-1}\bar{v}_m$ is also a leaf of $\mathbf{t}$. There is a largest $\ell < m$ such that $v_1 \dots v_\ell$ and $v_1 \dots v_{\ell-1}\bar{v}_\ell$ are both vertices of $\mathbf{t}$. In this case let $\kappa(\mathbf{t}, v) \in \mathbb{S}_n$ be the tree $\mathbf{t} \setminus (\{v_1 \dots v_p : \ell + 1 \le p \le m\} \cup \{v_1 \dots v_{m-1}\bar{v}_m\})$. That is, $\kappa(\mathbf{t}, v)$ is the tree with the same leaf set as $\mathbf{t}$ except that the leaf v and its sibling leaf $v_1 \dots v_{m-1}\bar{v}_m$ have both been removed and replaced by the single leaf $v_1 \dots v_\ell$.

Remark 3.3: If $\mathbf{t} = \mathbf{R}(z_1, \dots, z_{n+1})$ for distinct $z_1, \dots, z_{n+1}$ and y_{n+1} is the leaf of $\mathbf{t}$ corresponding to the input z_{n+1}, then $\kappa(\mathbf{t}, y_{n+1}) = \mathbf{R}(z_1, \dots, z_n)$.

Notation 3.4: Consider $\bar{\mathbf{t}} \in \bar{\mathbb{S}}_{n+1}$ and let $v = v_1 \dots v_m$ be a leaf of $\bar{\mathbf{t}}$. Let $\bar{\kappa}(\bar{\mathbf{t}}, v) \in \bar{\mathbb{S}}_n$ be the tree $(\bar{\mathbf{t}} \setminus \{w \in \bar{\mathbf{t}} : v_1 \dots v_{m-1} < w\}) \cup \{w_1 \dots w_{m-1} w_{m+1} \dots w_p : w \in \bar{\mathbf{t}},\ v_1 \dots v_{m-1}\bar{v}_m \le w_1 \dots w_p\}$.

Remark 3.5: If $\bar{\mathbf{t}} = \Phi \circ \mathbf{R}(z_1, \dots, z_{n+1})$ for distinct $z_1, \dots, z_{n+1}$ and y_{n+1} is the leaf of $\bar{\mathbf{t}}$ corresponding to the input z_{n+1}, then $\bar{\kappa}(\bar{\mathbf{t}}, y_{n+1}) = \Phi \circ \mathbf{R}(z_1, \dots, z_n)$.

Remark 3.6: Note that $\bar{\kappa}(\bar{\mathbf{t}}, v)$ is the tree obtained from $\bar{\mathbf{t}}$ by deleting v and its sibling and closing up the resulting gap if there is one (there will be a gap if and only if the sibling of v is not a leaf). This operation is the same as one that appears in the backward transition of the Rémy chain, and indeed, as part (ii) of the next proposition shows, the common backward transition probabilities of the PATRICIA chains are the same as that of the Rémy chain described at the beginning of Section 4 of [7].

Proposition 3.7:

(i) The process $({}^\nu R_n)_{n\in\mathbb{N}}$ is Markov. For $\mathbf{s} \in \mathbb{S}_n$ such that $\mathbb{P}\{{}^\nu R_n = \mathbf{s}\} > 0$ and $\mathbf{t} \in \mathbb{S}_{n+1}$ such that $\mathbb{P}\{{}^\nu R_{n+1} = \mathbf{t}\} > 0$ the associated backward transition probability is

$$\mathbb{P}\{{}^\nu R_n = \mathbf{s} \,|\, {}^\nu R_{n+1} = \mathbf{t}\} = \begin{cases} \frac{1}{n+1}, & \text{if } \mathbf{s} = \kappa(\mathbf{t}, v) \text{ for some } v \in \mathbf{L}(\mathbf{t}), \\ 0, & \text{otherwise.} \end{cases}$$

(ii) The process $({}^\nu \bar{R}_n)_{n\in\mathbb{N}}$ is Markov. For $\bar{\mathbf{s}} \in \bar{\mathbb{S}}_n$ such that $\mathbb{P}\{{}^\nu \bar{R}_n = \bar{\mathbf{s}}\} > 0$ and $\bar{\mathbf{t}} \in \bar{\mathbb{S}}_{n+1}$ such that $\mathbb{P}\{{}^\nu \bar{R}_{n+1} = \bar{\mathbf{t}}\} > 0$ the associated backward transition probability is

$$\mathbb{P}\{{}^\nu \bar{R}_n = \bar{\mathbf{s}} \,|\, {}^\nu \bar{R}_{n+1} = \bar{\mathbf{t}}\} = \begin{cases} \frac{1}{n+1}, & \text{if } \bar{\mathbf{s}} = \bar{\kappa}(\bar{\mathbf{t}}, v) \text{ for some } v \in \mathbf{L}(\bar{\mathbf{t}}), \\ 0, & \text{otherwise.} \end{cases}$$

Proof: Suppose that $\mathbf{s} \in \mathbb{S}_n$ and $\mathbf{t} \in \mathbb{S}_{n+1}$ is such that $\mathbb{P}\{{}^\nu R_{n+1} = \mathbf{t}\} > 0$. It is clear from Remark 2.1 and the assumption that $(Z_n)_{n\in\mathbb{N}}$ are independent that $\mathbb{P}\{{}^\nu R_n = \mathbf{s} \,|\, {}^\nu R_{n+1} = \mathbf{t}, {}^\nu R_{n+2}, \ldots\}$ is $\frac{1}{n+1}$ if $\mathbf{s} = \kappa(\mathbf{t}, v)$ for some $v \in \mathbf{L}(\mathbf{t})$ and 0 otherwise. In particular, the time-reversal of $({}^\nu R_n)_{n\in\mathbb{N}}$ is Markovian and hence the same is true of $({}^\nu R_n)_{n\in\mathbb{N}}$ itself. This establishes (i).

The proof of (ii) is similar. □

Notation 3.8: For $y \in \{0,1\}^\star$, let $\tau(y) := \{z \in \{0,1\}^\infty : y < z\}$; that is, $\tau(y)$ is the set of infinite extensions of the finite word y.

Remark 3.9: An alternative route to establishing the Markov property of $({}^\nu R_n)_{n\in\mathbb{N}}$ is to note that the random set $\{Z_1, \ldots, Z_n\}$ is conditionally independent of ${}^\nu R_1, \ldots, {}^\nu R_{n-1}$ given ${}^\nu R_n$. The conditional distribution of the random set $\{Z_1, \ldots, Z_n\}$ given $\{{}^\nu R_n = \mathbf{s}\}$, where $\mathbf{L}(\mathbf{s}) = \{y_1, \ldots, y_n\}$, is that of the random set $\{\hat{Z}_1, \ldots, \hat{Z}_n\}$, where $\hat{Z}_1, \ldots, \hat{Z}_n$ are independent and $\hat{Z}_k$ is distributed according to $\nu(\cdot \cap \tau(y_k))/\nu(\tau(y_k))$ (that is, according to ν conditioned on $\tau(y_k)$) for $1 \le k \le n$. Thus,

$$\mathbb{P}\{{}^\nu R_{n+1} = \mathbf{t} \mid {}^\nu R_1, \dots, {}^\nu R_{n-1}, {}^\nu R_n = \mathbf{s}\} = \mathbb{P}\{\mathbf{R}(\hat{Z}_1, \dots, \hat{Z}_n, Z_{n+1}) = \mathbf{t}\},$$

where $\hat{Z}_1, \dots, \hat{Z}_n$ are constructed to be independent of Z_{n+1}. This observation leads readily to an explicit calculation of the (forward) transition probabilities of $({}^\nu R_n)_{n\in\mathbb{N}}$ – see Section 2 of [8]. The backward transition probabilities may then be derived as in Section 3 of [8], giving a complete proof of part (i) of Proposition 3.7.

Remark 3.10: Once part (i) of Proposition 3.7 is known, an alternative derivation of part (ii) is to use the Markov property of the time-reversal of $({}^\nu R_n)_{n\in\mathbb{N}}$ and apply Dynkin's classical criterion for the composition of a function with a Markov process to be Markovian (see, for instance, p. 575 of [25]). Specifically, one checks that if $\bar{\mathbf{t}} \in \bar{\mathbb{S}}_{n+1}$ is such that $\mathbb{P}\{{}^\nu \bar{R}_{n+1} = \bar{\mathbf{t}}\} > 0$, $\mathbf{t} \in \mathbb{S}_{n+1}$ is such that $\mathbb{P}\{{}^\nu R_{n+1} = \mathbf{t}\} > 0$ and $\bar{\mathbf{t}} = \Phi(\mathbf{t})$, and $\bar{\mathbf{s}} \in \bar{\mathbb{S}}_n$, then $\mathbb{P}\{\Phi({}^\nu R_n) = \bar{\mathbf{s}} \mid {}^\nu R_{n+1} = \mathbf{t}\}$ is $\frac{1}{n+1}$ for $\bar{\mathbf{s}}$ of the form $\bar{\kappa}(\bar{\mathbf{t}}, v)$ when $v \in \mathbf{L}(\bar{\mathbf{t}})$ and 0 otherwise. The fact that this conditional probability is the same for all $\mathbf{t} \in \mathbb{S}_{n+1}$ such that $\mathbb{P}\{{}^\nu R_{n+1} = \mathbf{t}\} > 0$ and $\bar{\mathbf{t}} = \Phi(\mathbf{t})$ gives that the time-reversal of $({}^\nu \bar{R}_n)_{n\in\mathbb{N}}$ is Markov and hence the same is true of the process $({}^\nu \bar{R}_n)_{n\in\mathbb{N}}$ itself. Moreover, this same observation leads readily to the claimed backward transition probabilities.

Remark 3.11: Interestingly, Dynkin's criterion does not hold for the forwards in time processes $({}^\nu R_n)_{n\in\mathbb{N}}$ and the function Φ, as the following example with $\nu = \gamma$ and $R_n := {}^\gamma R_n$, $n \in \mathbb{N}$, shows. Define trees $\mathbf{s}', \mathbf{s}'', \bar{\mathbf{t}}$ by $\mathbf{L}(\mathbf{s}') = \{00, 01, 1\}$ $\mathbf{L}(\mathbf{s}'') = \{000, 001, 1\}$, and $\mathbf{L}(\bar{\mathbf{t}}) = \{000, 001, 01, 1\}$ (see Figure 2). Note that $\Phi(\mathbf{s}') = \Phi(\mathbf{s}'') = \mathbf{s}'$. If $R_3 = \mathbf{s}'$, then the only way that $\Phi(R_4)$ can be $\bar{\mathbf{t}}$ is if Z_4 is of the form $00\ldots$; thus, $\mathbb{P}\{\Phi(R_4) = \bar{\mathbf{t}} \mid R_3 = \mathbf{s}'\} = \frac{1}{4}$. Similarly, if $R_3 = \mathbf{s}''$, then the only way that $\Phi(R_4)$ can be $\bar{\mathbf{t}}$ is if Z_4 is of the form $000\ldots$ or $01\ldots$; thus $\mathbb{P}\{\Phi(R_4) = \bar{\mathbf{t}} \mid R_3 = \mathbf{s}''\} = \frac{1}{4} + \frac{1}{8}$. Since these two conditional probabilities are different, Dynkin's criterion does not hold.

As observed in [8], different choices of ν result in different distributions of the radix sort chains $({}^\nu R_n)$. By way of contrast, two different diffuse probability measures ν' and ν'' on $\{0,1\}^\infty$ may result in one and the same distribution of the PATRICIA chains $({}^{\nu'} \bar{R}_n)$ and $({}^{\nu''} \bar{R}_n)$. As a trivial example, consider $\nu' = \gamma$ and $\nu'' = \delta_0 \otimes \gamma$ – we may couple the two radix

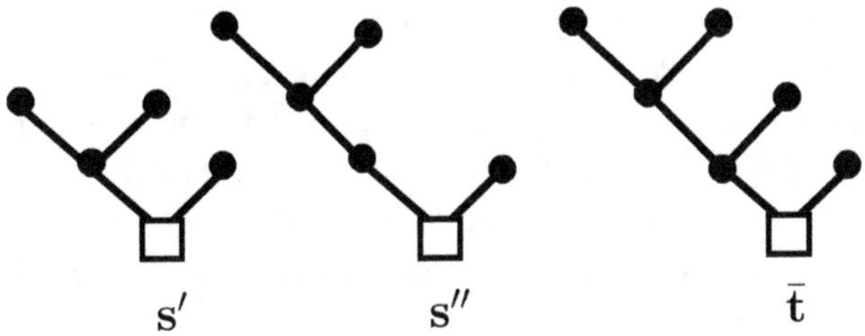

Fig. 2. Illustration of Remark 3.11: The trees $\mathbf{s}'$ and $\mathbf{s}''$ have the same PATRICIA contraction. However, there are more possibilities (and a higher probability) for a radix sort transition from $\mathbf{s}'$ than from $\mathbf{s}''$ to lead to a tree whose PATRICIA contraction is $\bar{\mathbf{t}}$.

sort chains together so that ${}^{\nu''}R_n = 0\,{}^{\nu'}R_n$ for all $n \in \mathbb{N}$, in which case $\Phi({}^{\nu'}\bar{R}_n) = \Phi({}^{\nu''}\bar{R}_n)$ for all $n \in \mathbb{N}$.

Definition 3.12: Declare that two probability measures ν' and ν'' are *PATRICIA-equivalent* when $({}^{\nu'}\bar{R}_n)_{n\in\mathbb{N}}$ and $({}^{\nu''}\bar{R}_n)_{n\in\mathbb{N}}$ have the same distribution.

For later use in Example 4.6 below we show the following.

Lemma 3.13: *For any diffuse probability measure ν on $\{0,1\}^\infty$ there exists a probability measure $\bar{\nu}$ on $\{0,1\}^\infty$ which is PATRICIA-equivalent to ν and assigns strictly positive probability to $\tau(y)$ for every $y \in \{0,1\}^\star$.*

Proof: We can write ν as $\delta_{u_1} \otimes \cdots \otimes \delta_{u_k} \otimes \nu^\emptyset$ for some finite, possibly empty, maximal sequence $u_1 \dots u_k$ and some probability measure $\nu^\emptyset$ on $\{0,1\}^\infty$. Set $\nu_1 := \nu^\emptyset$. Note by the assumed maximality of $u_1 \dots u_k$ that $\nu_1(\tau(0)) > 0$ and $\nu_1(\tau(1)) > 0$.

Suppose for some $n \in \mathbb{N}$ that we have built diffuse probability measures $\nu_1, \dots, \nu_n$ on $\{0,1\}^\infty$ such that:

- $\nu_1, \dots, \nu_n$ are PATRICIA-equivalent to ν,
- $\nu_n(\tau(y)) > 0$ for every $y \in \{0,1\}^n$,
- for $1 \le m < n$ the probability measures ν_m and ν_n agree on sets of the form $\tau(y)$, $y \in \{0,1\}^m$.

Conditioning the probability measure ν_n on the set $\tau(y)$, $y = y_1 \dots y_n \in \{0,1\}^n$, gives a probability measure of the form $\delta_{y_1} \otimes \cdots \otimes \delta_{y_n} \otimes \delta_{u_1} \otimes \cdots \otimes \delta_{u_k} \otimes \nu^y$ for some finite, possibly empty, maximal sequence $u_1 \dots u_k$ (we reuse notation from above and our notation doesn't record the dependence of this sequence on y) and some probability measure ν^y on $\{0,1\}^\infty$. Note that

$\nu^y(\tau(y0)) > 0$ and $\nu^y(\tau(y1)) > 0$. Put $\nu_{n+1} := \sum_{y \in \{0,1\}^n} \nu_n(\tau(y)) \delta_{y_1} \otimes \cdots \otimes \delta_{y_n} \otimes \nu^y$. It is clear that the probability measures ν_{n+1} and ν are PATRICIA-equivalent and that for $1 \le m < n+1$ the probability measures ν_m and ν_{n+1} agree on sets of the form $\tau(y)$, $y \in \{0,1\}^m$.

There is thus a unique diffuse probability measure $\bar{\nu}$ such that the probability measures ν_m and $\bar{\nu}$ agree on sets of the form $\tau(y)$, $y \in \{0,1\}^m$, for all $m \in \mathbb{N}$.

It is not difficult to see that ν and $\bar{\nu}$ are PATRICIA-equivalent: we can couple together two i.i.d. sequences of inputs distributed according to ν and $\bar{\nu}$ so that for each $n \in \mathbb{N}$ the tree ${}^{\bar{\nu}}R_n$ is obtained from the tree ${}^{\nu}R_n$ by a deterministic operation that removes certain vertices with out-degree 1, and hence $\Phi({}^{\bar{\nu}}R_n) = \Phi({}^{\nu}R_n)$ for all $n \in \mathbb{N}$. □

Remark 3.14: We leave to the reader the proof of the fact (which we shall not use) that if ν' and ν'' are PATRICIA-equivalent, then, in the notation of Lemma 3.13, $\bar{\nu'} = \bar{\nu''}$.

It follows from Lemma 3.13 that for every diffuse probability measure ν on $\{0,1\}^\infty$ almost surely the successive states of the PATRICIA chain $({}^{\nu}\bar{R}_n)_{n\in\mathbb{N}}$ "fill out" the space $\{0,1\}^\star$ – see Corollary 3.15. This observation provides a useful sufficient condition for determining when an infinite PATRICIA bridge is of the form $({}^{\nu}\bar{R}_n)_{n\in\mathbb{N}}$ for some diffuse probability measure ν on $\{0,1\}^\infty$ – see Example 4.6.

Corollary 3.15: *For any diffuse probability measure ν on $\{0,1\}^\infty$ we have $\bigcup_{m\in\mathbb{N}} \bigcap_{n \ge m} {}^{\nu}\bar{R}_n = \{0,1\}^\star$ almost surely.*

Proof: Let $\bar{\nu}$ be the diffuse probability measure constructed in Lemma 3.13 that is PATRICIA-equivalent to ν and satisfies $\bar{\nu}(\tau(y)) > 0$ for all $y \in \{0,1\}^\star$. Almost surely, for any $k \in \mathbb{N}$ there exists an $N \in \mathbb{N}$ such that for all $n \ge N$ and for all $y \in \{0,1\}^k$ there is an $m \in [n]$ with $Z_m \in \tau(y)$. Therefore, almost surely, ${}^{\bar{\nu}}R_n \supseteq \{\emptyset\} \cup \bigsqcup_{\ell=1}^k \{0,1\}^\ell$ for $n \ge N$. Since $({}^{\nu}\bar{R}_n)_{n\in\mathbb{N}}$ and $({}^{\bar{\nu}}\bar{R}_n)_{n\in\mathbb{N}}$ have the same distribution, this establishes the result. □

4. Infinite bridges

An *infinite bridge* for $({}^{\nu}R_n)_{n\in\mathbb{N}}$ (resp. $({}^{\nu}\bar{R}_n)_{n\in\mathbb{N}}$) is a Markov chain $({}^{\nu}R_n^\infty)_{n\in\mathbb{N}}$ (resp. $({}^{\nu}\bar{R}_n^\infty)_{n\in\mathbb{N}}$) such that ${}^{\nu}R_1^\infty = \emptyset$ (resp. ${}^{\nu}\bar{R}_1^\infty = \emptyset$) and

$$\mathbb{P}\{{}^{\nu}R_n^\infty = \mathbf{s} \mid {}^{\nu}R_{n+1}^\infty = \mathbf{t}\} = \mathbb{P}\{{}^{\nu}R_n = \mathbf{s} \mid {}^{\nu}R_{n+1} = \mathbf{t}\}$$

for $\mathbf{s} \in \mathbb{S}_n$, $\mathbf{t} \in \mathbb{S}_{n+1}$, $n \in \mathbb{N}$ (resp.

$$\mathbb{P}\{{}^\nu \bar{R}_n^\infty = \bar{\mathbf{s}} \mid {}^\nu \bar{R}_{n+1}^\infty = \bar{\mathbf{t}}\} = \mathbb{P}\{{}^\nu \bar{R}_n = \bar{\mathbf{s}} \mid {}^\nu \bar{R}_{n+1} = \bar{\mathbf{t}}\}$$

for $\bar{\mathbf{s}} \in \bar{S}_n$, $\bar{\mathbf{t}} \in \bar{\mathbb{S}}_{n+1}$, $n \in \mathbb{N}$). The reason we use this terminology is the following. A finite bridge for $({}^\nu R_n)_{n\in\mathbb{N}}$ with end-point $\mathbf{t} \in \mathbb{S}_m$ for some $m \in \mathbb{N}$ is the Markov process $({}^\nu R_n^{\mathbf{t}})_{n\in[m]}$ obtained by conditioning $({}^\nu R_n)_{n\in[m]}$ on the event $\{{}^\nu R_m = \mathbf{t}\}$. A finite bridge has the same backward transition probabilities as $({}^\nu R_n)_{n\in\mathbb{N}}$, and, writing $M(\mathbf{t})$ for the number of leaves of the binary tree $\mathbf{t}$, one way to produce an infinite bridge is to find a sequence of trees $(\mathbf{t}_k)_{k\in\mathbb{N}}$ such that $M(\mathbf{t}_k) \to \infty$ and the finite-dimensional distributions of $({}^\nu R_n^{\mathbf{t}_k})_{n\in[M(\mathbf{t}_k)]}$ converge as $k \to \infty$: the limiting finite-dimensional distributions will be consistent and hence define a stochastic process $({}^\nu R_n^\infty)_{n\in\mathbb{N}}$ that is an infinite bridge for $({}^\nu R_n)_{n\in\mathbb{N}}$. Similar remarks hold for the infinite bridges of $({}^\nu \bar{R}_n)_{n\in\mathbb{N}}$. We call the infinite bridges for the PATRICIA chains *infinite PATRICIA bridges.*

It follows from Proposition 3.7 that for an arbitrary diffuse probability measure ν any infinite bridge for $({}^\nu R_n)_{n\in\mathbb{N}}$ is an infinite bridge for $(R_n)_{n\in\mathbb{N}}$. Conversely, an infinite bridge $(R_n^\infty)_{n\in\mathbb{N}}$ for $(R_n)_{n\in\mathbb{N}}$ is a bridge for $({}^\nu R_n)_{n\in\mathbb{N}}$ provided $\{\mathbf{t} \in \mathbb{S}_n : \mathbb{P}\{R_n^\infty = \mathbf{t}\} > 0\} \subseteq \{\mathbf{t} \in \mathbb{S}_n : \mathbb{P}\{{}^\nu R_n = \mathbf{t}\} > 0\}$ for all $n \in \mathbb{N}$. We may therefore restrict our attention to infinite bridges for $(R_n)_{n\in\mathbb{N}}$. Similar observations give that in investigating the infinite bridges of the processes $({}^\nu \bar{R}_n)_{n\in\mathbb{N}}$ we may restrict attention to those of $(\bar{R}_n)_{n\in\mathbb{N}}$.

It follows from the general theory of Doob–Martin compactifications that the family of distributions of infinite bridges is a compact convex set, every element of which is unique mixture of extremal distributions. Moreover, every extremal infinite bridge is a limit of finite bridges (but the converse is not true in general). Furthermore, we see from Proposition 3.7 that each Markov process of the form $({}^\nu R_n)_{n\in\mathbb{N}}$ (resp. $({}^\nu \bar{R}_n)_{n\in\mathbb{N}}$) is an infinite bridge for $(R_n)_{n\in\mathbb{N}}$ (resp. $(\bar{R}_n)_{n\in\mathbb{N}}$) and, as we shall see below, these infinite bridges are extremal. The interesting questions for the infinite bridges of $(R_n)_{n\in\mathbb{N}}$ and $(\bar{R}_n)_{n\in\mathbb{N}}$ are therefore:

- What are all the extremal infinite bridges?
- Is every limit of finite bridges extremal?
- Is every extremal infinite bridge of the form $({}^\nu R_n)_{n\in\mathbb{N}}$ (resp. $({}^\nu \bar{R}_n)_{n\in\mathbb{N}}$) for some ν?

In the case of $(R_n)_{n\in\mathbb{N}}$, these questions were answered in [8] as follows.

Theorem 4.1:

(i) *An infinite bridge for* $(R_n)_{n\in\mathbb{N}}$ *is extremal if and only if it has an almost surely trivial tail* σ*-field.*
(ii) *An infinite bridge for* $(R_n)_{n\in\mathbb{N}}$ *is extremal if and only if it is a weak limit of finite bridges.*
(iii) *The extremal infinite bridges for* $(R_n)_{n\in\mathbb{N}}$ *coincide with collection of Markov processes of the form* $({}^{\nu}R_n)_{n\in\mathbb{N}}$ *for some diffuse probability measure* ν.

By part (ii) of Proposition 3.7 the class of infinite PATRICIA bridges and the class of infinite Rémy bridges coincide. The next theorem shows that analogues of parts (i) and (ii) Theorem 4.1 hold for the infinite PATRICIA bridges.

Theorem 4.2:

(i) *An infinite bridge for* $(\bar{R}_n)_{n\in\mathbb{N}}$ *is extremal if and only if it has an almost surely trivial tail* σ*-field.*
(ii) *An infinite bridge for* $(\bar{R}_n)_{n\in\mathbb{N}}$ *is extremal if and only if it is a weak limit of finite bridges.*

Proof: See Proposition 5.19 and the proof of Corollary 5.21 in [7]. □

Remark 4.3: Let us comment briefly on the proof of Proposition 5.19 in [7]. Remark 5.20 in that paper says that this proposition can also be proved along the lines of Lemma 5.3 therein. As pointed out to us by Julian Gerstenberg and Rudolf Grübel, the argument in the proof of Lemma 5.3 and consequently Remark 5.20 is incorrect. However, this does not invalidate the proof of Proposition 5.19 given directly after the statement, as this proof is along completely different lines. Also, this does not invalidate any of the remainder of [7] since Lemma 5.3 and Remark 5.20 of that paper are not used further therein.

The following proposition gives a way of producing (extremal) infinite PATRICIA bridges from (extremal) infinite radix sort bridges.

Proposition 4.4: *If* $(R_n^\infty)_{n\in\mathbb{N}}$ *is an infinite bridge for* $(R_n)_{n\in\mathbb{N}}$, *then* $(\bar{R}_n^\infty)_{n\in\mathbb{N}} := (\Phi(R_n^\infty))_{n\in\mathbb{N}}$ *is an infinite bridge for* $(\bar{R}_n)_{n\in\mathbb{N}}$. *Moreover, if* $(R_n^\infty)_{n\in\mathbb{N}}$ *is extremal and thus of the form* $({}^{\nu}R_n)_{n\in\mathbb{N}}$ *for some diffuse probability measure* ν, *then* $(\bar{R}_n^\infty)_{n\in\mathbb{N}} = ({}^{\nu}\bar{R}_n)_{n\in\mathbb{N}}$ *is extremal.*

Proof: Consider the first claim. For $n \in \mathbb{N}$, $\mathbf{s} \in \mathbb{S}_n$, $\mathbf{t} \in \mathbb{S}_{n+1}$, $\bar{\mathbf{s}} \in \bar{\mathbb{S}}_n$, and $\bar{\mathbf{t}} \in \bar{\mathbb{S}}_{n+1}$ put $\Lambda_n(\mathbf{s}) := \mathbb{P}\{R_n^\infty = \mathbf{s}\}$, $\Lambda_{n+1}(\mathbf{t}) := \mathbb{P}\{R_{n+1}^\infty = \mathbf{t}\}$, $Q_{n+1,n}(\mathbf{t}, \mathbf{s}) := \mathbb{P}\{R_n^\infty = \mathbf{s} \,|\, R_{n+1}^\infty = \mathbf{t}\} = \mathbb{P}\{R_n = \mathbf{s} \,|\, R_{n+1} = \mathbf{t}\}$, $\bar{\Lambda}_n(\bar{\mathbf{s}}) := \mathbb{P}\{\bar{R}_n^\infty = \bar{\mathbf{s}}\}$, $\bar{\Lambda}_{n+1}(\bar{\mathbf{t}}) := \mathbb{P}\{\bar{R}_{n+1}^\infty = \bar{\mathbf{t}}\}$, and $\bar{Q}_{n+1,n}(\bar{\mathbf{t}}, \bar{\mathbf{s}}) := \mathbb{P}\{\bar{R}_n = \bar{\mathbf{s}} \,|\, \bar{R}_{n+1} = \bar{\mathbf{t}}\}$. Because $(R_n^\infty)_{n\in\mathbb{N}}$ is an infinite bridge for $(R_n)_{n\in\mathbb{N}}$ we know that

$$\sum_{\mathbf{t}\in\mathbb{S}_{n+1}} \Lambda_{n+1}(\mathbf{t}) Q_{n+1,n}(\mathbf{t}, \mathbf{s}) = \Lambda_n(\mathbf{s})$$

and in order to show that $(\bar{R}_n^\infty)_{n\in\mathbb{N}}$ is an infinite bridge for $(\bar{R}_n)_{n\in\mathbb{N}}$ we need to show that

$$\sum_{\bar{\mathbf{t}}\in\bar{\mathbb{S}}_{n+1}} \bar{\Lambda}_{n+1}(\bar{\mathbf{t}}) \bar{Q}_{n+1,n}(\bar{\mathbf{t}}, \bar{\mathbf{s}}) = \bar{\Lambda}_n(\bar{\mathbf{s}}).$$

Now

$$\bar{\Lambda}_n(\bar{\mathbf{s}}) = \sum_{\mathbf{s}\in\mathbb{S}_n : \Phi(\mathbf{s})=\bar{\mathbf{s}}} \Lambda_n(\mathbf{s}),$$

and

$$\bar{\Lambda}_{n+1}(\bar{\mathbf{t}}) = \sum_{\mathbf{t}\in\mathbb{S}_{n+1} : \Phi(\mathbf{t})=\bar{\mathbf{t}}} \Lambda_{n+1}(\mathbf{t}).$$

Furthermore, by Remark 3.10 we have

$$\bar{Q}_{n+1,n}(\bar{\mathbf{t}}, \bar{\mathbf{s}}) = \sum_{\mathbf{s}\in\mathbb{S}_n : \Phi(\mathbf{s})=\bar{\mathbf{s}}} Q_{n+1,n}(\mathbf{t}, \mathbf{s})$$

when $\Phi(\mathbf{t}) = \bar{\mathbf{t}}$. Therefore,

$$\begin{aligned}
&\sum_{\bar{\mathbf{t}}\in\bar{\mathbb{S}}_{n+1}} \bar{\Lambda}_{n+1}(\bar{\mathbf{t}}) \bar{Q}_{n+1,n}(\bar{\mathbf{t}}, \bar{\mathbf{s}}) \\
&\quad = \sum_{\bar{\mathbf{t}}\in\bar{\mathbb{S}}_{n+1}} \sum_{\mathbf{t}\in\mathbb{S}_{n+1} : \Phi(\mathbf{t})=\bar{\mathbf{t}}} \Lambda_{n+1}(\mathbf{t}) \sum_{\mathbf{s}\in\mathbb{S}_n : \Phi(\mathbf{s})=\bar{\mathbf{s}}} Q_{n+1,n}(\mathbf{t}, \mathbf{s}) \\
&\quad = \sum_{\mathbf{s}\in\mathbb{S}_n : \Phi(\mathbf{s})=\bar{\mathbf{s}}} \sum_{\bar{\mathbf{t}}\in\bar{\mathbb{S}}_{n+1}} \sum_{\mathbf{t}\in\mathbb{S}_{n+1} : \Phi(\mathbf{t})=\bar{\mathbf{t}}} \Lambda_{n+1}(\mathbf{t}) Q_{n+1,n}(\mathbf{t}, \mathbf{s}) \\
&\quad = \sum_{\mathbf{s}\in\mathbb{S}_n : \Phi(\mathbf{s})=\bar{\mathbf{s}}} \sum_{\mathbf{t}\in\mathbb{S}_{n+1}} \Lambda_{n+1}(\mathbf{t}) Q_{n+1,n}(\mathbf{t}, \mathbf{s}) \\
&\quad = \sum_{\mathbf{s}\in\mathbb{S}_n : \Phi(\mathbf{s})=\bar{\mathbf{s}}} \Lambda_n(\mathbf{s}) \quad = \quad \bar{\Lambda}_n(\bar{\mathbf{s}}),
\end{aligned}$$

as required.

Turning to the second claim, suppose that the infinite bridge $(R_n^\infty)_{n\in\mathbb{N}}$ is extremal (and hence, by part (iii) of Theorem 4.1 is of the form $({}^\nu R_n)_{n\in\mathbb{N}}$ for some diffuse probability measure ν). By part (i) of Theorem 4.1 the tail σ-field of $(R_n^\infty)_{n\in\mathbb{N}}$ is almost surely trivial. It follows that the tail σ-field of $(\bar{R}_n^\infty)_{n\in\mathbb{N}} := (\Phi(R_n^\infty))_{n\in\mathbb{N}}$ is almost surely trivial and hence, by part (i) of Theorem 4.2, the infinite bridge $(\bar{R}_n^\infty)_{n\in\mathbb{N}}$ is extremal. (Alternatively, it is immediate from Remark 2.1 and the Hewitt-Savage zero-one law that $(\bar{R}_n^\infty)_{n\in\mathbb{N}} = ({}^\nu\bar{R}_n)_{n\in\mathbb{N}}$ has an almost surely trivial tail σ-field.) □

Remark 4.5: Comparing Theorem 4.1 with Theorem 4.2 and taking Proposition 4.4 into account, it is natural to conjecture that the analogue of part (iii) of Theorem 4.1 holds for infinite PATRICIA bridges; that is, that the set of extremal PATRICIA bridges is exhausted by the PATRICIA chains $({}^\nu\bar{R}_n)_{n\in\mathbb{N}}$, with ν a diffuse probability measure on $\{0,1\}^\infty$. This, however, is not the case, as the next example shows. This example recalls the construction from Example 4.1 of [7] of a specific extremal infinite Rémy (and hence PATRICIA) bridge based on a random zig-zag path with leaves attached to it. We will use Corollary 3.15 to conclude that this extremal infinite PATRICIA is not of the form $({}^\nu\bar{R}_n)_{n\in\mathbb{N}}$ for any diffuse probability measure ν on $\{0,1\}^\infty$ A concrete description of the extremal infinite PATRICIA bridges comprises the remainder of the paper.

Example 4.6: Let $\epsilon_2, \epsilon_3, \ldots$ be a sequence of independent identically distributed $\{0,1\}$-valued random variables with common distribution $\mathbb{P}\{\epsilon_k = 0\} = \mathbb{P}\{\epsilon_k = 1\} = \frac{1}{2}$, $2 \le k < \infty$. For $N \in \mathbb{N}$, define $(\bar{R}_n^N)_{n\in[N]}$ with $\bar{R}_n^N \in \bar{\mathbb{S}}_n$ for $n \in [N]$ by requiring that $(\bar{R}_n^N)_{n\in[N]}$ is Markov, $\bar{R}_N^N$ has the same distribution as $\{\emptyset\} \cup \bigcup_{k=2}^N \{\epsilon_2 \ldots \epsilon_k, \, \epsilon_2 \ldots \epsilon_{k-1}\bar{\epsilon}_k\}$, and the backward transition probabilities of $(\bar{R}_n^N)_{n\in[N]}$ are the same as those of $(\bar{R}_n)_{n\in\mathbb{N}}$. We may suppose that $\bar{R}_N^N = \{\emptyset\} \cup \bigcup_{k=2}^N \{\epsilon_2 \ldots \epsilon_k, \, \epsilon_2 \ldots \epsilon_{k-1}\bar{\epsilon}_k\}$. It follows from the form of the backward transition probabilities that for $M \in [N]$, $\bar{R}_M^N = \{\emptyset\} \cup \bigcup_{k=2}^M \{\epsilon_{I_2} \ldots \epsilon_{I_k}, \, \epsilon_{I_2} \ldots \epsilon_{k-1}\bar{\epsilon}_{I_k}\}$, where $2 \le I_2 < I_3 < \cdots < I_M \le N$ is a certain uniform random subset of $\{2, \ldots, N\}$ of cardinality M that is independent of $\epsilon_2, \ldots, \epsilon_N$. Thus $\bar{R}_M^N$ has the same distribution as $\bar{R}_M^M$ and hence, by Kolmogorov's extension theorem, there exists a Markov process $(\bar{R}_n^\infty)_{n\in\mathbb{N}}$ such that $(\bar{R}_n^\infty)_{n\in[N]}$ has the same distribution as $(\bar{R}_n^N)_{n\in[N]}$ for any $N \in \mathbb{N}$. Therefore $(\bar{R}_n^\infty)_{n\in\mathbb{N}}$ is an infinite bridge for $(\bar{R}_n)_{n\in\mathbb{N}}$.

We can give a pathwise construction of $(\bar{R}_n^\infty)_{n\in\mathbb{N}}$ as follows. Let $((Y_n, \eta_n))_{n=2}^\infty$ be an infinite sequence of independent identically distributed

$[0,1] \times \{0,1\}$-valued random variables such that Y_n has the uniform distribution on $[0,1]$, $\mathbb{P}\{\eta_n = 0\} = \mathbb{P}\{\eta_n = 1\} = \frac{1}{2}$, and Y_n and η_n are independent. For $2 \le n < \infty$, let σ_n be the permutation of $\{2,\ldots,n\}$ such that $Y_{\sigma_n(2)} < Y_{\sigma_n(3)} < \cdots < Y_{\sigma_n(n)}$. For $2 \le k \le n$, put $\epsilon_{n,k} = \eta_{\sigma_n(k)}$. Setting $\bar{R}_1^\infty = \emptyset$ and $\bar{R}_n^\infty := \{\emptyset\} \cup \bigcup_{k=2}^n \{\epsilon_{n,2}\ldots\epsilon_{n,k}, \epsilon_{n,2}\ldots\epsilon_{n,(k-1)}\bar{\epsilon}_{n,k}\}$ for $n \ge 2$ produces a process with the desired distribution. Note that if π is a permutation of $\{2,3,\ldots\}$ that leaves every element of $\{N+1, N+2, \ldots\}$ fixed for some $N \ge 2$ and we replace $((Y_n,\eta_n))_{n=2}^\infty$ by $((Y_{\pi(n)}, \eta_{\pi(n)}))_{n=2}^\infty$ in this construction, then the values of $\bar{R}_n^\infty$, $n \ge N$, are left unchanged. It therefore follows from the Hewitt-Savage zero-one law that the tail σ-field of $(\bar{R}_n^\infty)_{n\in\mathbb{N}}$ is $\mathbb{P}$-a.s. trivial and thus, by part (i) of Theorem 4.2, this process is an extremal infinite bridge for $(\bar{R}_n)_{n\in\mathbb{N}}$.

Now, for any $n \in \mathbb{N}$ the random tree $\bar{R}_n^\infty$ is a zig-zag path of length $n-1$ with leaves attached to it. It follows that for any $m \in \mathbb{N}$ the random tree $\bigcap_{n\ge m} \bar{R}_n^\infty$ is a zig-zag path of length at most $m-1$ with leaves attached to it and hence $\bigcup_{m\in\mathbb{N}} \bigcap_{n\ge m} \bar{R}_n^\infty$ is a zig-zag path of possibly infinite length with leaves attached to it and is certainly not all of $\{0,1\}^\star$. It is thus clear from Corollary 3.15 that the extremal infinite bridge $(\bar{R}_n^\infty)_{n\in\mathbb{N}}$ is not of the form $({}^\nu\bar{R}_n)_{n\in\mathbb{N}}$ for any diffuse probability measure ν.

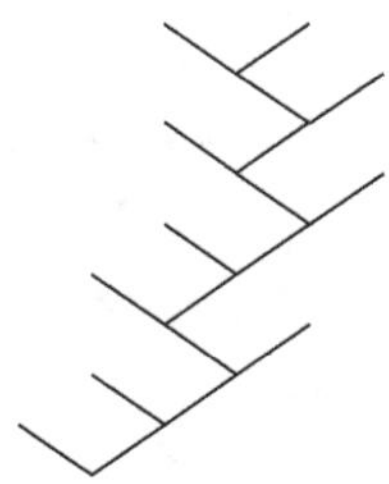

Fig. 3. A realization of $\bar{R}_9^\infty$.

For the sake of completeness, we note that if we fix $m \in \mathbb{N}$, then for $n \ge m$ the restriction of $\bar{R}_n^\infty$ to height $m-1$ is uniformly distributed over the set of zig-zag paths of length $m-1$ with leaves attached. Moreover, if $m \le N_1 < N_2 < \cdots < N_M$ for some fixed $M \in \mathbb{N}$, then the restrictions of $\bar{R}_{N_1}^\infty, \bar{R}_{N_2}^\infty, \ldots, \bar{R}_{N_M}^\infty$ to height $m-1$ become asymptotically independent as $N_2 - N_1, \ldots, N_M - N_{M-1} \to \infty$. It follows that almost surely $\bigcup_{m\in\mathbb{N}} \bigcap_{n\ge m} \bar{R}_n^\infty$ is just the set $\{\emptyset, 0, 1\}$. Incidentally, from these same observations it is clear that $\bigcap_{m\in\mathbb{N}} \bigcup_{n\ge m} \bar{R}_n^\infty = \{0,1\}^\star$ almost surely. This latter property would therefore not suffice to discriminate $(\bar{R}_n^\infty)_{n\in\mathbb{N}}$ from the extremal infinite bridges of the form $({}^\nu\bar{R}_n)_{n\in\mathbb{N}}$.

Our next goal is to introduce a class of extremal infinite bridges for $(\bar{R}_n)_{n\in\mathbb{N}}$ that subsumes both the class of Markov chains of the form $({}^\nu\bar{R}_n)_{n\in\mathbb{N}}$ and the Markov chain in Example 4.6. In fact, using Proposition 3.7 and

Remark 3.6 (showing that the common backward transitions of the PATRICIA chains coincide with those of the Rémy chain) we will be able to obtain a representation for *all* the extremal infinite PATRICIA bridges. This will be achieved through the concept of a *didendritic system* introduced in Definition 5.8 of [7] and revisited in Section 6.

Before moving on to this characterization of all the extremal infinite PATRICIA bridges, we include the following section which gives some idea of the divergent sample path behavior displayed by different infinite bridges.

5. Heights of trees in an infinite PATRICIA bridge

Given $\bar{\mathbf{t}} \in \bar{\mathbb{S}}_n$ for some $n \geq 0$ with leaves $y_1, \ldots, y_n$ let $\mathrm{ht}(\bar{\mathbf{t}})$ denote the height of $\bar{\mathbf{t}}$; that is, $\mathrm{ht}(\bar{\mathbf{t}}) := \max_{1 \leq k \leq n} |y_k|$. It is shown in [4] that

$$\lim_{n\to\infty} (\mathrm{ht}(\bar{R}_n) - \log_2 n)/\sqrt{2 \log_2 n} = 1$$

almost surely so that, in particular,

$$\lim_{n\to\infty} \mathrm{ht}(\bar{R}_n)/\log_2 n = 1$$

almost surely.

It is of interest to compare this asymptotic behavior with that of other infinite PATRICIA bridges. The infinite PATRICIA bridge $(\bar{R}_n^\infty)_{n\in\mathbb{N}}$ of Example 4.6 clearly has $\mathrm{ht}(\bar{R}_n^\infty) = n - 1$ for all $n \in \mathbb{N}$. On the other hand if $(\bar{R}_n^\infty)_{n\in\mathbb{N}}$ is the Rémy tree growth chain (which we have observed is an infinite PATRICIA bridge), then it follows from the results of [21] that

$$\lim_{n\to\infty} \mathrm{ht}(\bar{R}_n^\infty)/\sqrt{n}$$

exists and is strictly positive almost surely.

As one last example, it has been pointed out to us by Ralph Neininger that if $\nu = \bigotimes_{j=1}^\infty \mathrm{Ber}\left(\frac{1}{j+1}\right)$, where $\mathrm{Ber}(p)$ is the Bernoulli probability measure on $\{0,1\}$ that assigns mass $1-p$ to 0 and mass p to 1, then

$$\limsup_{n\to\infty} \mathrm{ht}({}^\nu \bar{R}_n)/\sqrt{2n/\log n} \geq 1$$

almost surely. For the sake of completeness, we sketch Neininger's argument.

Suppose that $Z_1, Z_2, \ldots$ is an infinite i.i.d. sequence of random elements of $\{0,1\}^\infty$ where $Z_n = (Z_{n,1}, Z_{n,2}, \ldots)$, $n \in \mathbb{N}$, with $Z_{n,1}, Z_{n,2}, \ldots$ independent, $\mathbb{P}\{Z_{n,j} = 0\} = \frac{j}{j+1}$, and $\mathbb{P}\{Z_{n,j} = 1\} = \frac{1}{j+1}$, $j \in \mathbb{N}$. Let ${}^\nu \bar{R}_n$, $n \in \mathbb{N}$, be the PATRICIA tree constructed from the inputs $Z_1, \ldots, Z_n$.

Consider the event

$$A_{n,t} := \bigcap_{\ell=1}^{t} \bigcup_{m=1}^{n} \{(Z_{m,1}, \ldots, Z_{m,\ell}) = (0, \ldots, 0, 1)\}.$$

Observe that on the event $A_{n,t}$ the tree ${}^{\nu}\bar{R}_n$ contains the vertices $\emptyset, 1, 01, 001, \ldots, 0\ldots01$, where the last sequence has $t-1$ zeros. Consequently, $\mathrm{ht}({}^{\nu}\bar{R}_n) \geq t$ on the event $A_{n,t}$. It therefore suffices to show that $\lim_{n\to\infty} \mathbb{P}(A_{n,t(n)}) = 1$, where $t(n) = \lfloor \sqrt{2n/\log n} - 1 \rfloor$.

Now

$$\begin{aligned}
1 - \mathbb{P}(A_{n,t}) &= \mathbb{P}(A_{n,t}^c) \\
&= \mathbb{P}\left(\bigcup_{\ell=1}^{t} \bigcap_{m=1}^{n} \{(Z_{m,1}, \ldots, Z_{m,\ell}) \neq (0, \ldots, 0, 1)\}\right) \\
&\leq \sum_{\ell=1}^{t} \mathbb{P}\left(\bigcap_{m=1}^{n} \{(Z_{m,1}, \ldots, Z_{m,\ell}) \neq (0, \ldots, 0, 1)\}\right) \\
&= \sum_{\ell=1}^{t} \left(1 - \frac{1}{\ell+1} \prod_{j=1}^{\ell-1} \frac{j}{j+1}\right)^n \\
&= \sum_{\ell=1}^{t} \left(1 - \frac{1}{\ell(\ell+1)}\right)^n \\
&\leq t \left(1 - \frac{1}{t(t+1)}\right)^n \\
&\leq (t+1) \left(1 - \frac{1}{(t+1)^2}\right)^n \leq (t+1) \exp\left(-\frac{n}{(t+1)^2}\right).
\end{aligned}$$

Thus

$$\begin{aligned}
1 - \mathbb{P}(A_{n,t(n)}) &\leq \sqrt{2n/\log n} \exp\left(-\frac{n}{(\sqrt{2n/\log n})^2}\right) \\
&= \sqrt{2n/\log n} \exp\left(-\frac{\log n}{2}\right) \\
&= \sqrt{2/\log n} \\
&\to 0 \quad \text{as } n \to \infty,
\end{aligned}$$

as required.

6. Didendritic systems

We now prepare some notation and definitions that serve to introduce the concept of a didendritic system. As announced at the end of Section 4, this concept will help to obtain a representation for *all* the extremal infinite PATRICIA bridges.

With the partial order $<$ defined on the set $\{0,1\}^\star \sqcup \{0,1\}^\infty$, we define, for any two elements u, v of this set, the *most recent predecessor* $u \wedge v$ of u and v. More precisely, for each such u, v there is a unique $w \in \{0,1\}^\star \sqcup \{0,1\}^\infty$ with the properties $w \le u$, $w \le v$, and if x is any element of $\{0,1\}^\star \sqcup \{0,1\}^\infty$ with $x \le u$ and $x \le v$, then $x \le w$. Denote this w by $u \wedge v$. For example, if $u = u_1 \dots u_m$ and $v = v_1 \dots v_n$ are elements of $\{0,1\}^\star$, then $u \wedge v = u_1 \dots u_p = v_1 \dots v_p$, where $p := \max\{q : 1 \le q \le m \wedge n,\ u_1 \dots u_q = v_1 \dots v_q\}$ and the maximum of the empty set is defined to be zero.

We define two further partial orders $<_L$ and $<_R$ on $\{0,1\}^\star$ by declaring that $u_1 \dots u_m <_L v_1 \dots v_n$ (resp. $u_1 \dots u_m <_R v_1 \dots v_n$) if $m < n$ and $v_1 \dots v_{m+1} = u_1 \dots u_m 0$ (resp. $m < n$ and $v_1 \dots v_{m+1} = u_1 \dots u_m 1$).

Definition 6.1: Given $\bar{\mathbf{t}} \in \bar{\mathbb{S}}_n$, consider a bijective labeling of the leaves of $\bar{\mathbf{t}}$ by a set $\mathcal{N}$ with $\#\mathcal{N} = n$ (that is, each leaf receives a distinct label). Use the notation $\tilde{\mathbf{t}}$ to denote the labeled object. Define an equivalence relation $\equiv$ on $\mathcal{N} \times \mathcal{N}$ by declaring that (g, h) and (i, j) are equivalent if g, h, i, j label respectively leaves u, v, w, x such that $u \wedge v = w \wedge x$. Denote by $\langle i, j \rangle$ the equivalence class containing $(i, j) \in \mathcal{N} \times \mathcal{N}$. Note that (i, i) is the only pair in the equivalence class $\langle i, i \rangle$ so we will usually denote this equivalence class more simply by i. If k, ℓ label respectively the leaves y, z, then label the vertex $y \wedge z$ with $\langle k, \ell \rangle$.

With a slight abuse of notation, define a partial order $<_L$ (resp. $<_R$) on $\{\langle i, j \rangle : i, j \in \mathcal{N}\}$ by declaring that $\langle g, h \rangle <_L \langle i, j \rangle$ if $\langle g, h \rangle$ labels a vertex u and $\langle i, j \rangle$ labels a vertex v such that $u <_L v$ (resp. $u <_R v$). Similarly, define a third partial order $<$ by declaring that $\langle g, h \rangle < \langle i, j \rangle$ if $\langle g, h \rangle$ labels a vertex u and $\langle i, j \rangle$ labels a vertex v with $u < v$.

The equivalence relation $\equiv$ and the partial orders $<_L$, $<_R$, and $<$ have the following properties.

(A) For $i, j \in \mathcal{N}$, $(i, j) \equiv (j, i)$.

(B) For distinct $i, j \in \mathcal{N}$, either $\langle i, j \rangle <_L \langle i, i \rangle$ and $\langle i, j \rangle <_R \langle j, j \rangle$, or $\langle i, j \rangle <_R \langle i, i \rangle$ and $\langle i, j \rangle <_L \langle j, j \rangle$.

(C) *"Triplet property"* For distinct i, j, k, exactly one of

$$\langle i, j \rangle = \langle i, k \rangle < \langle j, k \rangle$$

$$\langle j,k\rangle = \langle j,i\rangle < \langle k,i\rangle$$

or

$$\langle k,i\rangle = \langle k,j\rangle < \langle i,j\rangle$$

is valid.

(D) For $i, j, k, \ell \in \mathcal{N}$, at most one of the relations $\langle i,j\rangle <_L \langle k,\ell\rangle$ and $\langle i,j\rangle <_R \langle k,\ell\rangle$ can hold and $\langle i,j\rangle < \langle k,\ell\rangle$ if and only if either $\langle i,j\rangle <_L \langle k,\ell\rangle$ or $\langle i,j\rangle <_R \langle k,\ell\rangle$.

(E) Fix $f, g, h, i, j, k \in \mathcal{N}$. If $\langle f,g\rangle <_L \langle h,i\rangle < \langle j,k\rangle$, then $\langle f,g\rangle <_L \langle j,k\rangle$. Similarly, if $\langle f,g\rangle <_R \langle h,i\rangle < \langle j,k\rangle$, then $\langle f,g\rangle <_R \langle j,k\rangle$.

Definition 6.2: A *didendritic system* $\mathcal{D} = (\mathcal{N}, \equiv, \langle\cdot,\cdot\rangle, <_L, <_R, <)$ with the non-empty (possibly infinite) *label set* $\mathcal{N}$ is the set $\mathcal{N} \times \mathcal{N}$ equipped with an equivalence relation $\equiv$, equivalence classes $\langle\cdot,\cdot\rangle$, and partial orders $<_L$, $<_R$ and $<$ on the set of equivalence classes such that the above stated properties (A)-(E) hold.

Remark 6.3: We show in Proposition 6.6 that any didendritic system with a finite label set may be thought of as a leaf-labeled full binary tree. This claim was made in Remark 5.10 of [7]. The axioms in Definition 6.2 differ from those in [7]: the former are equivalent to the latter plus the triplet property (C). The Example 6.4 shows that the conclusion of Proposition 6.6 does not hold under the axioms of [7], contrary to the assertion of Remark 5.10 of [7]. However, the omission of axiom (C) in [7] is not critical: the development there depends solely on the conclusion of Remark 5.10 therein rather than the specific axioms that one uses to characterize a didendritic system. In brief, the arguments in [7] become correct once one adds axiom (C) to the definition there of a didendritic system.

Example 6.4: Consider the equivalence relation on $[3] \times [3]$ defined by

$$(h,i) \equiv (j,k) \text{ if an only if } h = j \text{ and } i = k, \text{ or } h = k \text{ and } i = j.$$

For $(i,j) \in \{(1,2),(1,3),(2,3)\}$ say that $\langle i,j\rangle <_L \langle i,i\rangle$ and $\langle i,j\rangle <_R \langle j,j\rangle$. Moreover, say that $\langle h,i\rangle < \langle j,k\rangle$ if $\langle h,i\rangle <_L \langle j,k\rangle$ or $\langle h,i\rangle <_R \langle j,k\rangle$. Then $([3], \equiv, <_L, <_R, <)$ meets the axioms of Definition 5.8 of [7]. However, this system does not correspond to a binary tree with 3 leaves, since this would require exactly 5 equivalence classes (each one corresponding to a vertex of the tree), whereas this system has $3 + 3 = 6$ equivalence classes.

Lemma 6.5: Any didendritic system with $\#\mathcal{N} \geq 3$ has the following property

(F) For distinct $i, j, k \in \mathcal{N}$, there exists a bijective mapping from $\{\langle g, h\rangle : g, h \in \{i, j, k\}\}$ to the set of vertices of a full binary tree $\tilde{\mathbf{t}}$ with three leaves labeled by i, j, k which preserves the three partial orders and, for $g \in \{i, j, k\}$, maps $\langle g, g\rangle$ to the leaf of $\tilde{\mathbf{t}}$ labeled g.

Proof: Assume that the first of the three possible set of relations in (C) is valid, i.e. $\langle i, j\rangle = \langle i, k\rangle < \langle j, k\rangle$. Then we must have by (B) that $\langle i, j\rangle = \langle i, k\rangle < \langle j, k\rangle < j$, and $\langle i, j\rangle < i$.

Applying (B) to i and j, we have either $\langle i, j\rangle <_L i$ and $\langle i, j\rangle <_R j$ or $\langle i, j\rangle <_L j$ and $\langle i, j\rangle <_R i$. Assume the former. Then (E) (together with (B)) enforces $\langle i, j\rangle <_R \langle j, k\rangle$.

Now applying (B) to j and k, we have either $\langle j, k\rangle <_L j$ and $\langle j, k\rangle <_R k$ or $\langle j, k\rangle <_L k$ and $\langle j, k\rangle <_R j$. Assume the former. This then results in the full binary tree $\tilde{\mathbf{t}}$ with vertex set $\{\emptyset, 0, 1, 10, 11\}$, where $\emptyset$ is labeled by $\langle i, j\rangle = \langle i, k\rangle$, 1 is labeled by $\langle j, k\rangle$, and the three leaves 0, 10 and 11 are labeled by i, j and k, respectively.

Combining all the ways allowed by (C) and (B), we arrive at the $3 \times 2 \times 2 = 12$ possible leaf-labeled full binary trees with three leaves and label set $\{i, j, k\}$. □

Let $\tilde{\mathbf{t}}$ be a leaf-labeled full binary tree, $\mathcal{D}$ be a didendritic system with finite label set $\mathcal{N}$, and ψ be a mapping from the set of equivalence classes of $\mathcal{D}$ to the set of vertices of $\tilde{\mathbf{t}}$. We call ψ an *isomorphism* from $\mathcal{D}$ to $\tilde{\mathbf{t}}$ if it is bijective and preserves the three partial orders. Necessarily, ψ then maps the label set $\mathcal{N}$ bijectively to the set of leaves of $\tilde{\mathbf{t}}$.

Proposition 6.6: *Let $\mathcal{D} = (\mathcal{N}, \equiv, \langle\cdot,\cdot\rangle, <_L, <_R, <)$ be a finite didendritic system. Then $\mathcal{D}$ is isomorphic to some leaf-labeled full binary tree $\tilde{\mathbf{t}}$. Moreover, both $\tilde{\mathbf{t}}$ and the isomorphism are unique.*

Proof: Denote the cardinality of $\mathcal{N}$ by n.

For $n = 2$ the assertion follows directly from properties (A)-(C).

For $n \geq 3$ we will use that $\mathcal{D}$ also has property (F) because of Lemma 6.5. Indeed, for $n = 3$ the assertion is immediate from that property.

We now proceed by induction and assume that the assertion is true for some $n \geq 3$.

For the induction step, assume without loss of generality that $\mathcal{N}_{n+1} = [n+1]$, and let $\mathcal{D}_{n+1}$ obey the assumptions of the proposition with label set $[n+1]$. We are going to construct a unique leaf-labeled full binary tree $\tilde{\mathbf{t}}_{n+1}$ and isomorphism ψ from $\mathcal{D}_{n+1}$ to $\tilde{\mathbf{t}}_{n+1}$.

To this end we denote the restriction of $\mathcal{D}_{n+1}$ to $[n]$ by $\mathcal{D}_n$. Since $\mathcal{D}_n$ meets the conditions of the induction hypothesis, there exists a unique leaf-labeled full binary tree $\tilde{\mathbf{t}}_n$ and isomorphism ϕ from $\mathcal{D}_n$ to $\tilde{\mathbf{t}}_n$. By a slight abuse of notation we will use the symbol $\langle i,j\rangle$ both for the equivalence classes in $\mathcal{D}_n$ and for the equivalence classes in $\mathcal{D}_{n+1}$.

Now choose $a, b \in [n]$ such that $\phi(\langle a,b\rangle) = \emptyset \in \tilde{\mathbf{t}}_n$. Focusing on $a, b, n+1$, the triplet property (C) allows to distinguish between two cases. The first case is when $\langle n+1, a\rangle = \langle n+1,\, b\rangle < \langle a,b\rangle$ and the second case is when $\langle a,b\rangle = \langle n+1,\, b\rangle < \langle n+1,a\rangle$ or $\langle a,\, b\rangle = \langle n+1,a\rangle < \langle n+1,b\rangle$.

Case 1: $\langle n+1,a\rangle = \langle n+1,b\rangle < \langle a,b\rangle$.
Arguing by the isomorphy of $\mathcal{D}_n$ and $\tilde{\mathbf{t}}_n$, and recalling that $\phi(\langle a,b\rangle) = \emptyset$, we see that for all $h \in [n]$ we have that either $\langle a,b\rangle = \langle a,h\rangle < \langle a,h\rangle$ or $\langle a,b\rangle = \langle a,h\rangle < \langle b,h\rangle$. In the former case, we obtain from $\langle n+1,a\rangle = \langle n+1,\, b\rangle < \langle a,b\rangle$, $\langle a,b\rangle = \langle a,h\rangle < \langle a,h\rangle$, and the transitivity of $<$ that $\langle n+1,a\rangle < \langle a,h\rangle$ and thence, from the triplet property (C) applied to the triplet $a, h, n+1$, that $\langle n+1,h\rangle = \langle n+1,a\rangle < \langle a,h\rangle$. In the latter case, we obtain by a similar argument that $\langle n+1,h\rangle = \langle n+1,b\rangle < \langle b,h\rangle$. Thus, in both cases we have for all $h \in [n]$ that $\langle n+1,h\rangle = \langle n+1,a\rangle = \langle n+1,b\rangle$. Since for all $i,j \in [n]$ with $\langle i,j\rangle \neq \langle a,b\rangle$ we have $\langle a,b\rangle < \langle i,j\rangle$ by the induction assumption, the transitivity of $<$ implies that $\langle n+1,h\rangle = \langle n+1,a\rangle = \langle n+1,b\rangle < \langle i,j\rangle$ for all $i,j \in [n]$. Thus, we are led to define

$$\psi(\langle n+1,h\rangle) := \emptyset \text{ for } h \in [n].$$

To define the other values of ψ, we consider two subcases:

Case 1a: $\langle n+1,\, a\rangle = \langle n+1,b\rangle <_L \langle a,b\rangle$.
Recall from above that $\langle a,b\rangle \leq \langle i,j\rangle$ for all $i,j \in [n]$. Property (E) then ensures that $\langle n+1,a\rangle = \langle n+1,b\rangle <_L \langle i,j\rangle$ for all $i,j \in [n]$. Moreover, the triplet property (C) applied to the triplet $n+1, a, b$ combined with property (D) guarantees that $\langle n+1,a\rangle = \langle n+1,b\rangle <_R \langle n+1,n+1\rangle$. Consequently, we are led to set

$$\begin{aligned} \psi(\langle n+1,n+1\rangle) &:= 0, \\ \psi(\langle i,j\rangle) &:= 1\phi(\langle i,j\rangle) \text{ for } i,j \in [n]. \end{aligned}$$

Case 1b: $\langle n+1,a\rangle = \langle n+1,b\rangle <_R \langle a,b\rangle$.
By an argument similar to that in Case 1b, we are led to set

$$\begin{aligned} \psi(\langle n+1,n+1\rangle) &:= 1, \\ \psi(\langle i,j\rangle) &:= 0\phi(\langle i,j\rangle) \text{ for } i,j \in [n]. \end{aligned}$$

Case 2: $\langle a,b\rangle = \langle n+1,b\rangle < \langle n+1,a\rangle$ or $\langle a,b\rangle = \langle n+1,a\rangle < \langle n+1,b\rangle$. From property (F) for $a, b, n+1$, we see that either of the two above subcases implies that $\langle a,b\rangle < n+1$. Also, we may assume without loss of generality from property (B) (by exchanging the roles of a and b if necessary) that $\langle a,b\rangle <_L a$ and $\langle a,b\rangle <_R b$.

Put $\mathcal{L} := \{g \in [n+1] : \langle a,b\rangle <_L g\}$, $\mathcal{R} := \{h \in [n+1] : \langle a,b\rangle <_R h\}$, and note from property (D) that $n+1 \in \mathcal{L} \cup \mathcal{R}$. Moreover, $a \in \mathcal{L}$ and $b \in \mathcal{R}$, hence $\#\mathcal{L} \leq n$ and $\#\mathcal{R} \leq n$. We may thus apply the induction assumption to $\mathcal{D}_{\mathcal{L}}$ and $\mathcal{D}_{\mathcal{R}}$, defined to be the restrictions of $\mathcal{D}_{n+1}$ to $\mathcal{L}$ and $\mathcal{R}$, respectively. Let $\tilde{\mathbf{t}}_{\mathcal{L}}$ and $\tilde{\mathbf{t}}_{\mathcal{R}}$ be the corresponding leaf-labeled full binary trees, and let $\chi_{\mathcal{L}}$ and $\chi_{\mathcal{R}}$ be the corresponding isomorphisms. The desired leaf-labeled full binary tree $\tilde{\mathbf{t}}_{n+1}$ has the vertex set $\{\emptyset\} \cup \{0v : v \in \tilde{\mathbf{t}}_{\mathcal{D}}\} \cup \{1v : v \in \tilde{\mathbf{t}}_{\mathcal{R}}\}$, and the isomorphism ψ from $\mathcal{D}_{n+1}$ to $\tilde{\mathbf{t}}_{n+1}$ is given by

- $\psi(\langle g,h\rangle) := \emptyset$ for $g \in \mathcal{L}$, $h \in \mathcal{R}$,
- $\psi(\langle g,h\rangle) := 0v$ for $g,h \in \mathcal{L}$, with $v := \chi_{\mathcal{L}}(\langle g,h\rangle)$
- $\psi(\langle g,h\rangle) := 1v$ for $g,h \in \mathcal{R}$, with $v := \chi_{\mathcal{R}}(\langle g,h\rangle)$. □

The next result shows that a finite didendritic system can be constructed in two stages. The first stage determines the partial order $<$ while the second stage resolves each instance of $<$ as either $<_L$ or $<_R$ in a consistent manner. A randomized version of this construction appears in the statement of the main representation result, Theorem 8.2 of [7]. In the next proposition we elaborate on the "deterministic heart" of that construction.

Proposition 6.7: *Let $\mathcal{N}$ be a finite set, $\equiv$ be an equivalence relation on $\mathcal{N} \times \mathcal{N}$ and $<$ be a partial order on the set of equivalence classes. Suppose also that for distinct $i,j \in \mathcal{N}$ there are elements $w(i,j)$ of the set $\{\curvearrowright, \curvearrowleft\}$. Assume properties (A) and (C) from Definition 6.2, as well as the following properties*

(B') For distinct $i,j \in \mathcal{N}$, $\langle i,j\rangle < i$ and $\langle i,j\rangle < j$.
(B") For $i \neq j$, $w(i,j) = \curvearrowright$ if and only if $w(j,i) = \curvearrowleft$.
(E') For distinct $i,j,k \in \mathcal{N}$, if $\langle i,j\rangle = \langle i,k\rangle < \langle j,k\rangle$, then $w(i,j) = w(i,k)$.

Then there is a unique pair of partial orders $<_L$ and $<_R$ on $\{\langle i,j\rangle : i,j \in \mathcal{N}\}$ such that

$$\langle i,j\rangle <_L i \text{ and } \langle i,j\rangle <_R j \iff w(i,j) = \curvearrowright$$

and the ensemble $(\mathcal{N}, \equiv, \langle\cdot,\cdot\rangle, <_L, <_R, <)$ is a didendritic system.

Proof: We may assume without loss of generality that $\mathcal{N} = [n]$ for some $n \in \mathbb{N}$.

Step 1. For $n = 1$ there is nothing to prove.

Step 2. For $n = 2$ we have two cases according to property (B'):
If $w(1,2) = \curvearrowright$ then $\langle 1,2\rangle <_L 1$ and $\langle 1,2\rangle <_R 2$. If $w(1,2) = \curvearrowleft$ then $\langle 1,2\rangle <_L 2$ and $\langle 1,2\rangle <_R 1$. In both cases, $(\{1,2\}, \equiv, \langle\cdot,\cdot\rangle, <_L, <_R, <)$ is a didendritic system.

Step 3. Now assume $n \geq 3$.

To show property (B) in the definition of a didendritic system we can argue as in the case $n = 2$.

Next we show for distinct $i, j, k \in [n]$ that the assumptions of the proposition define partial orders $<_L, <_R$ on $\{\langle g,h\rangle : g,h \in \{i,j,k\}\}$ which meet condition (F) formulated in Lemma 6.5.

Indeed, because of condition (C) we have one of the following cases:

Case 1: $\langle i,j\rangle = \langle i,k\rangle < \langle j,k\rangle$,
Case 2: $\langle j,k\rangle = \langle j,i\rangle < \langle k,i\rangle$,
Case 3: $\langle k,i\rangle = \langle k,j\rangle < \langle i,j\rangle$.

Because of property (E'), in Case 1 we have $w(i,j) = w(i,k)$, in Case 2 we have $w(j,k) = w(j,i)$, and in Case 3 we have $w(k,i) = w(k,j)$. Thus, in the three cases, we have the four different choices $(\curvearrowright, \curvearrowright)$, $(\curvearrowright, \curvearrowleft)$, $(\curvearrowleft, \curvearrowright)$, $(\curvearrowleft, \curvearrowleft)$ for the pair $(w(i,j), w(j,k))$ in Case 1, for the pair $(w(j,i), w(k,i))$ in Case 2, and for the pair $(w(k,i), w(i,j))$ in Case 3. Any of these $3 \times 4 = 12$ subcases 1a), ..., 3d) leads to partial orders $<_L, <_R$ on $\{\langle g,h\rangle : g,h \in \{i,j,k\}\}$ and to a (distinct) leaf-labeled full binary tree $\tilde{\mathbf{t}}$ whose five vertices are bijectively labeled by the elements of $\{\langle g,h\rangle : g,h \in \{i,j,k\}\}$ in an order preserving way. For example, in Case 1a) $\tilde{\mathbf{t}}$ consists of the vertex set $\{\emptyset, 0, 1, 10, 11\}$, with 0 labeled i, 10 labeled j and 11 labeled k. In Case 3c), $\tilde{\mathbf{t}}$ consists of the vertex set $\{\emptyset, 0, 00, 01, 1\}$, with 00 labeled i, 01 labeled j and 1 labeled k.

Step 4. We now proceed inductively to show also the remaining properties (D) and (E) in the definition of a didendritic system.

Let $\mathcal{N}_{n+1} = [n+1]$ and consider the restrictions of $\equiv$ and $<$ to $[n] \times [n]$. By the induction hypothesis there is a unique didendritic system $\mathcal{D}_n$

with label set $[n]$ that has the properties stated in the proposition. By Proposition 6.6 there is a unique leaf-labeled full binary tree $\tilde{\mathbf{t}}_n$ whose n vertices are bijectively labeled by the elements of $[n]$ such that the partial orders $<_L$, $<_R$ and $<$ are preserved. Again by Proposition 6.6, for the induction step it suffices to construct out of $\tilde{\mathbf{t}}_n$ a leaf-labeled full binary tree $\tilde{\mathbf{t}}_{n+1}$ whose $2n+1$ vertices are bijectively labeled by the equivalence classes $\langle i,j\rangle$, $i,j \in [n+1]$, such that the partial orders $<_L$, $<_R$ and $<$ are preserved, and to show that the construction of $\tilde{\mathbf{t}}_{n+1}$ is unique.

Let $a,b \in [n]$ be such that the root $\emptyset$ of $\tilde{\mathbf{t}}_n$ is labeled by $\langle a,b\rangle$. Because of condition (C) applied to the triplet $n+1, a, b$, one of the following three cases applies:

Case 1: $\langle n+1,a\rangle = \langle n+1,b\rangle < \langle a,b\rangle$,
Case 2a: $\langle a,b\rangle = \langle a,n+1\rangle < \langle b,n+1\rangle$,
Case 2b: $\langle b,n+1\rangle = \langle b,a\rangle < \langle n+1,a\rangle$.

In Case 1, because of the transitivity of $<$ and the isomorphy between $\mathcal{D}_n$ and $\tilde{\mathbf{t}}_n$, we have $\langle n+1,a\rangle = \langle n+1,b\rangle < \langle i,j\rangle$ for all $i,j \in [n]$. Therefore, within $\tilde{\mathbf{t}}_{n+1}$, the root $\emptyset$ must be labeled by $\langle n+1,a\rangle = \langle n+1,b\rangle$, and either the vertex 1 or the vertex 0 must be a leaf labeled by $n+1$. The former is the case if $w(n+1,a) = \curvearrowleft$, the latter if $w(n+1,a) = \curvearrowright$. In the former case we have $\langle n+1,a\rangle = \langle n+1,b\rangle <_L \langle a,b\rangle$, and the addresses in $\tilde{\mathbf{t}}_{n+1}$ of the equivalence classes $\langle i,j\rangle$, $i,j \in [n+1]$, are then given just as in Case 1a in the proof of Proposition 6.6. In the latter case we have $\langle n+1,a\rangle = \langle n+1,b\rangle <_R \langle a,b\rangle$, and the addresses in $\tilde{\mathbf{t}}_{n+1}$ of the equivalence classes $\langle i,j\rangle$, $i,j \in [n+1]$, are then given just as in Case 1b in the proof of Proposition 6.6.

Now we assume Case 2a and we will parallel the reasoning in Case 2 in the proof of Proposition 6.6. From Step 3 we have the property (F) in hand; this applied to the triplet $\{i,j,k\} = \{a,b,n+1\}$ ensures that the set of equivalence classes $\{\langle g,h\rangle : g,h \in \{a,b,n+1\}$ is isomorphic to a labeled binary tree $\tilde{\mathbf{t}}$ with 3 leaves, for which either 0 or 1 is a leaf. Assuming without loss of generality that $w(a,b) = \curvearrowright$ we see that the vertex 0 in $\tilde{\mathbf{t}}$ must be labeled with a and the vertex 1 in $\tilde{\mathbf{t}}$ must be labeled with $\langle b,n+1\rangle$. Thus, we arrive that $\langle a,b\rangle = \langle a,n+1\rangle <_R \langle b,n+1\rangle$ and $\langle a,b\rangle <_L a$.

Put $\mathcal{L} := \{h \in [n] : \langle a,b\rangle <_L h\}$ and $\mathcal{R} := \{h \in [n] : \langle a,b\rangle <_R h\} \cup \{n+1\}$. Note that $\#\mathcal{L} \le n$ and $\#\mathcal{R} \le n$ because $a \in \mathcal{L}$ and $b \in \mathcal{R}$. We may thus apply the induction assumption to the restrictions of $\equiv$, $<$ and w to the two set $\mathcal{L}$ and $\mathcal{R}$, respectively. Let $\tilde{\mathbf{t}}_{\mathcal{L}}$ and $\tilde{\mathbf{t}}_{\mathcal{R}}$ be the corresponding leaf-labeled full binary trees, and let $\chi_{\mathcal{L}}$ and $\chi_{\mathcal{R}}$ be the corresponding isomorphisms.

The desired leaf-labeled full binary tree $\tilde{\mathbf{t}}_{n+1}$ has the vertex set $\{\emptyset\} \cup \{0v : v \in \tilde{\mathbf{t}}_{\mathcal{L}}\} \cup \{1v : v \in \tilde{\mathbf{t}}_{\mathcal{R}}\}$, and the addresses of the equivalence classes $\langle g, h\rangle$, $g, h \in [n+1]$, in the tree $\tilde{\mathbf{t}}_{n+1}$ are given by

- $\psi(\langle g, h\rangle) := \emptyset$ for $g \in \mathcal{L}$, $h \in \mathcal{R}$,
- $\psi(\langle g, h\rangle) := 0v$ for $g, h \in \mathcal{L}$, with $v := \chi_{\mathcal{L}}(\langle g, h\rangle)$,
- $\psi(\langle g, h\rangle) := 1v$ for $g, h \in \mathcal{R}$, with $v := \chi_{\mathcal{R}}(\langle g, h\rangle)$.

The partial orders $<_L$ and $<_R$ are then simply read off from the corresponding orders in the leaf-labeled full binary tree $\tilde{\mathbf{t}}_{n+1}$, and clearly also the properties (D) and (E) are inherited.

It remains to deal with Case 2b. This is, however, completely analogous to Case 2a, and we refrain from giving the parallel arguments here. □

We have seen that finite didendritic systems are essentially finite leaf-labeled full binary trees. One way to produce didendritic systems with infinite label sets is via a projective limit construction as detailed in the following two lemmas. We omit the (simple) proofs. The first lemma says that the natural "projection" of a didendritic system to a subset of its label set is again a didendritic system.

Lemma 6.8: *Consider a didendritic system $(\mathcal{N}, \equiv, \langle\cdot,\cdot\rangle, <_L, <_R, <)$. Let $\mathcal{N}'$ be a non-empty subset of $\mathcal{N}$. Define an equivalence relation $\equiv'$ on $\mathcal{N}' \times \mathcal{N}'$ by declaring that $(i,j) \equiv' (k,\ell)$ if and only if $(i,j) \equiv (k,\ell)$. Write $\langle i,j\rangle'$ for the equivalence class of $(i,j) \in \mathcal{N}' \times \mathcal{N}'$. Define a partial order $<'_L$ on the equivalence classes of $\equiv'$ by declaring that $\langle i,j\rangle' <'_L \langle k,\ell\rangle'$ if and only if $\langle i,j\rangle <_L \langle k,\ell\rangle$. Define partial orders $<'_R$ and $<'$ analogously. Then $(\mathcal{N}', \equiv', \langle\cdot,\cdot\rangle', <'_L, <'_R, <')$ is a didendritic system.*

Definition 6.9: Suppose that the didendritic systems $(\mathcal{N}, \equiv, \langle\cdot,\cdot\rangle, <_L, <_R, <)$ and $(\mathcal{N}', \equiv', \langle\cdot,\cdot\rangle', <'_L, <'_R, <')$ are as in Lemma 6.8. We then say that the latter didendritic system is the *restriction* of the former to $\mathcal{N}'$.

The following lemma asserts the existence of a projective limit for a projective family of didendritic systems.

Lemma 6.10: *Suppose that $(\mathcal{N}^n, \equiv^n, \langle\cdot,\cdot\rangle^n, <^n_L, <^n_R, <^n)$, $n \in \mathbb{N}$, are didendritic systems such that for $m < n$ we have $\mathcal{N}^m \subseteq \mathcal{N}^n$ and $(\mathcal{N}^m, \equiv^m, \langle\cdot,\cdot\rangle^m, <^m_L, <^m_R, <^m)$ is the restriction of $(\mathcal{N}^n, \equiv^n, \langle\cdot,\cdot\rangle^n, <^n_L, <^n_R, <^n)$ to $\mathcal{N}^m$. Put $\mathcal{N}^\infty := \bigcup_{n\in\mathbb{N}} \mathcal{N}^n$. Then there is a unique didendritic system $(\mathcal{N}^\infty, \equiv^\infty, \langle\cdot,\cdot\rangle^\infty, <^\infty_L, <^\infty_R, <^\infty)$ such that for each $n \in \mathbb{N}$ the didendritic system $(\mathcal{N}^n, \equiv^n, \langle\cdot,\cdot\rangle^n, <^n_L, <^n_R, <^n)$ is the restriction of $(\mathcal{N}^\infty, \equiv^\infty, \langle\cdot,\cdot\rangle^\infty, <^\infty_L, <^\infty_R, <^\infty)$ to $\mathcal{N}^n$.*

7. Infinite PATRICIA bridges and exchangeable random didendritic systems

It is shown in Section 5 of [7] that if $(\bar{R}_n^\infty)_{n\in\mathbb{N}}$ is an infinite PATRICIA bridge, then there is a Markov chain $(\tilde{R}_n^\infty)_{n\in\mathbb{N}}$ such that for each $n \in \mathbb{N}$ the random element $\tilde{R}_n^\infty$ is a leaf-labeled full binary tree with n leaves labeled by $[n]$ and the following hold.

- The full binary tree obtained by removing the labels of $\tilde{R}_n^\infty$ is $\bar{R}_n^\infty$.
- For every $n \in \mathbb{N}$, the conditional distribution of $\tilde{R}_n^\infty$ given $\bar{R}_n^\infty$ is uniform over the $n!$ possible leaf-labelings of $\bar{R}_n^\infty$ by [n].
- In going backward from time $n+1$ to time n, $\tilde{R}_{n+1}^\infty$ is transformed into $\tilde{R}_n^\infty$ according to the following deterministic procedure:
  - Delete the leaf labeled $n+1$, along with its sibling (which may or may not be a leaf).
  - If the sibling of the leaf labeled $n+1$ is also a leaf, then assign the sibling's label to the common parent (which is now a leaf).
  - If the sibling of the leaf labeled $n+1$ is not a leaf, then attach the subtree below the sibling to the common parent with its leaf labels unchanged and leave all other leaf labels unchanged.

As we saw in Definition 6.1, the leaf-labeled full binary tree $\tilde{R}_n^\infty$ defines a random didendritic system with label set $[n]$ for any $n \in \mathbb{N}$. Moreover, for any $n \in \mathbb{N}$ the didendritic system defined by $\tilde{R}_n^\infty$ is the restriction to $[n]$ of the didendritic system defined by $\tilde{R}_{n+1}^\infty$. It follows from Lemma 6.10 that there is a random didendritic system $(\mathbb{N}, \equiv, \langle\cdot,\cdot\rangle, <_L, <_R, <)$ such that the restriction of this random didendritic system to $[n]$ is the random didendritic system defined by $\tilde{R}_n^\infty$ for all $n \in \mathbb{N}$. Because of Proposition 6.6 we can recover $\tilde{R}_n^\infty$ and hence $\bar{R}_n^\infty$ from this restriction and therefore we can recover $(\tilde{R}_n^\infty)_{n\in\mathbb{N}}$ and $(\bar{R}_n^\infty)_{n\in\mathbb{N}}$ from $(\mathbb{N}, \equiv, \langle\cdot,\cdot\rangle, <_L, <_R, <)$.

The random didendritic system defined by $(\tilde{R}_n^\infty)_{n\in\mathbb{N}}$ is not arbitrary: it inherits distributional symmetries from the uniform labeling in the construction of $(\tilde{R}_n^\infty)_{n\in\mathbb{N}}$ from $(\bar{R}_n^\infty)_{n\in\mathbb{N}}$. We now develop some terminology to describe these symmetries.

Definition 7.1: Given a didendritic system $\mathcal{D} = (\mathbb{N}, \equiv, \langle\cdot,\cdot\rangle, <_L, <_R, <)$ with label set $\mathbb{N}$ and a permutation σ of $\mathbb{N}$ such that $\sigma(i) = i$ for all but finitely many $i \in \mathbb{N}$, the didendritic system $\mathcal{D}^\sigma = (\mathbb{N}, \equiv^\sigma, \langle\cdot,\cdot\rangle^\sigma, <_L^\sigma, <_R^\sigma, <^\sigma)$ is defined by

- $(i',j') \equiv^\sigma (i'',j'')$ if and only if $(\sigma(i'),\sigma(j')) \equiv (\sigma(i''),\sigma(j''))$,

- $\langle i,j\rangle^\sigma$ is the equivalence class of the pair (i,j) for the equivalence relation $\equiv^\sigma$,
- $\langle h,i\rangle^\sigma <_L^\sigma \langle j,k\rangle^\sigma$ if and only if $\langle\sigma(h),\sigma(i)\rangle <_L \langle\sigma(j),\sigma(k)\rangle$,
- $\langle h,i\rangle^\sigma <_R^\sigma \langle j,k\rangle^\sigma$ if and only if $\langle\sigma(h),\sigma(i)\rangle <_R \langle\sigma(j),\sigma(k)\rangle$,
- $\langle h,i\rangle^\sigma <^\sigma \langle j,k\rangle^\sigma$ if and only if $\langle\sigma(h),\sigma(i)\rangle < R\langle\sigma(j),\sigma(k)\rangle$.

A random didendritic system $\mathcal{D} = (\mathbb{N}, \equiv, \langle\cdot,\cdot\rangle, <_L, <_R, <)$ with label set $\mathbb{N}$ is *exchangeable* if for each permutation σ of $\mathbb{N}$ such that $\sigma(i) = i$ for all but finitely many $i \in \mathbb{N}$ the random didendritic system $\mathcal{D}^\sigma$ has the same distribution as $\mathcal{D}$.

The following result is Lemma 5.12 of [7].

Lemma 7.2: *The random didendritic system on $\mathbb{N}$ corresponding to the labeled version of an infinite PATRICIA bridge is exchangeable. Conversely, if we have an exchangeable random didendritic system (on $\mathbb{N}$), apply Lemma 6.8 to restrict it to a sequence of random didendritic systems on $[n]$ for $n \in \mathbb{N}$, and apply Proposition 6.6 to construct a full binary tree that is leaf-labeled by $[n]$, then the resulting sequence of leaf-labeled full binary trees is a PATRICIA bridge.*

Our aim is to find concrete representations of the extremal infinite Rémy bridges (recall that an infinite Rémy bridge is extremal if it has a trivial tail σ-field). To this end, it will be useful to relate the extremality of an infinite Rémy bridge to properties of the associated exchangeable random didendritic system. We say that an exchangeable random didendritic system $\mathcal{D}$ is *ergodic* if

$$\mathbb{P}(\{\mathcal{D} \in A\} \triangle \{\mathcal{D}^\sigma \in A\}) = 0$$

for all permutations σ of $\mathbb{N}$ such that $\sigma(i) = i$ for all but finitely many $i \in \mathbb{N}$ implies that

$$\mathbb{P}\{\mathcal{D} \in A\} \in \{0,1\}.$$

The following result is Proposition 5.19 of [7].

Proposition 7.3: *An infinite Rémy bridge is extremal if and only if the associated exchangeable random didendritic system is ergodic.*

Example 7.4: We return to Example 4.6 and give a concrete representation of the corresponding random didendritic system. Let $U_1, U_2, \ldots$ be independent and uniformly distributed on $[0,1]$, and let $\varepsilon_1, \varepsilon_2, \ldots$ be i.i.d.

with values in $\{\curvearrowright, \curvearrowleft\}$ and with $\mathbb{P}\{\varepsilon_i = \curvearrowright\} = \mathbb{P}\{\varepsilon_i = \curvearrowleft\} = \frac{1}{2}$. We define the equivalence relation $\equiv$ on $\mathbb{N} \times \mathbb{N}$ by declaring that

$$(i, j) \equiv (k, k) \iff i = j = k$$

and if $i \neq j$ and $k \neq \ell$ then

$$(i, j) \equiv (k, \ell) \iff U_i \wedge U_j = U_k \wedge U_\ell.$$

The partial order $<$ is defined by declaring that

- for $i, j, k \in \mathbb{N}$, $\langle i, j \rangle < \langle k, k \rangle \iff i \neq j$ and $U_i \wedge U_j \leq U_k$,
- for $i \neq j$, $k \neq \ell$, $\langle i, j \rangle < \langle k, \ell \rangle \iff i \neq j$ and $U_i \wedge U_j < U_k \wedge U_\ell$.

We say that $\langle i, j \rangle <_L \langle k, \ell \rangle$ (respectively, $\langle i, j \rangle <_R \langle k, \ell \rangle$) if and only if $\langle i, j \rangle < \langle k, \ell \rangle$ and $\varepsilon_h = \curvearrowleft$ (respectively, $\varepsilon_h = \curvearrowright$), where

$$h = \begin{cases} i, & \text{if } U_i < U_j, \\ j, & \text{if } U_j < U_i. \end{cases}$$

One can check that the restriction of this random didendritic system to the label set $[n]$ defines a random full binary tree which has the same distribution as the random full binary tree $\bar{R}_n^\infty$ in 4.6. The labeling of this tree by $[n]$ is clearly uniform, and the result of passing from the restriction to $[n+1]$ to the restriction to $[n]$ is given by the deterministic transformation described at the beginning of this section. Together this shows that the corresponding full binary tree-valued process is distributed as the labeled version of the PATRICIA bridge from Example 4.6.

It is clear by the symmetry inherent in the construction that the above random didendritic system is exchangeable. We already saw in Example 4.6 that the above PATRICIA bridge is extremal, and hence by Proposition 7.3 the exchangeable random didendritic system is ergodic. The ergodicity can also be seen directly from the observation that the restrictions of the random didendritic system to disjoint finite subsets of $\mathbb{N}$ are independent (see Remark 5.18 of [7]).

8. Infinite PATRICIA bridges and real trees

The first part of Theorem 8.2 of [7] constructs an ergodic exchangeable random didendritic system $\mathcal{D}$ (and hence an extremal infinite PATRICIA bridge) from an $\mathbb{R}$-tree $\mathbf{S}$ with root ρ, a probability measure μ on $\mathbf{S}$, and a function $W : \mathbf{S} \times [0, 1] \times \mathbf{S} \times [0, 1] \to \{\curvearrowright, \curvearrowleft\}$. For the purposes of this construction, we first fix some notation. For $x, y \in \mathbf{S}$ let $x \curlywedge y$ denote that element in the segment $[x, y]$ that is closest to the root ρ of $\mathbf{S}$ (equivalently, $[\rho, x \curlywedge y] = [\rho, x] \cap [\rho, y]$). We say that $x \prec y$ if $x \in [\rho, y)$.

The following properties of $\mathbf{S}$, ρ, μ, and W are essential:

(T) Let $(\xi_n)_{n \in \mathbb{N}}$ be i.i.d. with common distribution μ. Then almost surely for distinct $i, j, k \in \mathbb{N}$, one of

$$\xi_i \curlywedge \xi_j = \xi_i \curlywedge \xi_k \prec \xi_j \curlywedge \xi_k,$$

$$\xi_j \curlywedge \xi_k = \xi_j \curlywedge \xi_i \prec \xi_k \curlywedge \xi_i,$$

or

$$\xi_k \curlywedge \xi_i = \xi_k \curlywedge \xi_j \prec \xi_i \curlywedge \xi_j$$

holds.

(LR) For an independent sequence of i.i.d. $[0, 1]$-valued random variables $(\vartheta_n)_{n \in \mathbb{N}}$ with common uniform distribution one has almost surely

- *for $i \neq j$, $W(\xi_i, \vartheta_i, \xi_j, \vartheta_j) = \curvearrowright$ if and only if $W(\xi_j, \vartheta_j, \xi_i, \vartheta_i) = \curvearrowleft$;*
- *for distinct i, j, k, if $\xi_i \curlywedge \xi_j = \xi_i \curlywedge \xi_k \prec \xi_j \curlywedge \xi_k$, then $W(\xi_i, \vartheta_i, \xi_j, \vartheta_j) = W(\xi_i, \vartheta_i, \xi_k, \vartheta_k)$.*

Assume (T) and (LR) hold. Let (ξ_i, ϑ_i) be i.i.d. copies of a random variable with distribution $\mu \otimes \lambda$, where λ is the uniform distribution on $[0, 1]$. Using the random input $(\xi_i, \vartheta_i)_{i \in \mathbb{N}}$, we define

- the equivalence relation $\equiv^{\mathbf{S}}$ on $\mathbb{N} \times \mathbb{N}$ by declaring, for $i, k, \ell \in \mathbb{N}$ that $(i, i) \equiv^{\mathbf{S}} (k, \ell)$ if and only if $i = k = \ell$, and for $i \neq j$, $k \neq \ell$, that $(i, j) \equiv^{\mathbf{S}} (k, \ell)$ if and only if $\xi_i \curlywedge \xi_j = \xi_k \curlywedge \xi_\ell$,
- the partial order $<^{\mathbf{S}}$ on the equivalence classes $\langle \cdot, \cdot \rangle^{\mathbf{S}}$ of $\equiv^{\mathbf{S}}$ by declaring, for $i, k, \ell \in \mathbb{N}$ that $\langle i, j \rangle^{\mathbf{S}} <^{\mathbf{S}} \langle k, k \rangle^{\mathbf{S}}$ if and only if $i \neq j$ and $\xi \curlywedge \xi_j \preccurlyeq \xi_k$, and for $i \neq j$, $k \neq \ell$, that $\langle i, j \rangle <^{\mathbf{S}} \langle k, \ell \rangle$ if and only if $\xi_i \curlywedge \xi_j \prec \xi_k \curlywedge \xi_\ell$,
- the mappings $w : \{(i, j) : i, j \in \mathbb{N}, i \neq j\} \to \{\curvearrowright, \curvearrowleft\}$ by putting $w(i, j) := W(\xi_i, \vartheta_i, \xi_j, \vartheta_j)$.

Proposition 6.7 then extends $(\mathbb{N}, \equiv^{\mathbf{S}}, <^{\mathbf{S}}, w)$ into a random didendritic system $\mathcal{D}^{\mathbf{S}} = (\mathbb{N}, \equiv^{\mathbf{S}}, \langle \cdot, \cdot \rangle^{\mathbf{S}}, <_L{}^{\mathbf{S}}, <_R{}^{\mathbf{S}}, <^{\mathbf{S}})$. The exchangeability of $\mathcal{D}^{\mathbf{S}}$ is clear because the random variables (ξ_i, ϑ_i), $i \in \mathbb{N}$, are independent and identically distributed. The ergodicity of $\mathcal{D}$ holds because the restrictions of the random didendritic system to disjoint finite subsets of $\mathbb{N}$ are independent (see Remark 5.18 of [7]).

We can now formulate the assertion of Theorem 8.2 of [7] in a still more explicit and "constructive" manner.

Theorem 8.1: *Let* $\mathbf{S}$ *be a complete separable* $\mathbb{R}$*-tree with root* ρ, μ *be a probability measure on* $\mathbf{S}$, *and* W *be a Borel measurable function from* $\mathbf{S} \times [0,1] \times \mathbf{S} \times [0,1]$ *to* $\{\curvearrowright, \curvearrowleft\}$. *Suppose that the properties (T) and (LR) hold. Then* $\mathcal{D}^{\mathbf{S}} = (\mathbb{N}, \equiv^{\mathbf{S}}, \langle \cdot, \cdot \rangle^{\mathbf{S}}, <_L{}^{\mathbf{S}}, <_R{}^{\mathbf{S}}, <^{\mathbf{S}})$ *is an ergodic exchangeable random didendritic system. Conversely, for any ergodic exchangeable random didendritic system* $\mathcal{D}$ *there exists a 4-tuple* $(\mathbf{S}, \rho, \mu, W)$ *with the abovementioned properties such that* $\mathcal{D}^{\mathbf{S}}$ *has the same distribution as* $\mathcal{D}$.

In short, the construction described at the beginning of the section builds an ergodic exchangeable random didendritic system (and hence an extremal infinite PATRICIA bridge) from a rooted $\mathbb{R}$-tree endowed with a sampling measure μ and a "left-right prescription" W that obey the consistency properties (T) and (LR). Conversely, any ergodic exchangeable random didendritic system (and hence any extremal infinite PATRICIA bridge) arises from such a construction.

The first part of Theorem 8.1 and its proof has already been explained. We now briefly review the proof of the second part. The arguments in Section 6 of [7] construct from a given ergodic exchangeable random didendritic system $\mathcal{D} = (\mathbb{N}, \equiv, \langle \cdot, \cdot \rangle, <_L, <_R, <)$ a complete separable (ultrametric) $\mathbb{R}$-tree $\mathbf{T}$ with a distinguished point $\rho \in \mathbf{T}$, and with an injective mapping from the set of equivalence classes $\{\langle i, j \rangle : i, j \in \mathbb{N}\}$ into $\mathbf{T}$ such that the partial order on $\mathbf{T}$ defined by the root ρ extends the partial order $<$ on $\{\langle i, j \rangle : i, j \in \mathbb{N}\}$. The *core* of $\mathbf{T}$, denoted by $\Gamma(\mathbf{T})$, is the closure of the set of points of attachment the leaves of $\mathbf{T}$. Here, $\Pi(x)$, the point of attachment of a leaf x, equals x if the leaf is not isolated, whereas for an isolated leaf it is such that the line segment $[x, \Pi(x))$ is the maximal one among all line segments $[x, y)$ that arise as intersections of $\mathbf{T}$ with open balls centered at x. Now Proposition 7.4 of [7] constructs on the complete separable $\mathbb{R}$-tree $\mathbf{S} := \Gamma(\mathbf{T})$ with root ρ a diffuse probability measure μ having property (T), and such that $(\mathbb{N}, \equiv, \langle \cdot, \cdot \rangle \;\; <)$ has the same distribution as $(\mathbb{N}, \equiv^{\mathbf{S}}, \langle \cdot, \cdot \rangle^{\mathbf{S}}, <^{\mathbf{S}})$.

In Section 8 of [7], using the Aldous-Hoover-Kallenberg theory on exchangeable random arrays, a Borel measurable function $W : \mathbf{S} \times [0,1] \times \mathbf{S} \times [0,1] \to \{\curvearrowright, \curvearrowleft\}$ is constructed which has the above property (LR), and which is such that the resulting ergodic exchangeable random didendritic system $\mathcal{D}$ has the same distribution as the resulting ergodic exchangeable random didendritic system $\mathcal{D}^{\mathbf{S}}$.

Example 8.2: We continue Examples 4.6 and 7.4. Here, the ultrametric $\mathbb{R}$-tree $\mathbf{T}$ may be taken to be the interval $[0, \frac{1}{2}]$, along with disjoint segments of length $\frac{1}{2}(1-U_i)$ attached to the points $\xi_i = \frac{1}{2}U_i$, and with the root $\rho = 0$. This is the same description of $\mathbf{T}$ as in Example 6.7 of [7], except that the roles of 0 and $\frac{1}{2}$ have been interchanged, in order to tie in with the way the order $<$ is constructed in Example 7.4. The core $\mathbf{S} = \Gamma(\mathbf{T})$ is then the interval $[0, \frac{1}{2}]$, and the sampling measure μ is the uniform distribution on this interval. The prescription of "left versus right" is then determined by the function

$$W(x,s,y,t) = \begin{cases} \curvearrowright, & \text{if } x < y \text{ and } s < \frac{1}{2}, \\ \curvearrowleft, & \text{if } x < y \text{ and } s > \frac{1}{2}, \\ \curvearrowright, & \text{if } y < x \text{ and } t < \frac{1}{2}, \\ \curvearrowleft, & \text{if } y < x \text{ and } t > \frac{1}{2}, \\ \curvearrowleft, & \text{otherwise.} \end{cases}$$

Example 8.3: We know from Remark 4.5 that $({}^{\nu}\bar{R}_n)_{n\in\mathbb{N}}$ is an extremal infinite PATRICIA bridge for each diffuse probability measure ν on $\{0,1\}^\infty$. Here the ultrametric $\mathbb{R}$-tree $\mathbf{T}$ may be taken as follows (cf. Examples 6.8 & 8.4 of [7]): Take the complete binary tree $\{0,1\}^*$, join two elements of the form $v_1 \dots v_k$ and $v_1 \dots v_k v_{k+1}$ with a segment of length $\frac{1}{2^{k+2}}$, and let $\mathbf{T}$ be the completion of this $\mathbb{R}$-tree. The root $\rho \in \mathbf{T}$ is the point corresponding to the root $\emptyset \in \{0,1\}^*$. The core $\mathbf{S} = \Gamma(\mathbf{T})$ is just $\mathbf{T}$ itself. There is a bijective correspondence between $\{0,1\}^\infty$ and the points "added" in passing to the completion. The sampling measure μ on $\mathbf{S}$ is identified via this correspondence with the probability measure ν on $\{0,1\}^\infty$. Let x and y be two points in the support of μ that correspond to the points u and v in $\{0,1\}^\infty$. The most recent common ancestor of x and y in $\mathbf{T}$ (that is, the point z such that $[\rho, z] = [\rho, x] \cap [\rho, y]$) is the point in $\mathbf{T}$ corresponding to the most recent common ancestor of u and v in $\{0,1\}^*$. The "left versus right" rule is then given by

$$W(x,s,y,t) = \begin{cases} \curvearrowright, & \text{if } u \wedge v <_L u \text{ and } u \wedge v <_R v, \\ \curvearrowleft, & \text{if } u \wedge v <_L v \text{ and } u \wedge v <_R u. \end{cases}$$

Acknowledgement

We thank the referee of [8] for suggesting that we extend our analysis of the radix sort chain to the PATRICIA chain and the referee of this paper for helpful and stimulating comments. We are grateful to Ralph Neininger for sharing his example in Section 5. We acknowledge the hospitality of the Institute for Mathematical Sciences, National University of Singapore, where part of this paper was written.

References

1. L. Alonso and R. Schott, *Random generation of trees.* Kluwer Academic Publishers, Boston, MA, 1995.
2. A. Andersson, Comments on: "On the balance property of Patricia tries: external path length viewpoint" [Theoret. Comput. Sci. **68** (1989), no. 1, 1–17] by P. Kirschenhofer, H. Prodinger and W. Szpankowski. With a reply by Kirschenhofer, Prodinger and Szpankowski. *Theoret. Comput. Sci.*, **106**(2):391–400, 1992.
3. J. Bourdon, Size and path length of Patricia tries: dynamical sources context. *Random Structures Algorithms*, **19**(3-4):289–315, 2001.
4. L. Devroye, A note on the probabilistic analysis of Patricia trees. *Random Structures Algorithms*, **3**(2):203–214, 1992.
5. L. Devroye, Laws of large numbers and tail inequalities for random tries and PATRICIA trees. *J. Comput. Appl. Math.*, **142**(1):27–37, 2002.
6. L. Devroye, Universal asymptotics for random tries and PATRICIA trees. *Algorithmica*, **42**(1):11–29, 2005.
7. S. N. Evans, R. Grübel, and A. Wakolbinger, Doob-Martin boundary of Rémy's tree growth chain. *Ann. Probab.*, **45**(1):225–277, 2017.
8. S. N. Evans and A. Wakolbinger, Radix sort trees in the large. *Electron. Commun. Probab.*, **22**:Paper No. 68, 13pp, 2017.
9. N. Forman, C. Haulk, and J. Pitman, A representation of exchangeable hierarchies by sampling from random real trees. *Probab. Theory Related Fields*, **172**(1-2):1–29, 2018.
10. M. Fuchs and C.-K. Lee, A general central limit theorem for shape parameters of m-ary tries and PATRICIA tries. *Electron. J. Combin.*, **21**(1):Paper 1.68, 26pp, 2014.
11. J. Gerstenberg, Exchangeable interval hypergraphs and limits of ordered discrete structures, 2018. eprint arXiv:1802.09015 [math.PR], to appear in *Ann. Probab.*
12. G. Gwehenberger, Anwendung einer binären Verweiskettenmethode beim Aufbau von Listen. *Elektronische Rechenanlagen*, **10**:223–226, 1968.
13. P. Kirschenhofer, H. Prodinger, and W. Szpankowski, On the balance property of Patricia tries: external path length viewpoint. *Theoret. Comput. Sci.*, **68**(1):1–17, 1989.

14. C. Knessl and W. Szpankowski, Limit laws for the height in PATRICIA tries. *J. Algorithms*, **44**(1):63–97, 2002.
15. D. E. Knuth, *The art of computer programming. Volume 3.* Addison-Wesley Publishing Co., Reading, Mass.-London-Don Mills, Ont., 1973.
16. J.-F. Le Gall, *Spatial branching processes, random snakes and partial differential equations.* Lectures in Mathematics ETH Zürich. Birkhäuser Verlag, Basel, 1999.
17. K. Leckey, *Probabilistic Analysis of Radix Algorithms on Markov Sources.* PhD thesis, Johann Wolfgang Goethe-Universität in Frankfurt am Main, 2015.
18. A. Magner, C. Knessl, and W. Szpankowski, Expected external profile of PATRICIA tries. In *ANALCO14—Meeting on Analytic Algorithmics and Combinatorics*, pp. 16–24. SIAM, Philadelphia, PA, 2014.
19. A. Magner and W. Szpankowski, Profiles of PATRICIA tries. *Algorithmica*, **80**(1):331–397, 2018.
20. A. N. Magner, *Profiles of Patricia Tries.* Thesis (Ph.D.)–Purdue University. ProQuest LLC, Ann Arbor, MI, 2015.
21. P. Marchal, Constructing a sequence of random walks strongly converging to Brownian motion. In *Discrete random walks (Paris, 2003)*, pp. 181–190. Assoc. Discrete Math. Theor. Comput. Sci., Nancy, 2003.
22. D. R. Morrison, Patricia—Practical Algorithm To Retrieve Information Coded in Alphanumeric. *Journal of the ACM*, **15**:514–534, 1968.
23. B. Rais, P. Jacquet, and W. Szpankowski, Limiting distribution for the depth in PATRICIA tries. *SIAM J. Discrete Math.*, **6**(2):197–213, 1993.
24. J.-L. Rémy, Un procédé itératif de dénombrement d'arbres binaires et son application à leur génération aléatoire. *RAIRO Inform. Théor.*, **19**(2):179–195, 1985.
25. L. C. G. Rogers and J. W. Pitman, Markov functions. *Ann. Probab.*, **9**(4):573–582, 1981.
26. R. P. Stanley, *Catalan numbers.* Cambridge University Press, New York, 2015.
27. W. Szpankowski, Patricia tries again revisited. *J. Assoc. Comput. Mach.*, **37**(4):691–711, 1990.
28. W. Tong, R. Goebel, and G. Lin, On the smoothed heights of Trie and Patricia index trees. In Z. Cai, A. Zelikovsky and A. Bourgeois (eds), *Computing and combinatorics.* COCOON 2014. *Lecture Notes in Comput. Sci.*, vol. 8591, pp. 94–103. Springer, Cham, 2014.
29. W. Tong, R. Goebel, and G. Lin, Smoothed heights of tries and patricia tries. *Theoret. Comput. Sci.*, **609**(part 3):620–626, 2016.

Collision Times of Random Walks and Applications to the Brownian Web

David Coupier*, Kumarjit Saha†, Anish Sarkar‡ and Viet Chi Tran§

*Univ. de Valenciennes, CNRS, EA 4015 - LAMAV
F-59313 Valenciennes Cedex 9, France
david.coupier@univ-valenciennes.fr

†Ashoka University
Rai - 131029, Haryana, India
kumarjitsaha@gmail.com

‡Indian Statistical Institute
7 S. J. S. Sansanwal Marg, New Delhi - 110016, India
anish@isid.ac.in

§LAMA, Univ. Gustave Eiffel, UPEM
Univ. Paris Est Creteil, CNRS
F-77447, Marne-la-Vallée, France
chi.tran@u-pem.fr

Convergence of directed forests, spanning on random subsets of lattices or on point processes, towards the Brownian web has made the subject of an abundant literature, a large part of which relies on a criterion proposed by Fontes, Isopi, Newman and Ravishankar (2004). One of their convergence condition, called (B2), states that the probability of the event that there exist three distinct paths for a time interval of length t (> 0), all starting within a segment of length ε, is of small order of ε. This condition is often verified by applying an FKG type correlation inequality together with a coalescing time tail estimate for two paths. For many models where paths have complex interactions, it is hard to establish FKG type inequalities. In this article, we show that for a non-crossing path model, with certain assumptions, a suitable upper bound on expected first collision time among three paths can be obtained directly using Lyapunov functions. This, in turn, provides an alternate verification of Condition (B2). We further show that in case of independent simple symmetric one-dimensional random walks or in case of independent Brownian motions, the expected value can be computed explicitly. We apply this alternate method of verification of (B2) to several models in the basin of attraction of the Brownian web studied earlier in the literature ([22], [14], [11]).

Contents

1. Introduction

The Brownian web can be described as a collection of coalescing one-dimensional Brownian motions starting from everywhere on a space-time plane $\mathbb{R}^2$. The origin of this random object goes back to the work of Arratia (see [1]), where he studied the diffusive scaling limit of coalescing simple symmetric random walk paths starting from every point of $2\mathbb{Z}$ at time 0 and showed that it converges to a collection of coalescing Brownian motions starting from every point of $\mathbb{R}$ at time 0. The construction of a system of coalescing Brownian motions starting from every point in space and time was first given by Tóth and Werner in [26] where they used it to construct what they called "the true self-repelling motions". In [10], Fontes et al. finally introduced a suitable topology so that the Brownian web can be viewed as a random variable taking values in a Polish space. In particular, they developed a criterion for convergence of rescaled forests to the Brownian web, that we will recall later (Theorem 2.2).

The convergence conditions include two criteria about the number of distinct paths for a time interval of length t (> 0). In particular, their condition (B2) states that the probability of the event that there exists three distinct paths for a time interval of length t (> 0), all starting within a segment of length ε, is of small order of ε. For non-crossing path models, this condition is often verified by applying FKG type correlation inequality together with a bound on the distribution of the coalescence time between *two paths*. However, FKG is a strong property that is difficult to verify (even to expect) for many models, especially for models where paths have complex interactions.

The purpose of this article is to propose another method of verification of (B2) which is applicable for a non-crossing path models with certain assumptions. We show that under these assumptions, it is enough to have suitable estimate on the tail probability of the *first collision time among three paths* (see Proposition 3.2). Using homogeneity and Markovian properties of the model and using a Lyapunov function type technique, we prove a bound on the expected first collision time among three paths (Theorem 3.1). This allows us to use simple Markov inequality to obtain the tail estimate and verify condition (B2) without invoking FKG type inequalities.

Further, we show that part of this method, can be extended to explicitly calculate the expected first collision time among three paths in case of independent simple symmetric random walks and in case of independent Brownian motions (Theorem 3.3). Although, these results seem to be known, the only reference that we found was [24] [Lemma 12]. In Section 4 we apply our method to verify (B2) for two discrete drainage network models studied earlier in the literature: Scheidegger model [22] and Howard model [14]. Our domination condition (Condition (iii) of Proposition 2.3) is difficult to verify for forests in continuum spaces; however, it might be still possible to apply the core ideas of our method. As an example, the Poisson tree of Ferrari et al. [11, 12] is treated in Section 4.3.

It should be mentioned here that there are other approaches for studying convergence to the Brownian web which do not require Condition (B2) to check. Schertzer et al. provided such an approach applicable for non-crossing path model (Theorem 6.6 of [23]). Slightly stronger version of this approach appear in Theorem 26 of [7]. Both these approaches applicable for non-crossing path models and do not require Markovian properties unless a suitable bound on the distribution of the coalescing time between two paths is available. Long before this, Newman et al. developed an alternate condition replacing (B2) (Theorem 1.4 of [17]), which is applicable for crossing path models also.

2. Convergence to the Brownian web

2.1. *The Brownian web*

We describe the Brownian web and the topology introduced in Fontes et al. [10], that we are going to work with. This subsection ends by recalling the convergence criterion proposed by these authors.

Let $\mathbb{R}_c^2$ denote the completion of the space-time plane $\mathbb{R}^2$ with respect to the metric

$$\rho((x_1,t_1),(x_2,t_2)) := |\tanh(t_1)-\tanh(t_2)| \vee \left|\frac{\tanh(x_1)}{1+|t_1|}-\frac{\tanh(x_2)}{1+|t_2|}\right|.$$

As a topological space $\mathbb{R}_c^2$ can be identified with the continuous image of $[-\infty,\infty]^2$ under a map that identifies the line $[-\infty,\infty]\times\{\infty\}$ with the point $(*,\infty)$, and the line $[-\infty,\infty]\times\{-\infty\}$ with the point $(*,-\infty)$. A path π in $\mathbb{R}_c^2$ with starting time $\sigma_\pi\in[-\infty,\infty]$ is a mapping $\pi:[\sigma_\pi,\infty]\to[-\infty,\infty]\cup\{*\}$ such that $\pi(\infty)=*$ and, when $\sigma_\pi=-\infty$, $\pi(-\infty)=*$. Also $t\mapsto(\pi(t),t)$ is a continuous map from $[\sigma_\pi,\infty]$ to $(\mathbb{R}_c^2,\rho)$. We then define Π to be the space of all paths in $\mathbb{R}_c^2$ with all possible starting times in $[-\infty,\infty]$.

The following metric, for $\pi_1,\pi_2\in\Pi$

$$d_\Pi(\pi_1,\pi_2) := \max\Big\{|\tanh(\sigma_{\pi_1})-\tanh(\sigma_{\pi_2})|, \\ \sup_{t\geq\sigma_{\pi_1}\wedge\sigma_{\pi_2}}\left|\frac{\tanh(\pi_1(t\vee\sigma_{\pi_1}))}{1+|t|}-\frac{\tanh(\pi_2(t\vee\sigma_{\pi_2}))}{1+|t|}\right|\Big\}$$

makes Π a complete, separable metric space. The metric d_Π is slightly different from the original choice in [10] which is somewhat less natural as explained in [25]. Convergence in this metric can be described as locally uniform convergence of paths as well as convergence of starting times. In what follows, for $\mathbf{x}=(\mathbf{x}(1),\mathbf{x}(2))\in\mathbb{R}^2$ the notation $\pi^{\mathbf{x}}$ is used to denote a path starting from $\mathbf{x}$, i.e., $\pi^{\mathbf{x}}:[\mathbf{x}(2),\infty)\mapsto\mathbb{R}$.

Let $\mathcal{H}$ be the space of compact subsets of (Π,d_Π) equipped with the Hausdorff metric $d_{\mathcal{H}}$ given by,

$$d_{\mathcal{H}}(K_1,K_2) := \sup_{\pi_1\in K_1}\inf_{\pi_2\in K_2} d_\Pi(\pi_1,\pi_2)\vee\sup_{\pi_2\in K_2}\inf_{\pi_1\in K_1} d_\Pi(\pi_1,\pi_2).$$

Since (Π,d_Π) is Polish, $(\mathcal{H},d_{\mathcal{H}})$ is also Polish. Let $B_{\mathcal{H}}$ be the Borel $\sigma-$algebra on the metric space $(\mathcal{H},d_{\mathcal{H}})$. In the following, for any $\mathbf{x}^1,\dots,\mathbf{x}^k\in\mathbb{R}^2$, the notation $(W^{\mathbf{x}^1},\dots,W^{\mathbf{x}^k})$ represents coalescing Brownian motions starting from $\mathbf{x}^1,\dots,\mathbf{x}^k$ respectively. The Brownian web $\mathcal{W}$ is characterized as (Theorem 2.1 of [10]):

Theorem 2.1: *There exists an $(\mathcal{H}, \mathcal{B}_{\mathcal{H}})$ valued random variable $\mathcal{W}$ such that whose distribution is uniquely determined by the following properties:*

(a) *for each deterministic point $\mathbf{x} \in \mathbb{R}^2$ there is a unique path $\pi^{\mathbf{x}} \in \mathcal{W}$ almost surely;*
(b) *for a finite set of deterministic points $\mathbf{x}^1, \ldots, \mathbf{x}^k \in \mathbb{R}^2$, the collection $(\pi^{\mathbf{x}^1}, \ldots, \pi^{\mathbf{x}^k})$ is distributed as $(W^{\mathbf{x}^1}, \ldots, W^{\mathbf{x}^k})$;*
(c) *for any countable deterministic dense set $\mathcal{D} \subset \mathbb{R}^2$, $\mathcal{W}$ is the closure of $\{\pi^{\mathbf{x}} : \mathbf{x} \in \mathcal{D}\}$ in (Π, d_Π) almost surely.*

Note that the above theorem shows how to work with uncountably many Brownian motions as the collection is almost surely determined by only countably many coalescing Brownian motions.

2.2. *Convergence criterion of Fontes et al.*

Since the random variable $\mathcal{W}$ takes values in a Polish space, we can talk about weak convergence. Fontes et al. [10] provided the criteria to study weak convergence to the Brownian web $\mathcal{W}$. We present their convergence criteria for models with non-crossing paths only. A subset Γ of Π is said to be consists of non-crossing paths if there exist no $\pi_1, \pi_2 \in \Gamma$ such that

$$(\pi_1(t) - \pi_2(t))(\pi_1(s) - \pi_2(s)) < 0 \text{ for some } t, s \geq \max\{\sigma_{\pi_1}, \sigma_{\pi_2}\}. \quad (2.1)$$

For a subset $\Gamma \subseteq \Pi$ of paths and for $t \in \mathbb{R}$ let $\Gamma^t := \{\pi \in \Gamma : \sigma_\pi \leq t\}$ denote the set of paths which start 'before' time t. For $t > 0$ and $t_0, a, b \in \mathbb{R}$ with $a < b$, we define a counting random variable as follows

$$\eta_\Gamma(t_0, t; a, b) := \#\{\pi(t_0 + t) : \pi \in \Gamma^{t_0} \text{ and } \pi(t_0) \in [a, b]\}. \quad (2.2)$$

The following theorem of [10] provides convergence criteria to the Brownian web for non-crossing path models.

Theorem 2.2: *Let $\mathcal{X}_n$, $n \in \mathbb{N}$ be $(\mathcal{H}, B_{\mathcal{H}})$ valued random variables with non-crossing paths. Assume that the following conditions hold:*

(I1) *Let $\mathcal{D}$ be a deterministic countable dense set of $\mathbb{R}^2$. For each $\mathbf{x} \in \mathcal{D}$, there exists $\pi_n^{\mathbf{x}} \in \mathcal{X}_n$ such that for any finite set of points $\mathbf{x}^1, \ldots, \mathbf{x}^k \in \mathcal{D}$, as $n \to \infty$, we have $(\pi_n^{\mathbf{x}^1}, \ldots, \pi_n^{\mathbf{x}^k})$ converges in distribution to $(W^{\mathbf{x}^1}, \ldots, W^{\mathbf{x}^k})$.*
(B1) *For all $t > 0$ and $a, t_0 \in \mathbb{R}$,*

$$\limsup_{n\to\infty} \sup_{(a,t_0)\in\mathbb{R}^2} \mathbb{P}(\eta_{\mathcal{X}_n}(t_0, t; a, a + \epsilon) \geq 2) \to 0 \text{ as } \epsilon \downarrow 0.$$

(B2) *For all* $t > 0$ *and* $a, t_0 \in \mathbb{R}$,

$$\frac{1}{\epsilon} \limsup_{n\to\infty} \sup_{(a,t_0)\in\mathbb{R}^2} \mathbb{P}(\eta_{\mathcal{X}_n}(t_0, t; a, a+\epsilon) \geq 3) \to 0 \text{ as } \epsilon \downarrow 0.$$

Then $\mathcal{X}_n$ *converges in distribution to the standard Brownian web* $\mathcal{W}$ *as* $n \to \infty$.

Fontes et al. [10] used this convergence criteria to prove that the collection of diffusively scaled coalescing one-dimensional simple symmetric random walk paths converges to the Brownian web. This model is described in more detail in Subsection 4.1. Since then, the convergence of coalescing path families (crossing as well as non-crossing) to the Brownian web has made the subject of an abundant literature ([3, 5, 6, 7, 12, 10, 17, 20, 21, 27]). We refer to [23] for a review.

2.3. *Main result: An alternative verification of* (B2)

As mentioned in the introduction, for non-crossing path models in order to verify (B2) it is enough to have sufficient tail decay for the first collision time among three paths and often it is verified using an FKG type correlation inequality. However, FKG property is difficult to verify for many models where paths have complex interactions, e.g., Howard's model [14], discrete directed spanning forest [20], directed spanning forest [2, 8] etc. Among these, for Howard's model, only a partial version of the FKG property is proved till now [5]. The main contribution of this paper is that, we propose an alternate method for verification of (B2) based on a bound for expected first collision time and show that this method is applicable for many non-crossing path models with certain Markovian properties and homogeneity assumptions (see Proposition 3.2 for detail).

To fix ideas let us introduce some notations. Suppose $\mathcal{V}$ is a locally finite random subset of $\mathbb{R}^2$ such that $\mathcal{V} \cap \{(x,s) : x \in \mathbb{R}, s \geq t\}$ is nonempty for all $t \in \mathbb{R}$. Let $h : \mathbb{R}^2 \to \mathcal{V}$ is a random map such that for each $(x,t) \in \mathbb{R}^2$, almost surely we have $h(x,t)(2) > t$ where $h(x,t)(i)$ denotes the i-th coordinate of $h(x,t)$ for $i = 1, 2$. The vertex $h(x,t)$ is the ancestor of (x,t). We can define recursively the next ancestors, for $k \geq 1$, by $h^k(x,t) := h(h^{k-1}(x,t))$, with $h^0(x,t) = (x,t)$. For each $(x,t) \in \mathbb{R}^2$, the path $\pi^{(x,t)} : [t,\infty) \to \mathbb{R}$ starting from (x,t) is obtained by joining the successive vertices $h^{k-1}(x,t), h^k(x,t),\ k \geq 1$ linearly. It is useful to observe that both $h(x,t)$ and $\pi^{(x,t)}$ are defined for any $(x,t) \in \mathbb{R}^2$.

The collection of all the paths starting from the vertices in $\mathcal{V}$ is given as $\mathcal{X} := \{\pi^{(x,t)} : (x,t) \in \mathcal{V}\}$. For a path $\pi \in \Pi$, for $n \geq 1$ and for normalization constants $\gamma, \sigma > 0$, the n-th order diffusively scaled path is given by $\{\pi_n(t) = \pi_n(\gamma,\sigma)(t) = \frac{1}{\sqrt{n}\sigma}\pi(n\gamma t)\}$. Corresponding collection of scaled paths is denoted by $\mathcal{X}_n := \{\pi_n^{(x,t)} : (x,t) \in \mathcal{V}\}$. Let $\overline{\mathcal{X}}_n$ denote the closure of $\mathcal{X}_n$ in (Π, d_Π). We assume that for each $n \geq 1$, $\overline{\mathcal{X}}_n$ is a $(\mathcal{H}, B_{\mathcal{H}})$ valued random variable. Set $-\infty < x < y < z < \infty$ and consider the three paths $\pi^{(x,0)}, \pi^{(y,0)}, \pi^{(z,0)}$ starting from the points $(x,0), (y,0)$ and $(z,0)$ respectively. The *first collision time* of these three paths is defined as

$$T_{(x,y,z)} := \inf\{t \geq 0 : \text{ either } \pi^{(y,0)}(t) = \pi^{(x,0)}(t) \text{ or } \pi^{(z,0)}(t) = \pi^{(y,0)}(t)\}. \tag{2.3}$$

The next proposition is motivated from the argument of Fontes et al. [12] and uses Markov inequality in place of FKG inequality.

Proposition 2.3: *Suppose $(\mathcal{V}, h)$ satisfies the following conditions:*

(i) [Non-crossing] For any (x_1,t_1) and (x_2,t_2) in $\mathbb{R}^2$, the paths $\pi^{(x_1,t_1)}, \pi^{(x_2,t_2)}$ are non-crossing.

(ii) [Homogeneity] For every $a, t_0 \in \mathbb{R}$ and $t, \epsilon > 0$, the distribution of the counting random variable $\eta_{\mathcal{X}_n}(t_0, t_0+t; a, a+\epsilon)$ does not depend on a, t_0.

(iii) For each $t > 0$, we have

$$\frac{1}{\epsilon}\limsup_{n\to\infty}\mathbb{P}\Big(\{\pi(nt) : \pi \in \mathcal{X}, \sigma_\pi \leq 0, \pi(0) \in [0, \epsilon\sqrt{n}]\} \not\subseteq \{\pi^{(j,0)}(nt) : j \in [0, \lfloor\epsilon\sqrt{n}\rfloor + 1] \cap \mathbb{Z}\}\Big) \to 0$$

as $\epsilon \downarrow 0$.

(iv) For $x, y, z \in \mathbb{R}$ with $x < y < z$ and constants $C_1, C_2 > 0$ (independent of x, y and z), we have

$$\mathbb{E}\big(T_{(x,y,z)}\big) \leq C_1 + C_2(y-x)(z-y). \tag{2.4}$$

Then $\{\overline{\mathcal{X}}_n : n \geq 1\}$ satisfies (B2).

Proof: We prove the result for $\sigma = \gamma = 1$, the general case being similar. From Condition (ii) [homogeneity], we observe that, for any $a, t_0 \in \mathbb{R}$ and $t, \epsilon > 0$,

$$\mathbb{P}(\eta_{\mathcal{X}_n}(t_0, t_0+t; a, a+\epsilon) \geq 3) = \mathbb{P}(\eta_{\mathcal{X}_n}(0, t; 0, \epsilon) \geq 3).$$

Hence, it suffices to show that for any $t > 0$,

$$\frac{1}{\epsilon}\limsup_{n\to\infty}\mathbb{P}(\eta_{\mathcal{X}_n}(0,t;0,\epsilon)\geq 3)=\frac{1}{\epsilon}\limsup_{n\to\infty}\mathbb{P}(\eta_{\mathcal{X}}(0,nt;0,\epsilon\sqrt{n})\geq 3)\to 0, \tag{2.5}$$

as $\epsilon\downarrow 0$. Equation (2.5) means, for $\mathcal{X}$, that there exist three distinct paths starting between 0 and $\epsilon\sqrt{n}$ that do not coalesce before nt. We prove (2.5) for $t = 1$. The argument for general $t > 0$ is similar.

Let us define the set of paths starting from vertices $\{(j,0) : 0 \leq j \leq \lfloor\epsilon\sqrt{n}\rfloor + 1\}$ by $\mathcal{P}$. Let $B_n(\epsilon)$ denote the event $\{\{\pi(n) : \pi \in \mathcal{X}_1, \sigma_\pi \leq 0, \pi(0) \in [0, \epsilon\sqrt{n}]\} \nsubseteq \{\pi(n) : \pi \in \mathcal{P}\}\}$. On the complement of the event $B_n(\epsilon)$, there must be at least 3 distinct paths in $\mathcal{P}$ till time n, the event which we denote by $A_{\mathcal{P}}(3;n)$. Therefore, we have,

$$\{\eta_{\mathcal{X}_1}(0,n;0,\epsilon\sqrt{n})\geq 3\}\subseteq A_{\mathcal{P}}(3;n)\cup B_n(\epsilon).$$

Next, we estimate $\mathbb{P}(A_{\mathcal{P}}(3;n))$. We follow a modification of the argument of Fontes et al. [12] and use the straightforward Markov inequality. For any $i \in \mathbb{Z}$ and $j \geq 2$, define the event where no coalescence occur before time n between the paths starting from the vertices $(i,0)$, $(i+1,0)$ and $(i+j,0)$, i.e.,

$$E(i,j;n) := \{\pi^{(i,0)}(n) < \pi^{(i+1,0)}(n) < \pi^{(i+j,0)}(n)\}. \tag{2.6}$$

Note that the paths $\pi^{(i,0)}$, $\pi^{(i+1,0)}$ and $\pi^{(i+j,0)}$ are in $\mathcal{P}$ but need not be in $\mathcal{X}$. From non-crossing nature of paths (Condition (i)) it follows, if $A_{\mathcal{P}}(3;n)$ occurs then at least one of $E(i,\lfloor\epsilon\sqrt{n}\rfloor + 1 - i;n)$ must occur for some $i = 0,1,\ldots,\lfloor\epsilon\sqrt{n}\rfloor - 1$. Hence we have:

$$A_{\mathcal{P}}(3;n)\subseteq\cup_{i=0}^{\lfloor\epsilon\sqrt{n}\rfloor-1}E(i,\lfloor\epsilon\sqrt{n}\rfloor+1-i;n). \tag{2.7}$$

Using Condition (iv) and the Markov inequality,

$$\begin{aligned}&\mathbb{P}(E(i,\lfloor\epsilon\sqrt{n}\rfloor+1-i;n))=\mathbb{P}(T_{(i,i+1,\lfloor\epsilon\sqrt{n}\rfloor+1)}>n)\\&\leq\frac{1}{n}\mathbb{E}\big(T_{(i,i+1,\lfloor\epsilon\sqrt{n}\rfloor+1)}\big)\leq\frac{1}{n}\big[C_1+C_2\big(\lfloor\epsilon\sqrt{n}\rfloor-i\big)\big]\\&\leq\frac{1}{n}\big[C_1+C_2\epsilon\sqrt{n}\big].\end{aligned}$$

Thus, from (2.7),

$$\begin{aligned}
&\limsup_{n\to\infty} \mathbb{P}(\eta_{\mathcal{X}_n}(0,1;0,\epsilon) \geq 3) \\
&\leq \limsup_{n\to\infty} \Big[\frac{1}{n} \sum_{i=0}^{\lfloor \epsilon\sqrt{n}\rfloor - 1} [C_1 + C_2\epsilon\sqrt{n}] + \mathbb{P}(B_n(\epsilon))\Big] \\
&\leq \limsup_{n\to\infty} \frac{1}{n}\left[C_1\epsilon\sqrt{n} + C_2\left(\epsilon\sqrt{n}\right)^2\right] + \limsup_{n\to\infty} \mathbb{P}(B_n(\epsilon)) \\
&\leq C_2\epsilon^2 + \limsup_{n\to\infty} \mathbb{P}(B_n(\epsilon)).
\end{aligned}$$

Hence, using condition (iii), $\frac{1}{\epsilon}\limsup_{n\to\infty} \mathbb{P}(\eta_{\mathcal{X}_n}(0,1;0,\epsilon) \geq 3) \to 0$ as $\epsilon \downarrow 0$. □

3. First collision time of three paths

Conditions (i), (ii) and (iii) of Propositions 2.3 are usually easy to read from the model. We focus in this section on Condition (iv). We first show that when the paths of $\mathcal{X}_n$ are non-crossing and exhibit certain Markov properties, then (2.4) can be obtained by Lyapunov techniques. This is done in Section 3.1. In case of coalescing simple symmetric one-dimensional random walks or in case of Brownian motions, we calculate the expectation of the first collision time explicitly. This is presented in Sections 3.2 and 3.3.

3.1. *Control of the collision time expectation*

We start with a preliminary result on entrance time of a Markov chain with countable state space.

3.1.1. *An entrance time result for a Markov chain*

Let $\{Y_j : j \geq 0\}$ be such a Markov chain with countable state space $\mathcal{M}$. For any subset $M \subseteq \mathcal{M}$, let $\tau(M) := \inf\{n \geq 1 : Y_n \in M\}$ be the first entrance time to M.

Theorem 3.1: *Suppose there exist a function $V : \mathcal{M} \to [0,\infty)$, $b \geq 0$, $p_0 > 0$ and two disjoint subsets M_0, M_1 of $\mathcal{M}$ such that*

(i) $\mathbb{E}[V(Y_1) - V(Y_0)|Y_0 = x] \leq -1 + b\mathbb{I}\{x \in M_1\}$ *for all* $x \in M_0^c$;
(ii) $\mathbb{P}(Y_1 \in M_0|Y_0 = x) \geq p_0$ *for all* $x \in M_1$

where $\mathbb{I}(A)$ denote the indicator of the set A. Then for $x \in M_0^c$ we have

$$\mathbb{E}[\tau(M_0)|Y_0 = x] \leq V(x) + b/p_0. \tag{3.1}$$

Note here that $M_1 \cup M_0$ need not be finite. In our applications, we would have $V(x) = 0$ for $x \in M_0$ and the class of states in M_0 is closed, i.e., if we enter this class, we would not get out of it.

Let us briefly explain the intuition behind this theorem. Condition (i) is essentially a Lyapunov type condition, where we have a strictly positive drift towards $M_1 \cup M_0$, which by standard arguments would imply that the expected entrance time into $M_1 \cup M_0$ for any $x \in (M_1 \cup M_0)^c$ is bounded by $V(x)$. Condition (ii) ensures that each time the chain is in M_1, there is a strictly positive probability (bounded below) to hit M_0 at the next step and the chain should enter the set M_0 after a geometric number of time points. If M_1 is the empty set, the theorem is a known Lyapunov result. When $M_1 \neq \emptyset$, this theorem shows that it is sufficient to focus on the entrance time to $M_0 \cup M_1$.

Proof: [Proof of Theorem 3.1] Let us define $W_0 = 0$ and for $n \geq 1$,

$$W_n = V(Y_n) - V(Y_0) - \sum_{i=1}^{n} \mathbb{E}\big[V(Y_i) - V(Y_{i-1}) \mid Y_{i-1}\big]. \tag{3.2}$$

Letting $\mathcal{F}_n = \sigma(Y_0, Y_1, \ldots, Y_n)$, it is easy to check that $\{W_n : n \geq 0\}$ is $\mathcal{F}_n$-adapted martingale.

Let us set $Y_0 = x \in M_0^c$ and $T_0 := \tau(M_0)$. Clearly T_0 is $\mathcal{F}_n$-measurable stopping time. Therefore, we have that $\{W_{n\wedge T_0} : n \geq 0\}$ is also $\mathcal{F}_n$-adapted martingale. Let $\mathbb{P}_x(\cdot) := \mathbb{P}(\cdot \mid Y_0 = x)$ and $\mathbb{E}_x(\cdot) := \mathbb{E}(\cdot|Y_0 = x)$ denote respectively the law and the expectation conditionally to the initial condition. Hence, we have $\mathbb{E}_x(W_{n\wedge T_0}) = \mathbb{E}_x(W_{0\wedge T_0}) = 0$ for all $n \geq 1$. Thus, using condition (i), we obtain that

$$\begin{aligned}
\mathbb{E}_x\big[V(Y_{n\wedge T_0})\big] &= V(x) + \mathbb{E}_x\Big[\sum_{i=1}^{n\wedge T_0} \mathbb{E}_x\big[V(Y_i) - V(Y_{i-1}) \mid Y_{i-1}\big]\Big] \\
&\leq V(x) + \mathbb{E}_x\Big[\sum_{i=1}^{n\wedge T_0} -1 + b\mathbb{I}\{Y_{i-1} \in M_1\}\Big] \\
&= V(x) - \mathbb{E}_x(n \wedge T_0) + b\mathbb{E}_x\Big[\sum_{i=1}^{n\wedge T_0} \mathbb{I}\{Y_{i-1} \in M_1\}\Big] \qquad (3.3)
\end{aligned}$$

since for any $n \geq 1$ and $i = 0, 1, \ldots, n\wedge T_0 - 1$, $Y_i \in M_0^c$. Hence, transferring $\mathbb{E}_x(n\wedge T_0)$ to the other side and observing that V is a non-negative function, we have

$$\mathbb{E}_x(n\wedge T_0) \le \mathbb{E}_x(n\wedge T_0)+\mathbb{E}_x\big[V(Y_{n\wedge T_0})\big] \le V(x)+b\mathbb{E}_x\Big[\sum_{i=1}^{n\wedge T_0}\mathbb{I}\{Y_{i-1}\in M_1\}\Big].$$

We argue below that $T_0 < \infty$ almost surely and $\sum_{i=1}^{n\wedge T_0}\mathbb{I}\{Y_{i-1}\in M_1\}$ is stochastically dominated by a geometric random variable (the total number of trials before a success) with expectation $1/p_0$. Then letting $n \to +\infty$ in the left hand side of above equation and using the monotone convergence theorem gives the required result.

Let us now prove that $T_0 < \infty$ almost surely. Set $T_1 := \tau(M_0 \cup M_1)$. We consider first the case $x \notin M_0 \cup M_1$. From Theorem 11.3.4 in Meyn and Tweedie [15] and assumption (i), $\mathbb{E}_x(T_1) \le V(x)$ so that the chain will almost surely enter the set $M_0 \cup M_1$ in finite time. When it enters $M_0 \cup M_1$, either it hits the set M_0 and we are done, or it hits the set M_1. In this case, the chain has a strictly positive probability p_0 of hitting the set M_0 at the next step, by assumption (ii). So, in finitely many steps, the chain will either hit M_0, which again finishes the proof, or will go out to $(M_0 \cup M_1)^c$. In the latter case, we are back to the starting situation. So, again it will hit the set $M_0 \cup M_1$ in finite time and so on. Since $p_0 > 0$, the chain can only go out of M_1 to $(M_0 \cup M_1)^c$ finitely many times, before it hits the set M_0. Thus, the chain will hit the set M_0 in finite time almost surely. When the chain starts from $x \in M_1$, the situation is as above without having first to hit the set $M_0 \cup M_1$. So, in all cases, $T_0 < \infty$ almost surely.

For every time $i < n \wedge T_0$ before hitting M_0 (at time T_0) where $Y_i \in M_1$, the probability that $Y_{i+1} \in M_0$ is lower bounded by p_0. Thus, taking the event that the chain hits M_0 at the next time as a success, it is like conducting a geometric trial with success probability at least p_0. Hence the number of times $i < n \wedge T_0$ such that $Y_i \in M_1$ is stochastically dominated by a geometric random variable with expectation $1/p_0$. □

3.1.2. *Application to the expectation of first collision times*

Let us construct a discrete drainage network model where the vertex set $\mathcal{V}$ is a random subset of $\mathbb{Z}^2$ satisfying some homogeneity and Markovian properties (as mentioned in Proposition 3.2) and apply Theorem 3.1 to show that the Condition (iv) of Proposition 2.3 holds. We assume that for any $(x,t) \in \mathbb{Z}^2$, it's ancestor $h(x,t)$ sits on the next level, i.e., $h(x,t)(2) = t+1$ where $h(x,t)(i)$ is the i-th coordinate of $h(x,t)$ for $i = 1,2$. As in Proposition 2.3, $\mathcal{X}$ denotes the collection of all paths starting from $\mathcal{V}$ and

$\mathcal{X}_n$ is the corresponding collection of n-th order diffusively scaled paths. Define $\mathcal{S} = \{(u, v) \in \mathbb{Z}^2 : u, v \geq 0\}$ and $\mathcal{S}_0 = \{(u, v) \in \mathcal{S} : uv = 0\}$. $\mathcal{S}$ (resp. $\mathcal{S}_0$) plays the role of $\mathcal{M}$ (resp. $\mathcal{M}_0$) in Theorem 3.1. We are now ready to state our Proposition regarding alternate verification of (B2).

Proposition 3.2: *Suppose $(\mathcal{V}, h)$ satisfies the following conditions:*

(i) [Markov property] For $x, y, z \in \mathbb{Z}$ with $x < y < z$, the process $\{X_j := (\pi^{(y,0)}(j) - \pi^{(x,0)}(j), \pi^{(z,0)}(j) - \pi^{(y,0)}(j)) : j \geq 0\}$ forms a Markov chain with state space $\mathcal{S}$.

(ii) [Control of the collision time] There exist a subset $\mathcal{S}_1$ of $\mathcal{S} \setminus \mathcal{S}_0$ and $p_0 > 0$, such that

(a) there exist positive constants d_1 and d_2 such that for all $(l, m) \notin \mathcal{S}_0$,

$$\begin{aligned}\mathbb{E}\big(X_1(1)X_1(2) - X_0(1)X_0(2) \mid X_0 = (l, m)\big) \\ \leq -d_1 + d_2 \mathbf{1}_{\{(l,m) \in \mathcal{S}_1\}}\end{aligned} \tag{3.4}$$

where $X_n = (X_n(1), X_n(2))$.

(b) For all $(l, m) \in \mathcal{S}_1$, $\mathbb{P}(X_1 \in \mathcal{S}_0 | X_0 = (l, m)) \geq p_0$.

Then for any $-\infty < x < y < z < \infty$, there exist positive constants C_1, C_2 (independent of x, y, z) such that

$$\mathbb{E}(T_{(x,y,z)}) \leq C_1 + C_2(y - x)(z - y).$$

In addition, if $(\mathcal{V}, h)$ satisfies Condition (ii) of Proposition 2.3, then $\{\overline{\mathcal{X}}_n : n \geq 1\}$ satisfies (B2).

Note that we have considered the state space of the Markov chain as the first quadrant, which implies that the paths are automatically non-crossing. Since the vertex set $\mathcal{V}$ is a random subset of $\mathbb{Z}^2$, and for any $(x, t) \in \mathbb{Z}^2$ we have $h(x, t)(2) = t + 1$ almost surely, hence Condition (iii) of Proposition 2.3 automatically holds.

Proof: Consider the function $V : \mathcal{S} \to [0, \infty)$ defined by

$$V(u, v) = \frac{uv}{d_1}.$$

Part (a) and part (b) of Assumption (ii) of Proposition 3.2 ensure that Conditions (i) and (ii) of Theorem 3.1 hold for the markov chain $\{X_j : j \geq 0\}$.

For any $-\infty < x < y < z < \infty$, first collision time for the paths $\pi^{(x,0)}, \pi^{(y,0)}$ and $\pi^{(z,0)}$ is same as the entrance time to the set $\mathcal{S}_0$ for the

Markov chain $\{X_j : j \geq 0\}$. Thus, the bound for $\mathbb{E}(T_{(x,y,z)})$ follows from Theorem 3.1. Finally the proof follows from Proposition 2.3. □

3.2. *First collision time of three random walks*

The calculations leading to equation (3.3) yields a surprising exact computational result for collision time among three simple symmetric random walks. With $V(\cdot)$ being defined as the product of the distances between the pair of the left and the middle random walk and the pair of middle and right random walks (see below for exact definition), it turns out that the condition (i) in Theorem 3.1 is an equality with M_1 being the empty set. Though the exact computation is not required for the convergence to the Brownian web, we give it here for the sake of completeness.

Let us consider three symmetric nearest neighbour independent random walks, which we denote by $S_n^{(L)}, S_n^{(M)}$ and $S_n^{(R)}$, starting from $-2i, 0$ and $2j$ respectively with $i, j \geq 1$. Given three independent sequences of **i.i.d** increment random variables $\{I_k^{(L)} : k \geq 1\}, \{I_k^{(M)} : k \geq 1\}$ and $\{I_k^{(R)} : k \geq 1\}$ with Rademacher distributions (each of these random variables take values $+1$ and -1 with probability $1/2$), the random walks are represented by

$$S_n^{(L)} = -2i + \sum_{k=1}^{n} I_k^{(L)}, \quad S_n^{(M)} = \sum_{k=1}^{n} I_k^{(M)} \quad \text{and} \quad S_n^{(R)} = 2j + \sum_{k=1}^{n} I_k^{(R)}.$$

Let us define the associated filtration $\{\mathcal{F}_n : n \geq 0\}$ where $\mathcal{F}_n = \sigma(I_k^{(L)}, I_k^{(M)}, I_k^{(R)} : k \leq n)$ is the σ-algebra generated by the increment random variables up to time n.

We define the meeting times of these random walks by

$$\tau_{(L,M)} := \inf\{n \geq 1 : S_n^{(L)} = S_n^{(M)}\} \quad \text{and}$$
$$\tau_{(M,R)} := \inf\{n \geq 1 : S_n^{(M)} = S_n^{(R)}\}.$$

We know that $\tau_{(L,M)}$ and $\tau_{(M,R)}$ are almost surely finite, but that their expectations are infinite. Let us define the first collision time of these random walks by

$$\tau_C := \min\{\tau_{(L,M)}, \tau_{(M,R)}\}. \tag{3.5}$$

Theorem 3.3: *With the notation as above, we have* $\mathbb{E}(\tau_C) = 4ij$.

Although this theorem seems to be known, the only reference that we found was [24]. We give a proof of the theorem for the sake of completeness. Before we proceed further, let us set up the notation. For all $n \geq 0$, we denote the distances between the paths by

$$D_n^{(L,M)} = S_n^{(M)} - S_n^{(L)} \quad \text{and} \quad D_n^{(M,R)} = S_n^{(R)} - S_n^{(M)}.$$

We have $D_0^{(L,M)} = 2i$ and $D_0^{(M,R)} = 2j$. Also set the maximum distance between the pairs as

$$M_n = \max\{D_n^{(L,M)}, D_n^{(M,R)}\}.$$

Remark 3.4: It is obvious that at the first collision time, only one pair of random walks can meet. Assuming Theorem 3.3, we can also compute that the other pair of paths is at an expected distance of $2(i+j)$ at the collision time. Since at τ_C, at least one of $D_{\tau_C}^{(L,M)}$ or $D_{\tau_C}^{(M,R)}$ is zero, we have $M_{\tau_C} = D_{\tau_C}^{(L,M)} + D_{\tau_C}^{(M,R)}$. But we may write

$$D_n^{(L,M)} + D_n^{(M,R)} - 2(i+j) = \sum_{j=1}^{n}(I_j^{(R)} - I_j^{(L)}).$$

Therefore, using Wald's identity, we have

$$\mathbb{E}(M_{\tau_C} - 2(i+j)) = \mathbb{E}(D_{\tau_C}^{(L,M)} + D_{\tau_C}^{(M,R)} - 2(i+j)) = 0.$$

Thus, $\mathbb{E}(M_{\tau_C}) = 2(i+j)$.

The main observation is that we may rewrite τ_C as

$$\tau_C = \inf\{n \geq 1 : D_n^{(L,M)} D_n^{(M,R)} = 0\}. \tag{3.6}$$

Next we have two straightforward propositions which involve the product of the distances.

Proposition 3.5: *The family* $\{D_n^{(L,M)} D_n^{(M,R)} + n : n \geq 0\}$ *is an* $\mathcal{F}_n$*-adapted martingale.*

Proposition 3.6: *The family* $\{D_n^{(L,M)} D_n^{(M,R)} (D_n^{(L,M)} + D_n^{(M,R)}) : n \geq 0\}$ *is an* $\mathcal{F}_n$*-adapted martingale.*

Assuming the Propositions, we can now prove Theorem 3.3.

Proof: [Proof of Theorem 3.3] By Proposition 3.5, we have that, $\{D_{n\wedge\tau_C}^{(L,M)} D_{n\wedge\tau_C}^{(M,R)} + n\wedge\tau_C : n \geq 0\}$ is a martingale and hence for any $n \geq 1$,

$$\begin{aligned} \mathbb{E}(D_{n\wedge\tau_C}^{(L,M)} D_{n\wedge\tau_C}^{(M,R)} + n\wedge\tau_C) &= \mathbb{E}(D_{0\wedge\tau_C}^{(L,M)} D_{0\wedge\tau_C}^{(M,R)} + 0\wedge\tau_C) \\ &= \mathbb{E}(D_0^{(L,M)} D_0^{(M,R)}) = 4ij. \end{aligned} \tag{3.7}$$

This is the version of equation (3.3) in this case. Since $\tau_{(L,M)}, \tau_{(M,R)} < +\infty$ almost surely, we have that $\tau_C < +\infty$ almost surely. By the monotone convergence theorem, we have that, as $n \to \infty$, $\mathbb{E}(n \wedge \tau_C) \to \mathbb{E}(\tau_C)$. Since $\tau_C < +\infty$ almost surely, we have that, almost surely, as $n \to \infty$, $D^{(L,M)}_{n\wedge\tau_C} D^{(M,R)}_{n\wedge\tau_C} \to D^{(L,M)}_{\tau_C} D^{(M,R)}_{\tau_C} = 0$. To complete the proof, we show that, as $n \to \infty$,

$$\mathbb{E}(D^{(L,M)}_{n\wedge\tau_C} D^{(M,R)}_{n\wedge\tau_C}) \to 0. \tag{3.8}$$

From Proposition 3.6, we have that $\{D^{(L,M)}_{n\wedge\tau_C} D^{(M,R)}_{n\wedge\tau_C}(D^{(L,M)}_{n\wedge\tau_C} + D^{(M,R)}_{n\wedge\tau_C}) : n \geq 0\}$ is also a martingale and hence, for any $n \geq 1$,

$$\begin{aligned}
&\mathbb{E}\big[D^{(L,M)}_{n\wedge\tau_C} D^{(M,R)}_{n\wedge\tau_C}(D^{(L,M)}_{n\wedge\tau_C} + D^{(M,R)}_{n\wedge\tau_C})\big] \\
&= \mathbb{E}\big[D^{(L,M)}_{0\wedge\tau_C} D^{(M,R)}_{0\wedge\tau_C}(D^{(L,M)}_{0\wedge\tau_C} + D^{(M,R)}_{0\wedge\tau_C})\big] \\
&= \mathbb{E}\big[D^{(L,M)}_{0} D^{(M,R)}_{0}(D^{(L,M)}_{0} + D^{(M,R)}_{0})\big] = 8ij(i+j).
\end{aligned} \tag{3.9}$$

Therefore, observing that $D^{(L,M)}_{n\wedge\tau_C}$ and $D^{(M,R)}_{n\wedge\tau_C}$ are both non-negative, we have

$$\begin{aligned}
&\sup_{n\geq 1} \mathbb{E}\big[(D^{(L,M)}_{n\wedge\tau_C} D^{(M,R)}_{n\wedge\tau_C})^{3/2}\big] \\
&\leq \tfrac{1}{2} \sup_{n\geq 1} \mathbb{E}\big[D^{(L,M)}_{n\wedge\tau_C} D^{(M,R)}_{n\wedge\tau_C}(D^{(L,M)}_{n\wedge\tau_C} + D^{(M,R)}_{n\wedge\tau_C})\big] = 4ij(i+j)
\end{aligned}$$

and hence $\{D^{(L,M)}_{n\wedge\tau_C} D^{(M,R)}_{n\wedge\tau_C} : n \geq 1\}$ is a sequence of uniformly integrable random variables. We conclude (3.8) using Theorem 16.13. of Billingsley [4]. □

Next we prove the Propositions which are straightforward.

Proof: [Proof of Proposition 3.5] This proposition follows from straightforward calculation:

$$\begin{aligned}
\mathbb{E}\big[(&D^{(L,M)}_{n+1} D^{(M,R)}_{n+1} + n + 1) - (D^{(L,M)}_{n} D^{(M,R)}_{n} + n) \mid \mathcal{F}_n\big] \\
&= D^{(L,M)}_{n}\mathbb{E}\big[(I^{(R)}_{n+1} - I^{(M)}_{n+1}) \mid \mathcal{F}_n\big] + D^{(M,R)}_{n}\mathbb{E}\big[(I^{(M)}_{n+1} - I^{(L)}_{n+1}) \mid \mathcal{F}_n\big] \\
&\quad + \mathbb{E}\big[(I^{(M)}_{n+1} - I^{(L)}_{n+1})(I^{(R)}_{n+1} - I^{(M)}_{n+1}) \mid \mathcal{F}_n\big] + 1 \\
&= \mathbb{E}\big[I^{(M)}_{n+1}I^{(R)}_{n+1} - (I^{(M)}_{n+1})^2 - I^{(L)}_{n+1}I^{(R)}_{n+1} + I^{(L)}_{n+1}I^{(M)}_{n+1})\big] + 1 = 0.
\end{aligned}$$ □

Proof: [Proof of Proposition 3.6] Again straightforward calculations yield the result. We have

$$\begin{aligned}
&D_{n+1}^{(L,M)}D_{n+1}^{(M,R)}(D_{n+1}^{(L,M)}+D_{n+1}^{(M,R)})-D_n^{(L,M)}D_n^{(M,R)}(D_n^{(L,M)}+D_n^{(M,R)})\\
&=D_n^{(L,M)}D_n^{(M,R)}(I_{n+1}^{(R)}-I_{n+1}^{(L)})+D_n^{(L,M)}(D_n^{(L,M)}+D_n^{(M,R)})(I_{n+1}^{(R)}-I_{n+1}^{(M)})\\
&\quad+D_n^{(L,M)}(I_{n+1}^{(R)}-I_{n+1}^{(M)})(I_{n+1}^{(R)}-I_{n+1}^{(L)})\\
&\quad+D_n^{(M,R)}(D_n^{(L,M)}+D_n^{(M,R)})(I_{n+1}^{(M)}-I_{n+1}^{(L)})\\
&\quad+D_n^{(M,R)}(I_{n+1}^{(M)}-I_{n+1}^{(L)})(I_{n+1}^{(R)}-I_{n+1}^{(L)})\\
&\quad+(D_n^{(L,M)}+D_n^{(M,R)})(I_{n+1}^{(M)}-I_{n+1}^{(L)})(I_{n+1}^{(R)}-I_{n+1}^{(M)})\\
&\quad+(I_{n+1}^{(M)}-I_{n+1}^{(L)})(I_{n+1}^{(R)}-I_{n+1}^{(M)})(I_{n+1}^{(R)}-I_{n+1}^{(L)}).
\end{aligned}$$

The conditional expectation with respect to $\mathcal{F}_n$ of the first term, second term and the fourth term is clearly 0. We observe that the conditional expectation with respect to $\mathcal{F}_n$ of the third term is $D_n^{(L,M)}$, the fifth term is $D_n^{(M,R)}$ and the sixth term is $-(D_n^{(L,M)}+D_n^{(M,R)})$. The last term is independent of $\mathcal{F}_n$ and has expectation 0. This proves the result. □

3.3. *First collision time of three Brownian motions*

As one can expect, these results generalize to the Brownian motion set up in a straightforward way. Let us start with three independent Brownian motions $\{(B_t^{(L)}, B_t^{(M)}, B_t^{(R)}) : t \geq 0\}$ started from $-x$, 0 and y respectively, for $x, y > 0$. We also define the associated filtration $\{\mathcal{F}_t : t \geq 0\}$ where $\mathcal{F}_t = \sigma(B_s^{(L)}, B_s^{(M)}, B_s^{(R)} : s \leq t)$.

Again, let us define

$$\tau_C^{(B)} = \inf\{t > 0 : (B_t^{(M)} - B_t^{(L)})(B_t^{(R)} - B_t^{(M)}) = 0\}. \tag{3.10}$$

Then, we have the following result

Theorem 3.7: *With the notation as above, we have* $\mathbb{E}(\tau_C^{(B)}) = xy$.

The proof follows from the propositions below which are exact replica of Propositions 3.5 and 3.6, exactly as in the discrete case.

Proposition 3.8: *We have* $\{(B_t^{(M)} - B_t^{(L)})(B_t^{(R)} - B_t^{(M)}) + t : t \geq 0\}$ *is a* $(\mathcal{F}_t)$*-adapted martingale.*

Proposition 3.9: *We have* $\{(B_t^{(M)} - B_t^{(L)})(B_t^{(R)} - B_t^{(M)})(B_t^{(R)} - B_t^{(L)}) : t \geq 0\}$ *is a* $(\mathcal{F}_t)$*-adapted martingale.*

Remark 3.10: It should be observed that there is nothing surprising in the above propositions as well as Propositions 3.5 and 3.6. For example, we can write

$$\begin{aligned}(B_t^{(M)} - B_t^{(L)})(B_t^{(R)} - B_t^{(M)}) + t \\ = B_t^{(M)} B_t^{(R)} - B_t^{(L)} B_t^{(R)} + B_t^{(L)} B_t^{(M)} - \left[(B_t^{(M)})^2 - t\right].\end{aligned}$$

Since Brownian motions are themselves martingales and since products of independent martingales and sums of martingales are martingales, it is easy to observe that the first three terms are martingales. Since t is the quadratic variation process of the Brownian motion, the last term is also a martingale. With these arguments, other processes can also be shown to be martingales.

4. Applications to drainage network models

We now apply Propositions 2.3 and 3.2 to some directed forests of the literature and provide an alternate verification of (B2) for these models.

4.1. *Scheidegger's model*

We start with the system of coalescing simple symmetric random walks starting from every point on $\mathbb{Z}^2_{\text{even}} := \{(x,t) : x + t \text{ even}\}$.

Let $\{b_{(x,t)} : (x,t) \in \mathbb{Z}^2_{\text{even}}\}$ be an i.i.d. collection such that

$$b_{(x,t)} = \begin{cases} +1 & \text{with probability } 1/2 \\ -1 & \text{with probability } 1/2. \end{cases}$$

Define $h : \mathbb{Z}^2 \mapsto \mathbb{Z}^2_{\text{even}}$ as

$$h(x,t) - \begin{cases} (x + b_{(x,t)}, t+1) & \text{if } (x,t) \in \mathbb{Z}^2_{\text{even}} \\ (x, t+1) & \text{otherwise.} \end{cases}$$

For $k \geq 1$ let $h^k(x,t) := h(h^{k-1}(x,t))$ with $h^0(x,t) = (x,t)$. For $(x,t) \in \mathbb{Z}^2$ let $\pi^{(x,t)} : [t, \infty) \to \mathbb{R}$ be the path starting at (x,t) obtained by joining the successive vertices $h^{k-1}(x,t), h^k(x,t),\ k \geq 1$ by straight line segments. Let $\mathcal{X} := \{\pi^{(x,t)} : (x,t) \in \mathbb{Z}^2_{\text{even}}\}$, be the collection of all paths starting from $\mathbb{Z}^2_{\text{even}}$. For $n \geq 1$ and for normalization constants $\gamma, \sigma > 0$, with slight abuse of notation let $\mathcal{X}_n = \mathcal{X}_n(\gamma, \sigma)$ denote the collection of the scaled paths. For each $n \geq 1$, $\overline{\mathcal{X}}_n$, the closure of $\mathcal{X}_n$ in (Π, d_Π), is a $(\mathcal{H}, B_{\mathcal{H}})$ valued random variable.

Fontes et al. [10] proved that $\overline{\mathcal{X}}_n$ converges in distribution to the Brownian web with $\gamma = \sigma = 1$ (see Theorem 6.1 of Fontes et al. [10]). To verify the Condition (B2), they prove a version of FKG inequality for paths and using it, along with the coalescing time tail estimate for a pair of random walks, to derive suitable bounds. In this case it is not hard to obtain FKG inequalities as the paths are independent till the time of coalescence. On the other hand, we proved in Section 3 that the expected first collision time of three simple symmetric random walks starting at $2i$ and $2j$ is exactly $4ij$, which gives the same order bound (with known constant values) as obtained by Fontes et al. However, as a warmup, we show that Proposition 3.2 is applicable here. Note that the exact value of this expectation is not required for our calculations.

In this case, we have $\mathcal{V} = \mathbb{Z}^2_{\text{even}}$ and $h(\cdot)$ is defined as above. From the construction it is clear that the paths are non-crossing and Condition (i) of Proposition 2.3 is satisfied. Because of the i.i.d. nature of the collection of random variables $\{b_{(x,t)} : (x,t) \in \mathbb{Z}^2_{\text{even}}\}$, the homogeneity condition, i.e., Condition (ii) of Proposition 2.3 is also satisfied.

For any $(x,0) \in \mathbb{Z}^2_{\text{even}}$, $h^{j+1}(x,0)(1) = h^j(x,0)(1) + b_{(h^j(x,0))}$. Set x, y, $z \in 2\mathbb{Z}$ such that $x < y < z$. Then we have $(h^{j+1}(y,0)(1) - h^{j+1}(x,0)(1), h^{j+1}(z,0)(1) - h^{j+1}(y,0)(1)) = (h^j(y,0)(1) - h^j(x,0)(1), h^j(z,0)(1) - h^j(y,0)(1)) + (b_{(h^j(y,0))} - b_{(h^j(x,0))}, b_{(h^j(z,0))} - b_{(h^j(y,0))})$. By the independence of the family of random variables $\{b_{(x,t)} : (x,t) \in \mathbb{Z}^2_{\text{even}}\}$, it is clear that

$$\begin{aligned}\{X_j &= (\pi^{(y,0)}(j) - \pi^{(x,0)}(j), \pi^{(z,0)}(j) - \pi^{(y,0)}(j)) \\ &= (h^j(y,0)(1) - h^j(x,0)(1), h^j(z,0)(1) - h^j(y,0)(1)) : j \geq 0\}\end{aligned}$$

is a Markov chain with state space $\{(x,t) \in \mathbb{Z}^2 : x, t \geq 0\}$, which verifies Condition (i) of Proposition 3.2.

Finally, to verify Condition (ii), we take $\mathcal{S}_1 = \emptyset$ so that part (b) of Condition (ii) is trivially satisfied. For Condition (a) of Proposition 3.2 (ii), let us denote, $b_{(x,0)} = I_x, b_{(y,0)} = I_y$ and $b_{(z,0)} = I_z$. Hence,

$$\begin{aligned}&\mathbb{E}[X_1(1)X_1(2) - X_0(1)X_0(2) | X_0 = (y-x, z-y)] \\ &= \mathbb{E}\big[(y - x + (I_y - I_x))(z - y + (I_z - I_y)) \\ &\qquad - (y-x)(z-y) | X_0 = (y-x, z-y)\big] \\ &= \mathbb{E}\big[(y - x + (I_y - I_x))(z - y + (I_z - I_y)) - (y-x)(z-y)\big] \\ &= (z-y)\mathbb{E}(I_y - I_x)) + (y-x)\mathbb{E}(I_z - I_y) + \mathbb{E}[(I_y - I_x)(I_z - I_y)].\end{aligned}$$

Clearly from definition, we have $\mathbb{E}(I_x) = \mathbb{E}(I_y) = \mathbb{E}(I_z) = 0$ and $\mathbb{E}(I_y^2) = 1$. Further for $x < y < z$, the random variables I_x, I_y and I_z are independent.

Hence,

$$\mathbb{E}[X_1(1)X_1(2) - X_0(1)X_0(2)|X_0 = (y-x, z-y)] = -1.$$

If $x = y$ or $y = z$, i.e., at least one pair of paths have already coalesced, we have that $X_1(1)X_1(2) = 0$. Therefore, $\mathcal{X}$ satisfies the conditions in Proposition 3.2 with the choice of $\mathcal{S}_1 = \emptyset$, $d_1 = -1$. This allows to conclude that Condition (B2) of Theorem 2.2 holds for $\{\overline{\mathcal{X}}_n : n \in \mathbb{N}\}$.

4.2. *Howard's model*

We first describe a 2-dimensional drainage network proposed by Howard (see [14]). Fix $0 < p < 1$ and let $\{B_{(x,t)} : (x,t) \in \mathbb{Z}^2\}$ be an i.i.d. collection of Bernoulli random variables with success probability p. Set $\mathcal{V} = \{(x,t) \in \mathbb{Z}^2 : B_{(x,t)} = 1\}$. Let $\{U_{(x,t)} : (x,t) \in \mathbb{Z}^2\}$ be an i.i.d. collection of Rademacher random variables, independent of the collection $\{B_{(x,t)} : (x,t) \in \mathbb{Z}^2\}$: $\mathbb{P}(U_{(x,t)} = 1) = \mathbb{P}(U_{(x,t)} = -1) = 1/2$. For a vertex $(x,t) \in \mathbb{Z}^2$, we consider $k_0 = \min\{|k| : k \in \mathbb{Z}, B_{(x+k,t+1)} = 1\}$. Clearly, k_0 is almost surely finite. Now, we define,

$$h(x,t) := \begin{cases} (x+k_0, t+1) & \text{if } (x-k_0, t+1) \notin \mathcal{V} \\ (x-k_0, t+1) & \text{if } (x+k_0, t+1) \notin \mathcal{V} \\ (x+U_{(x,t)}k_0, t+1) & \text{otherwise.} \end{cases} \tag{4.1}$$

Here each open vertex $(x,t) \in \mathcal{V}$ represents a water source and the edge $\langle (x,t), h(x,t) \rangle$ represents the channel through which water can flow. Consider the random graph $G = (\mathcal{V}, E)$ where the random edge set E is given by $E = \{\langle (x,t), h(x,t) \rangle : (x,t) \in \mathcal{V}\}$. Gangopadhyay et al. [13] proved that this random graph G is connected almost surely. This result, shows that Howard's model generates a random directed tree on $\mathbb{Z}^2$. The authors of [13] actually used a general construction and studied this model for higher dimensions also. The construction that we presented here is applicable for $d = 2$ only and agrees with that of [13] for $d = 2$.

Let $\mathcal{X}$ be the collection of all paths obtained by following the edges and for $\gamma, \sigma > 0$, let $\mathcal{X}_n = \mathcal{X}_n(\gamma, \sigma)$ be the collection of scaled paths with normalization constants γ, σ. Coletti et al. [5, 6] proved that, for $\gamma = 1$ and $\sigma = \sigma_0(p)$ with $\sigma_0(p)$ as given by $\sigma_0(p) = \big(\frac{(1-p)(2-2p+p^2)}{p^2(2-p)^2}\big)^{1/2}$, $\overline{\mathcal{X}}_n$ converges in distribution to the Brownian web. Again, to verify Condition (B2) they proved a partial FKG inequality for the Howard's model. We show here that Howard's model satisfies Proposition 3.2, and thereby Condition (B2) is verified.

By the i.i.d. nature of the collections of random variables $\{B_{(x,t)} : (x,t) \in \mathbb{Z}^2\}, \{U_{(x,t)} : (x,t) \in \mathbb{Z}^2\}$, it follows that $\mathcal{X}$ satisfies Condition (ii) of Proposition 3.2. It is not difficult to see that $\mathcal{X}$ consists of non-crossing paths only and Condition (i) of Proposition 3.2 holds. From definition it follows that for any $(y,s) \in \mathbb{Z}^2$, $h(y,s)$ depends only on the random variables $\{B_{(x,s+1)} : x \in \mathbb{Z}\}$ and $U_{(y,s)}$. Thus from the i.i.d. nature of the random variables it follows that for $x, y, z \in \mathbb{Z}$ with $x < y < z$, the process

$$\begin{aligned}\{X_j &= (\pi^{(y,0)}(j) - \pi^{(x,0)}(j), \pi^{(z,0)}(j) - \pi^{(y,0)}(j)) \\ &= (h^j(y,0)(1) - h^j(x,0)(1), h^j(z,0)(1) - h^j(y,0)(1)) : j \geq 0\}\end{aligned}$$

is a Markov chain with state space $\{(x,t) \in \mathbb{Z}^2 : x, t \geq 0\}$, which verifies Condition (i) of Proposition 3.2.

To verify Condition (ii), we take $\mathcal{S}_1 = \{(l,m) \in \mathbb{Z}^2 : 1 \leq \min(l,m) \leq r_0\}$ where $r_0 \in \mathbb{N}$ will be chosen later. We observe that, for any $r_0 \geq 1$,

$$\mathbb{P}[X_1 \in \mathcal{S}_0 \mid X_0 = (y-x, z-y)] \geq (1-p)^{2r_0} p$$

for all $((y-x),(z-y)) \in \mathcal{S}_1$. For this assume that $y - x \leq r_0$ and consider the event $\{B_{(u,1)} = 0, \text{ for } |u - x| \leq r_0, u \neq y, B_{(y,1)} = 1\}$, i.e., the vertex above the second path is open and all other points closed which are within a distance r_0 (on both sides) from the vertex above the first path. This ensures that the first path and the second path coalesce. Similar arguments hold when $z - y \leq r_0$. This verifies part (b) of Condition (ii) Proposition 3.2 with $p_0 = (1-p)^{2r_0} p > 0$.

For part (a) of Condition (ii) in Proposition 3.2, let us denote, $I_x := h(x,0)(1) - x$, $I_y := h(y,0)(1) - y$ and $I_z := h(z,0)(1) - z$. As in the earlier model, we obtain,

$$\begin{aligned}&\mathbb{E}[X_1(1)X_1(2) - X_0(1)X_0(2) | X_0 = (y-x, z-y)] \\ &= (z-y)\mathbb{E}(I_y - I_x) + (y-x)\mathbb{E}(I_z - I_y) + \mathbb{E}[(I_y - I_x)(I_z - I_y)].\end{aligned}$$

Now, we use the fact that the marginal distributions of I_x, I_y and I_z are same as I (say) with $\mathbb{E}(I) = 0$. But as opposed to the earlier model, these random variables are no longer independent. Using Cauchy Schwartz inequality for any $x, y, z \in \mathbb{Z}$ with $x < y < z$ we have

$$\begin{aligned}&\mathbb{E}[X_1(1)X_1(2) - X_0(1)X_0(2) | X_0 = (y-x, z-y)] \\ &= \mathbb{E}[(I_y - I_x)(I_z - I_y)] \leq 4\mathbb{E}[I^2] < \infty. \qquad (4.2)\end{aligned}$$

To verify part (a) of the Condition (ii) of Proposition 3.2, we still need to show that for $((y-x),(z-y)) \in (\mathcal{S}_0 \cup \mathcal{S}_1)^c$ there exists $d_1 > 0$ such that

$\mathbb{E}[X_1(1)X_1(2) - X_0(1)X_0(2)|X_0 = (y-x, z-y)] \le -d_1$. For $x < y < z \in \mathbb{Z}$ with $\min(y-x, z-y) > r > 0$ we have

$$\begin{aligned}&\mathbb{E}[X_1(1)X_1(2) - X_0(1)X_0(2)|X_0 = (y-x, z-y)]\\&= \mathbb{E}[(I_y - I_x)(I_z - I_y)] = -\mathbb{E}[I_y^2] + \mathbb{E}[I_y I_z] - \mathbb{E}[I_x I_z] + \mathbb{E}[I_x I_y]. \quad (4.3)\end{aligned}$$

The last three terms converge to 0 as $r \to \infty$. We show it for $\mathbb{E}[I_y I_z]$, the others being exactly the same. Observe that $\mathbb{E}[(I_y I_z)\mathbb{I}(\max\{|I_y|, |I_z|\} < r/2)] = 0$ and hence we obtain

$$\begin{aligned}&|\mathbb{E}[I_y I_z]|\\&= |\mathbb{E}[(I_y I_z)\mathbb{I}(\max\{|I_y|, |I_z|\} < r/2)] + \mathbb{E}[(I_y I_z)\mathbb{I}(\max\{|I_y|, |I_z|\} \ge r/2)]|\\&\le \sqrt{\mathbb{E}[I_y^2 I_z^2]}\sqrt{\mathbb{P}(\max\{|I_y|, |I_z|\} \ge r/2)}\\&\le \sqrt[4]{\mathbb{E}[I_y^4]}\sqrt[4]{\mathbb{E}[I_z^4]}\sqrt{2}\sqrt{\mathbb{P}(|I_y| > r/2)}\\&\to 0\end{aligned}$$

as $r \to \infty$. Thus, we have $-\mathbb{E}[I_y^2] + \mathbb{E}[I_y I_z] - \mathbb{E}[I_x I_z] + \mathbb{E}[I_x I_y] \le -\mathbb{E}(I^2)/2$ for all $r > r_0$. Hence it follows that Howard's model satisfies Proposition 3.2 with $\mathcal{S}_1 = \{(l,m) \in \mathbb{Z}^2 : 1 \le \min\{l,m\} \le r_0\}$, $d_1 = \mathbb{E}(I^2)/2, d_2 = 4\mathbb{E}[I^2]$ and $p_0 = (1-p)^{2r_0}p$ for some r_0 and Condition (ii) of Proposition 2.3 holds for this discrete model. This shows that the condition (B2) is satisfied.

4.3. *Ferrari, Landim and Thorisson's model*

We are interested in a directed forest spanning on Poisson process on $\mathbb{R}^2$ introduced by Ferrari et al. [11]. The vertices of the forest are the atoms of a homogeneous Poisson point process N having unit intensity in $\mathbb{R}^2$, embedded with the Euclidean norm. With an abuse of notation, the Poisson point process N will be considered in the sequel sometimes as a point measure with Dirac masses on its atoms and sometimes as a union of points of $\mathbb{R}^2$. The ancestor of $(x,t) \in \mathbb{R}^2$, denoted by $h(x,t)$, is the closest point $(x', t') \in N$ such that

$$t' > t \qquad \text{and} \qquad |x' - x| < \frac{1}{2}.$$

This means that in order to find the ancestor of (x,t), we are growing a tube of width 1 in the upward direction whose base is centered around x. The first point of N encountered by the tube is the ancestor of (x,t).

The graph, considered by Ferrari et al. [11], with vertices as the atoms of N and edge set $\{\langle (x,t), h(x,t)\rangle,\ (x,t) \in N\}$, is of out-degree 1 almost surely and hence a forest. We should mention here that the authors of [11] considered this forest on $\mathbb{R}^d$ for general d and showed that for $d = 2$, the random forest is indeed a tree, i.e., connected almost surely. We concentrate only on $d = 2$. In another work, Ferrari, Fontes and Wu [12] showed that for $d = 2$ under diffusive scaling, as a collection of paths, this tree converges to the Brownian web.

It is not *a priori* clear that Condition (iii) of Proposition 2.3 holds in this model. However, we show that the main idea of Proposition 2.3 can be extended to give an alternate verification of (B2) for this model as well.

For any path $\pi \in \mathcal{X}$ emanating from (x, t), we define a càdlàg path, denoted by $\tilde{\pi}$ henceforth, emanating from (x, t) to its ancestor $h(x, t) = (x', t')$ which remains constant on $[t, t')$ with value x and then jumps to x'. It is easy to observe that π and $\tilde{\pi}$ meet at jump times of $\tilde{\pi}$ and hence the number of steps between two time segments are exactly the same for both types of paths. From now on, we will only *consider the càdlàg paths.*

In the sequel, we will deal with triplets $(x, y, z) \in \mathbb{R}^3$ with $x < y < z$ and $\min\{(y - x), (z - y)\} \geq 1/2$. Let $X_t = (\tilde{\pi}^{(x,0)}(t), \tilde{\pi}^{(y,0)}(t), \tilde{\pi}^{(z,0)}(t))$ denote the triplet of càdlàg paths starting from the points $(x, 0), (y, 0)$ and $(z, 0)$ respectively. Further, it is useful to observe that our choice of starting points ensures that either $\min\{(X_t(2) - X_t(1)), (X_t(3) - X_t(2))\} \geq 1/2$ or $(X_t(2) - X_t(1))(X_t(3) - X_t(2)) = 0$ for all $t \geq 0$ where $X_t(i)$ denotes the i-th coordinate of X_t. In other words, the paths are separated by distance $1/2$ or they have coalesced for all time points. Hence, the process $\{X_t : t \geq 0\}$, is a pure jump Markov process with state space $\mathcal{M} = \{(u, v, w) : u \leq v \leq w,\ \text{either } u = v \text{ or } v = w \text{ or } \min\{v - u, w - v\} \geq 1/2\}$.

This part of the argument is motivated from Ferrarri et al. [12]. We are interested in finding $\lim_{n\to\infty} \mathbb{P}(\eta_{\mathcal{X}}(0, n; 0, \epsilon\sqrt{n}) \geq 3)$. Set A_n as the (random) point set $\{\tilde{\pi}(0) : \pi \in \mathcal{X}, \sigma_\pi \leq 0\} \cap [-1/2, \epsilon\sqrt{n} + 1/2]$. It is easy to observe that for any $x, y \in A_n$ we must have $|x - y| \geq 1/2$. Hence elements of the set A_n can be enumerated as $\{x_1, x_2, \ldots, x_J\}$ where $x_1, \ldots, x_J$ and J are random variables such that $x_1 < x_2 < \cdots < x_J$ and $x_{i+1} - x_i \geq 1/2$ almost surely for $1 \leq i \leq J - 1$ (with convention that if $J = 0$, the point set is empty). If required, we consider an enlarged set B_n containing A_n such that $B_n = \{y_1, y_2, \ldots, y_{J'}\}$ where

$$y_1 \geq -1/2, y_{J'} \leq \epsilon\sqrt{n} + 1/2 \text{ and } 1/2 \leq y_{i+1} - y_i \leq 1 \text{ for } 1 \leq i \leq J' - 1.$$

Note that J' is random but that we necessarily have $J' \leq 2\epsilon\sqrt{n} + 2$.

For $1 \leq j \leq J' - 2$, let us define the event

$$E(j) = \{\tilde{\pi}^{(y_j,0)}(n) < \tilde{\pi}^{(y_{j+1},0)}(n) < \tilde{\pi}^{(y_{J'},0)}(n)\}.$$

On the event $\{\eta_{\mathcal{X}}(0,n;0,\epsilon\sqrt{n}) \geq 3\}$, there must exist $j = j(\omega) \in \{1,2,\ldots,J'-2\}$ such that $E(j)$ is satisfied. Let $\mathcal{F}_0$ denote the σ-field generated by the Poisson point process on the negative half-plane $\{(x,t) \in \mathbb{R}^2 : t \leq 0\}$. Clearly, A_n is $\mathcal{F}_0$ measurable. Given the set A_n, the set B_n can be chosen by a deterministic algorithm so that B_n, and hence J', are $\mathcal{F}_0$ measurable. On the other hand, given B_n, the event $E(j)$ depends only on Poisson process on the upper half-plane $\{(x,t) \in \mathbb{R}^2 : t > 0\}$ and hence $E(j)$ depends on $\mathcal{F}_0$ only through the positions of B_n.

We prove the following proposition.

Proposition 4.1: *For any set B_n having properties as above, we have, for $j \geq 1$,*

$$\mathbb{P}(E(j) \mid B_n) \leq \frac{1}{n}[C_1 + C_2\epsilon\sqrt{n}]$$

for constants $C_1, C_2 > 0$ independent of j, as well as B_n.

Assuming the Proposition 4.1, we have

$$\begin{aligned}
&\mathbb{P}(\eta_{\mathcal{X}}(0,n;0,\epsilon\sqrt{n}) \geq 3) = \mathbb{E}\big(\mathbb{P}((\eta_{\mathcal{X}}(0,n;0,\epsilon\sqrt{n}) \geq 3)|\mathcal{F}_0)\big)\\
&\leq \mathbb{E}\big(\mathbb{P}(\cup_{j=1}^{J'-2}E(j)|\mathcal{F}_0)\big) \leq \mathbb{E}\Big(\sum_{j=1}^{J'-2}\mathbb{P}(E(j)|\mathcal{F}_0)\Big)\\
&= \mathbb{E}\Big(\sum_{j=1}^{J'-2}\mathbb{P}(E(j)|B_n)\Big) \leq \mathbb{E}\Big(\sum_{j=1}^{J'-2}\frac{1}{n}[C_1 + C_2\epsilon\sqrt{n}]\Big)\\
&= \frac{1}{n}[C_1 + C_2\epsilon\sqrt{n}]\mathbb{E}(J'-2) \leq \frac{1}{n}[C_1 + C_2\epsilon\sqrt{n}]2\epsilon\sqrt{n}\\
&\to 2C_2\epsilon^2
\end{aligned}$$

as $n \to \infty$, which proves (B2).

4.3.1. *A Lyapunov condition for càdlàg branches*

It now remains to prove Proposition 4.1. We are inspired by a result due to Meyn and Tweedie [16] , which is based on establishing a Lyapunov control similar to Theorem 3.1 (i) on the infinitesimal generator of $(X_t)_{t\geq 0}$ (see e.g. [9] or [16] for definitions of these generators).

Let us define $\mathcal{M}_0 = \{(u,v,w) : u \leq v \leq w,\ \text{either } u = v \text{ or } v = w\}$ and $\tau = \tau(\mathcal{M}_0) = \inf\{t \geq 0,\ X_t \in \mathcal{M}_0\}$. We note that $\tau < \infty$ almost surely.

Theorem 4.2: *Assume that there exist $d > 0$ and a non-negative function $V : \mathcal{M} \to [0,\infty)$ such that the generator G of the Markov process $(X_t)_{t\geq 0}$*

satisfies

$$GV(u,v,w) \le -d\mathbf{1}_{(\mathcal{M}_0)^c}(u,v,w). \tag{4.4}$$

Then, for any $(u,v,w) \in (\mathcal{M}_0)^c$,

$$\mathbb{E}_{(u,v,w)}(\tau) \le \frac{1}{d} V(u,v,w).$$

Proof: [Theorem 4.2] The result is a particular case of [16] (Th. 4.3). Using Dynkin's formula:

$$\begin{aligned} 0 &\le \mathbb{E}_{(u,v,w)}\big(V(X_{\tau\wedge t})\big) \\ &\le V(u,v,w) + \mathbb{E}_{(u,v,w)}\Big(\int_0^{\tau\wedge t} \big(-d\mathbf{1}_{(\mathcal{M}_0)^c}(X_s)\big)ds\Big) \\ &\le V\big(u,v,w\big) - d\mathbb{E}_{(u,v,w)}\big(\tau\wedge t\big) \end{aligned}$$

from which we obtain the announced result by letting $t \to \infty$. □

We claim that for the function

$$V\big(u,v,w\big) := (v-u)(w-v) \tag{4.5}$$

the equation (4.4) is satisfied with $d = 1/12$. Assuming the claim, we first prove Proposition 4.1.

Proof: [Proposition 4.1] With τ being as defined before Theorem 4.2, for $j \in \{1, \ldots, J'-2\}$,

$$\begin{aligned} \mathbb{P}(E(j) \mid B_n) = \mathbb{P}_{(y_j,y_{j+1},y_{J'})}(\tau > n) &\le \frac{1}{n}\mathbb{E}_{(y_j,y_{j+1},y_{J'})}(\tau) \\ &\le \frac{12}{n}(y_{j+1}-y_j)(y_{J'}-y_{j+1}) \le \frac{12}{n}[\epsilon\sqrt{n}+1] \end{aligned}$$

using the fact that $y_{j+1} - y_j \le 1$ and $y_{J'} - y_{j+1} \le \epsilon\sqrt{n} + 1$. □

Now we verify the claim above. For this, we start with computing the generator of $(X_t)_{t\ge 0}$. The basic idea is the following: the generator computed on V can be interpreted as an expectation, which gives us a way to estimate it easily. Let us fix $(u,v,w) \in (\mathcal{M}_0)^c$. We need some notation. Let $\mathcal{J} = [-1/2, 1/2]$ and $A = (u+\mathcal{J}) \cup (v+\mathcal{J}) \cup (w+\mathcal{J})$. Note that these sets in A may overlap and the total length (Lebesgue measure) $|A|$ of A can be strictly less than 3. Consider a uniform random variable U on the set A. Now, for $s \in \{u,v,w\}$, define I_s as follows:

$$I_s := \begin{cases} U - s & \text{if } U \in s + \mathcal{J} \\ 0 & \text{otherwise.} \end{cases}$$

Note that I_s denotes the increment of the jump of the path starting from $(s,0)$ at the first jump time of the process $(X_t)_{t\geq 0}$. Then we have

$$GV(u,v,w) = |A|\mathbb{E}\big[V(u+I_u, v+I_v, w+I_w) - V(u,v,w)\big].$$

Using the form of V, we have

$$GV(u,v,w) = |A|\mathbb{E}\big[(v-u)(I_w-I_v)+(w-v)(I_v-I_u)+(I_w-I_v)(I_v-I_u)\big]. \tag{4.6}$$

First we make a few observations on I_s for $s \in \{u,v,w\}$. Clearly, I_s is symmetric and hence $\mathbb{E}(I_s) = 0$. Further, $I_u I_w = 0$ almost surely since $(u+\mathcal{J}) \cap (w+\mathcal{J}) = \emptyset$ as $(u,v,w) \in (\mathcal{M}_0)^c$. Also, we observe that $I_u I_v$ is non-zero only when $U \in (u+\mathcal{J})\cap(v+\mathcal{J})$. In such a case, $I_u > 0$ and $I_v < 0$. Thus, $I_u I_v \leq 0$ almost surely. Similar argument shows that $I_v I_w \leq 0$ almost surely. Now,

$$\mathbb{E}(I_v^2) = \frac{1}{|A|}\int_{-1/2}^{1/2} t^2 dt = \frac{1}{12|A|}.$$

Thus, putting back in (4.6), we have

$$GV(u,v,w) = |A|\mathbb{E}\big[(I_w - I_v)(I_v - I_u)\big] \leq -|A|\mathbb{E}(I_v^2) = -\frac{1}{12}.$$

This proves the claim that V satisfies the condition (4.4) with $d = 1/12$.

Acknowledgements

The authors thank Rongfeng Sun for his careful reading and valuable comments. This work was supported by the GdR GeoSto 3477 and Labex CEMPI (ANR-11-LABX-0007-01). D.C. is funded by ANR PPPP (ANR-16-CE40-0016). A.S. wishes to thank Science & Engineering Research Board (SERB), Department of Science & Technology for the grant (MTR/2017/000293). This work was partially done while K.S., A.S. and V.C.T. were visiting the Institute for Mathematical Sciences, National University of Singapore in 2017. The visit was supported by the Institute.

References

1. R. Arratia, "Coalescing Brownian motions on the line", Ph.D. thesis, University of Wisconsin, Madison, 1979.
2. F. Baccelli and C. Bordenave, "The radial spanning tree of a Poisson point process" *Ann. Appl. Probab.*, **17**(1): 305–359, 2007.
3. N. Berestycki, C. Garban and A. Sen, "Coalescing Brownian flows: a new approach". *Ann. Probab.*, **43**(6): 3177–3215, 2015.

4. P. Billingsley, "Probability and measure", Second Edition, John Wiley and Sons, 1986.
5. C.-F. Coletti, E. S. Dias and L. R. G. Fontes, "Scaling limit for a drainage network model", *J. Appl. Probab.*, **46**, 1184–1197, 2009.
6. C. Coletti and G. Valle, "Convergence to the Brownian web for a generalization of the drainage network model", *Ann. Inst. H. Poincaré Probab. Statist.*, **50**:(3), 899–919, 2014.
7. D. Coupier, K. Saha, A. Sarkar and V. C. Tran, "The 2d-directed spanning forest converges to the Brownian web", arXiv:1805.09399, 2018.
8. D. Coupier and V. C. Tran, "The 2d-directed spanning forest is almost surely a tree", *Random Structures and Algorithms*, **42**(1), 59–72, 2013.
9. S. N. Ethier and T. G. Kurtz, "Markov Processes: Characterization and Convergence", Wiley, 2009.
10. L. R. G. Fontes, M. Isopi, C. M. Newman and K. Ravishankar, "The Brownian web: characterization and convergence", *Ann. Probab.*, **32**, 2857–2883, 2004.
11. P. Ferrari, C. Landim and H. Thorisson, "Poisson trees, succession lines and coalescing random walks", *Ann. Inst. H. Poincaré Probab. Statist.*, **40**(2), 141–152, 2004.
12. P. A. Ferrari, L. R. G. Fontes and X. Y. Wu, "Two-dimensional Poisson Trees converge to the Brownian web", *Ann. Inst. H. Poincaré Probab. Statist.*, **41**, 851–858, 2005.
13. S. Gangopadhyay, R. Roy and A. Sarkar, "Random oriented trees: a model of drainage networks", *Ann. Appl. Probab.*, **14**, 1242–1266, 2004.
14. A. D. Howard, "Simulation of stream networks by headward growth and branching", *Geogr. Anal.*, **3**, 29–50, 1971.
15. S. P. Meyn and R. L. Tweedie, "Markov chains and stochastic stability", Springer-Verlag, London, 1993.
16. S. P. Meyn and R. L. Tweedie, "Stability of Markovian processes III: Foster-Lyapunov criteria for continuous-time processes", *Advances in Applied Probability*, **25**(3), 518–548, 1993.
17. C. M. Newman, K. Ravishankar and R. Sun, "Convergence of coalescing nonsimple random walks to the Brownian web", *Electron. J. Prob.*, **10**, 21–60, 2005.
18. I. Rodriguez-Iturbe and A. Rinaldo, "Fractal river basins: chance and self-organization", Cambridge University Press, New York (1997).
19. R. Roy, K. Saha and A. Sarkar, "Random directed forest and the Brownian web", *Ann. Inst. H. Poincaré Probab. Statist.*, **52**, 1106–1143., 2005.
20. R. Roy, K. Saha and A. Sarkar, "Hack's law in a drainage network model: a Brownian web approach", *Ann Appl. Probab.*, **26**:3, 1807–1836, 2016.
21. A. Sarkar and R. Sun, "Brownian web in the scaling limit of supercritical oriented percolation in dimension $1+1$", *Electron. J. of Probab.*, **18**, 1–23, 2013.
22. A. E. Scheidegger, "A stochastic model for drainage pattern into an intramontane trench", *Bull. Ass. Sci. Hydrol.*, **12**, 15–20, 1967.
23. E. Schertzer, R. Sun and J. Swart, "The Brownian web, the Brownian net,

and their universality", *Advances in Disordered Systems, Random Processes and Some Applications*, 270–368, Cambridge University Press, 2017.

24. A. Sturm and J. M. Swart, "A particle system with cooperative branching and coalescence", *Ann. of Appl. Probab.*, **15**:3, 1616–1649, 2015.
25. R. Sun and J. Swart, "The Brownian net", *Ann. Probab.*, **36**, 1153–1208, 2008.
26. B. Tóth and W. Werner, "The true self-repelling motion", *Probab. Theory Related Fields*, **111**, 375–452, 1998.
27. G. Valle and L. Zuaznabar, "A version of the random directed forest and its convergence to the Brownian web", arXiv:1704.05555, 2017.

A Note on Characterizing Tightness of Random Sets of Càdlàg Paths

Nic Freeman

School of Mathematics and Statistics, University of Sheffield, UK
n.p.freeman@sheffield.ac.uk

Sandra Palau

Department of Statistics and Probability
Instituto de Investigaciones en Matemáticas Aplicadas y en Sistemas
Universidad Nacional Autónoma de México, México
sandra@sigma.iimas.unam.mx

We record two natural tightness criteria for random sets of càdlàg paths, within the state space introduced by Etheridge et al. in [4]. This state space is the equivalent, for càdlàg paths, of the (continuous path) state space introduced for the Brownian web by Fontes et al. in [6].

We discuss some implications of these criteria, in particular for cases of coalescing random walkers with heavy tailed jumps, in which tightness fails.

1. Introduction

In this article, we focus on an extension of the "usual" state space for the Brownian web that was introduced by [6]; the extension, which was introduced by [4], allows for the possibility that paths may be càdlàg, but otherwise preserves direct analogy to the usual state space. We record two natural characterizations of tightness for random sets of càdlàg paths, in the style of the classical characterizations of tightness for (single) random càdlàg paths. Using these characterizations, we discuss a corresponding failure of tightness for coalescing random walk approximations to webs made up of α-stable Lévy process, which were introduced recently by [9].

This article developed out of discussions following a learning session, given by the present authors, on the Brownian web and net, at the Workshop on Genealogies of Interacting Particle Systems, held at the Institute for Mathematic Sciences in Singapore. The material used in our learning session, which provides a brief introduction to the theory of the Brownian web and net, are available online as [7]. The reader is referred to [11] for a comprehensive introduction to the Brownian web and related objects. However, in this article, we are interested only in their state spaces.

Let us begin by setting up key pieces of notation and briefly describing the state space introduced by [6]. For $s \in [-\infty, \infty]$, let

$$C[s] = \{f : [s,\infty] \to [-\infty,\infty] \,;\, f \text{ is continuous on } [s,\infty] \cap (-\infty,\infty)\}$$
$$D[s] = \{f : [s,\infty] \to [-\infty,\infty] \,;\, f \text{ is càdlàg on } [s,\infty] \cap (-\infty,\infty)\}.$$

For $f \in D[s]$ we write $\sigma_f = s$ for the first time at which f is defined. We set

$$\widetilde{\Pi} = \bigcup_{t\in[-\infty,\infty]} C[t], \qquad \Pi = \bigcup_{t\in[-\infty,\infty]} D[t].$$

For each $f \in D[s]$ we associate f to a function $\bar{f} : [\sigma_{\bar{f}}, 1] \to [-1,1]$, by applying the space-time transformation

$$(x,t) \mapsto \left(\frac{\tanh(x)}{1+|t|}, \tanh(t)\right) \tag{1.1}$$

which maps the space-time plane from $[-\infty,\infty]^2$ into a subset of $[-1,1]^2$ (see Figure 1). The choice of the function tanh is arbitrary: it could be replaced by any order preserving continuous bijection of $[-\infty,\infty]$ to $[-1,1]$.

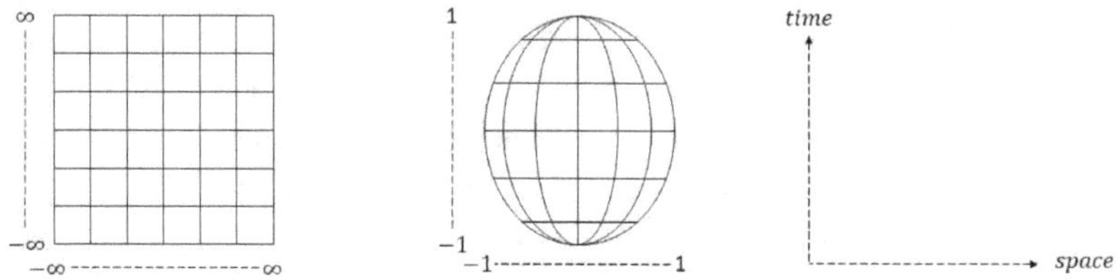

Fig. 1. The space-time transformation (1.1): before (left), after (mid), axis labels (right).

Formally, for each $s \in [-\infty,\infty]$ and $f \in D[s]$ we define $\bar{f}$ as follows. Let $\kappa_t = \tanh^{-1}(t)$ and note that κ is an order preserving homeomorphism between $[-1,1]$ and $[-\infty,\infty]$ (we use the symbol κ in place of $\tanh^{-1}$ to denote a change of time rather than rescaling of space). Then for $f \in \Pi$ we define

$$\bar{f}(t) = \frac{\tanh(f(\kappa_t))}{1 + |\kappa_t|} \tag{1.2}$$

for $t \in [\kappa^{-1}(\sigma_f), 1]$, with the convention $1/\infty = 0$. It follows immediately that $\bar{f}$ is càdlàg. Moreover, if $f \in \widetilde{\Pi}$, then $\bar{f}$ is continuous.

We use the map $f \mapsto \bar{f}$ to induce a pseudo-metric on $\widetilde{\Pi}$

$$d_{\widetilde{\Pi}}(f_1, f_2) = |\sigma_{\bar{f}_1} - \sigma_{\bar{f}_2}| \vee \sup_{t \in [-1,1]} \left|\bar{f}_1\left(t \vee \sigma_{\bar{f}_1}\right) - \bar{f}_2\left(t \vee \sigma_{\bar{f}_2}\right)\right|. \tag{1.3}$$

In standard fashion, we associate each $f \in \widetilde{\Pi}$ with its corresponding equivalence class and, with slight abuse of notation, regard $d_{\widetilde{\Pi}}$ as a metric on $\widetilde{\Pi}$. Essentially, $d_{\widetilde{\Pi}}$ is the usual supremum metric on continuous paths, ignoring behaviour near $\pm\infty$ in both space and time, and modified to handle paths with possibly different starting times.

The metric $d_{\widetilde{\Pi}}$ is that of [6], although their notation is a little different. It was shown in [6] that $(\widetilde{\Pi}, d_{\widetilde{\Pi}})$ is both complete and separable. The space $\mathcal{K}(\widetilde{\Pi})$ is defined to be the space of compact subsets of $\widetilde{\Pi}$, equipped with the induced Hausdorff metric, and including the empty set as an isolated point.

Thus, an element of $\mathcal{K}(\widetilde{\Pi})$ is a (compact) set of continuous paths. An analogous construction for càdlàg paths was introduced by [4]; it replaces the sup in (1.3) with a function based on the Skorohod metric, resulting in a space $\mathcal{K}(\Pi)$ whose elements are (suitably compact) sets of càdlàg paths. We will recall the detail of this construction in Section 1.1.

We now look to state our pair of characterizations of tightness in $\mathcal{K}(\Pi)$. In order to explain where they come from, we will first recall tightness characterizations for $\mathcal{K}(\widetilde{\Pi})$ and for spaces of (single) continuous/càdlàg paths. The reader who is already familiar with such results may wish to simply skip forwards to equation (1.12).

Let $\mathscr{C}$ denote the space of continuous functions $f : [-1,1] \to [-1,1]$, with the usual supremum metric. Theorem 7.3 of [3] provides a characterization of tightness in $\mathscr{C}$: a random sequence (X_n) with values in $\mathscr{C}$ is tight if and only if, for all $\eta > 0$,

$$\lim_{\delta \to 0} \limsup_{n \to \infty} \mathbb{P}\left[w(X_n, \delta) > \eta\right] = 0, \tag{1.4}$$

where

$$w(f, \delta) = \sup_{|s-t|<\delta} |f(s) - f(t)| \tag{1.5}$$

is the modulus of continuity of f. The supremum in (1.5) ranges over all s,t in the domain of f for which $|s-t|<\delta$: we adopt the convention that whenever an expression of the form $\sup_{\dots}(\dots)$ contains $f(\cdot)$, the argument of f is automatically restricted to its domain, and similarly for inf.

Let $\Lambda_{L,T}=\{(x,t)\,;\,|x|\le L,|t|\le T\}$. Extending the same ideas to $\mathcal{K}(\widetilde{\Pi})$, as in Section 6.1 of [11], it turns out that what is required to have tightness is uniform control of the modulus of continuity: a random sequence $\widetilde{\mathcal{X}}_n$ in $\mathcal{K}(\widetilde{\Pi})$ is tight if and only if, for all $\eta>0$, and for all bounded $\Lambda_{L,T}$,

$$\lim_{\delta\to 0}\limsup_{n\to\infty}\mathbb{P}\left[\sup_{f\in\widetilde{\mathcal{X}}_n} w(f,\delta,\Lambda_{L,T})>\eta\right]=0. \tag{1.6}$$

Here,

$$w(f,\delta,\Lambda_{L,T})=\sup_{\substack{|s-t|<\delta\\(f(t),t),(f(s),s)\in\Lambda_{L,T}}}|f(s)-f(t)|, \tag{1.7}$$

which one might think of (loosely) as a way of restricting $w(f,\delta)$ to only care about behaviour that originates from within the box $\Lambda_{L,T}$.

In another vein, if we let $\mathscr{D}$ denote the set of càdlàg functions $f:[-1,1]\to[-1,1]$, with the usual Skorohod metric, then it is well known that a sequence X_n of random elements of $\mathscr{D}$ is tight if and only if, for all $\eta>0$

$$\lim_{\delta\to 0}\limsup_{n\to\infty}\mathbb{P}\left[w'(X_n,\delta)>\eta\right]=0 \tag{1.8}$$

where

$$\begin{aligned} w'(f,\delta)&=\inf_{\{t_i\}}\max_{1\le i<m} w_i(f,\delta) \\ w_i(f,\delta)&=\sup_{\substack{|s-t|<\delta\\ s,t\in[t_i,t_{i+1})}}|f(s)-f(t)|. \end{aligned} \tag{1.9}$$

Here, the infimum ranges over all finite sequences $\{t_i\}_{i=1}^m$ such that $-1=t_1<t_2<\dots<t_m=1$ and $\min_i|t_{i+1}-t_i|\ge\delta$. We refer to such sequences as δ-sparse partitions (of $[-1,1]$). The function w' plays the role of the modulus of continuity for càdlàg paths. Essentially, w' differs from w by permitting f to make jumps (at times t_i) without these jumps increasing the value of w'.

It is natural to expect that the combination of these two extensions of (1.4) should result in a characterization of tightness for sequences of random sets of càdlàg paths. It does, but:

Remark 1.1: In (1.7), with continuous paths, we did not need to consider the possibility of paths making large jumps between $-\infty$ and ∞ (or points close to $\pm\infty$) without passing through $\Lambda_{L,T}$ on the way. With càdlàg paths, we must account for this possibility.

With the above remark in mind, for $f \in \Pi$, define $f_{L,T}$, with the same domain as f, by

$$f_{L,T}(t) = (-L) \vee f\Big((-T) \vee t \wedge T\Big) \wedge L. \tag{1.10}$$

We refer to $f_{L,T}$ as the restriction of f to $\Lambda_{L,T}$.

It is straightforward to check that $w(f, \delta, \Lambda_{L,T})$ in (1.7) could be replaced by

$$\hat{w}(f, \delta, \Lambda_{L,T}) = \sup_{|s-t|<\delta} |f_{L,T}(s) - f_{L,T}(t)|, \tag{1.11}$$

to characterize tightness in $\mathcal{K}(\widetilde{\Pi})$. It is this formulation which extends to $\mathcal{K}(\Pi)$.

For the space we are primarily interested in, $\mathcal{K}(\Pi)$, the natural equivalent of the modulus of continuity is

$$w'(f, \delta, \Lambda_{L,T}) = \inf_{\{t_i\}} \max_{1 \le i < m} w_i(f, \delta, \Lambda_{L,T}) \tag{1.12}$$

$$w_i(f, \delta, \Lambda_{L,T}) = \sup_{\substack{|s-t|<\delta \\ s,t \in [t_i, t_{i+1})}} |f_{L,T}(s) - f_{L,T}(t)|. \tag{1.13}$$

The infimum in (1.12) ranges over δ-sparse partitions of $[\sigma_f, T]$, i.e. $\sigma_f = t_1 < \cdots < t_m = T$ and $\min_i |t_{i+1} - t_i| \ge \delta$. Thus:

Proposition 1.2: *Let $(\mathcal{X}_n)_{n\in\mathbb{N}}$ be a sequence of random elements of $\mathcal{K}(\Pi)$. Then $(\mathcal{X}_n)_{n\in\mathbb{N}}$ is tight if and only if, for all $\eta > 0$ and all bounded $\Lambda_{L,T}$,*

$$\lim_{\delta\to 0} \limsup_{n\to\infty} \mathbb{P}\left[\sup_{f\in\mathcal{X}_n} w'(f, \delta, \Lambda_{L,T}) > \eta\right] = 0. \tag{1.14}$$

A second characterization of tightness in the Skorodhod space $\mathscr{D}$ is often useful, and provides a version of the modulus of continuity that does not rely on δ-sparse partitions. In particular, from Theorem 12.4 of [3], it can be seen that function w' in (1.9) can be replaced by

$$w''(f, \delta) = \sup_{\substack{s_1 \le t \le s_2 \\ |s_2 - s_1| < \delta}} |f(s_1) - f(t)| \wedge |f(t) - f(s_2)|, \tag{1.15}$$

plus the additional requirements of uniform right continuity at -1 and uniform left limits at 1.

When compared to (1.5), w'' has a natural interpretation: in the world of continuous paths, tightness fails if the sequence X_n contains a jump that does not vanish in the limit; in the world of càdlàg paths tightness fails if, in the limit, the sequence X_n contains two jumps, occurring arbitrarily close together, neither of which vanishes in the limit.

Extending to $\mathcal{K}(\Pi)$, the natural equivalent of (1.15) is

$$w''(f,\delta,\Lambda_{L,T}) = \sup_{\substack{s_1\le t\le s_2\\ |s_2-s_1|<\delta}} \big|f_{L,T}(s_1)-f_{L,T}(t)\big| \wedge \big|f_{L,T}(t)-f_{L,T}(s_2)\big|. \quad (1.16)$$

Additionally, we will need a way to require (uniform) right continuity at the starting time of f,

$$w^+(f,\delta,\Lambda_{L,T}) = \sup_{s\in[0,\delta)} \big|f_{L,T}(\sigma_f) - f_{L,T}(\sigma_f+s)\big|. \quad (1.17)$$

We won't need to consider the end time of f, because (1.1) pinches space to a point at time $+\infty$ and consequently enforces left continuity there.

We obtain:

Proposition 1.3: *Let $(\mathcal{X}_n)_{n\in\mathbb{N}}$ be a sequence of random elements of $\mathcal{K}(\Pi)$. Then $(\mathcal{X}_n)_{n\in\mathbb{N}}$ is tight if and only if, for all $\eta>0$ and all bounded $\Lambda_{L,T}$,*

$$\lim_{\delta\to 0}\limsup_{n\to\infty}\mathbb{P}\left[\sup_{f\in\mathcal{X}_n}\left(w''(f,\delta,\Lambda_{L,T})\vee w^+(f,\delta,\Lambda_{L,T})\right)>\eta\right]=0. \quad (1.18)$$

Since it does not use δ-sparse partitions, the characterization in Proposition 1.3 may be better suited for use in one type of argument that has often been used to prove tightness for random sets of paths i.e. cutting space-time up into small regions and looking to control macroscopic jumps originating from within each region – see e.g. Section 6.1 of [11] for details in the case of the Brownian web.

Propositions 1.2 and 1.3 are not deep results, but they are nonetheless useful. The proofs are closely connected to the classical theory of tightness in path-space, and we provide them in Section 2.

We note that the state space $\widetilde{\Pi}$ introduced by [6] has become widely used, but it is by no means the only suitable state space for random sets of paths such as the Brownian web. Several authors, including [12] and [10] have used state spaces based on stochastic flows, in which case special care is needed to handle the existence of points within the Brownian web with more than one (equivalence class of) outgoing paths. A second alternative, used in e.g. [8], is the 'marked measure space', which seeks to characterize a set of paths in terms of the times until meetings of paths emanating from finite, random, sets of points within the underlying space.

Recently, [2] proposed a new state space based on the concept of 'tubes'. In their topology, sets of paths are characterized by which (tube-shaped) regions of space they pass through. The resulting state space has many appealing properties, including compactness of the state space itself, which makes tightness automatic.

A further example is provided by a construction of the α-stable web in [9]. Although [9] does introduce a space that is essentially equivalent to our $\mathcal{K}(\Pi)$, using a metric analogous to that defined below in (1.20), their construction of the α-stable web then takes place under a *weaker* topology. Loosely speaking, they go on to introduce an auxiliary metric which, in a sense that we won't describe fully here, simply ignores the positions of paths until after they have coalesced with an older path. In this weakened topology, [9] proves tightness by constructing an explicit family of suitable relatively compact sets, rather than via the sort of general characterisation stated above.

In fact, for heavy tailed random walk approximations to the α-stable web, tightness will fail in $\mathcal{K}(\Pi)$. This failure is closely related to the $3-\epsilon$ moment condition that was discovered by [1] to be necessary for convergence in $\mathcal{K}(\widetilde{\Pi})$ of coalescing random walks to the Brownian web. We demonstrate this failure of tightness in Section 3, in which the discussion relies heavily on applying equation (1.6) and Proposition 1.3, but is otherwise self-contained.

1.1. *A state space for sets of càdlàg paths*

In this section we recall the spaces Π and $\mathcal{K}(\Pi)$ introduced by [4] (in which they were denoted M and $\mathcal{K}(M)$). We begin by considering the space

$$G = \{g : [\sigma_g, 2] \to [-1, 1]\,;\, g \text{ is càdlàg},\ \sigma_g \in [-1, 1],\ g \text{ is constant on } [1, 2]\}. \tag{1.19}$$

We wish to view G as a space of càdlàg paths, with possibly different starting times, and we now set about defining a metric which embodies this intuition.

For $g, h \in G$, let $\Lambda'[g, h]$ denote the set of strictly increasing bijections from $[\sigma_g, 2] \to [\sigma_h, 2]$. We define $\Lambda[g, h] \subseteq \Lambda'[g, h]$ to be the subset of

$\lambda \in \Lambda'[g,h]$ for which

$$\gamma_{g,h}(\lambda) = \sup_{\sigma_g \leq t < s \leq 2} \left| \log \frac{\lambda(s) - \lambda(t)}{s-t} \right| < \infty.$$

For such g, h, λ we define

$$d(g,h,\lambda) = \sup_{t \in [\sigma_g, 2]} |g(t) - h(\lambda(t))|,$$

and

$$\rho(g,h) = \inf_{\lambda \in \Lambda[g,h]} \Big(\gamma_{g,h}(\lambda) \vee d(g,h,\lambda) \Big). \tag{1.20}$$

Remark 1.4: If we used the domain $[\sigma_g, 1]$ for $g \in G$ in (1.19), instead of $[\sigma_g, 2]$, $\Lambda'[g,h]$ would be empty whenever $\sigma_h < \sigma_g = 1$. To avoid this technicality, we work over time $[-1, 2]$ and require constancy on $[1, 2]$.

It was shown in Section 5 of [4] that G is complete and separable under the metric

$$d_G(g,h) = \rho(g,h) \vee |\sigma_g - \sigma_h|.$$

Intuitively, this metric says that paths in G converge if their domains converge and, as the domains become close, the paths also become close (in the Skorohod sense). It is the direct equivalent, for càdlàg paths, to the metric (1.3) introduced by [6] for the "usual" state space of continuous paths used for the Brownian web. In fact, if we restrict to paths starting at some fixed time, d_G becomes the usual Skorohod metric. The metric d_G is the key ingredient for metrizing Π.

Recall $\bar{f}$ from (1.2). With slight abuse of notation, we extend each $\bar{f}$ to a function $\bar{f} \in G$ by setting $\bar{f}(t) = 0$ for all $t \in (1, 2]$. Then,

$$d_\Pi(f_1, f_2) = d_G(\bar{f}_1, \bar{f}_2) \tag{1.21}$$

is a pseudo-metric on Π. In standard fashion, from now on we implicitly work with equivalence classes of Π and, with mild abuse of notation, treat (Π, d_Π) as a metric space. Separability of G immediately implies the same for Π, and it was shown in Lemma 5.6 of [4] that completeness is also inherited by Π. Of course, convergence in (Π, d_Π) can be described as local Skorohod convergence of the paths plus convergence of the starting times. It was shown in Lemma 5.8 of [4] that $\widetilde{\Pi}$ is continuously embedded in Π.

Recall that $\mathcal{K}(\Pi)$ is the set of compact subsets of Π, equipped with the Hausdorff metric $d_{\mathcal{K}(\Pi)}$ and including the empty set as an isolated point.

2. Proofs

2.1. *On relative compactness in the Hausdorff metric*

In this section we record two general results, for later use, on the relationship between relative compactness and the Hausdorff metric. They are certainly known, but we were unable to locate suitable references within the literature. We make use of these results in the proof of Lemma 2.8.

Let M be any complete metric space and let $\mathcal{K}(M)$ be the set of compact subsets of M, equipped with the Hausdorff metric. Since both M and (consequently, also) $\mathcal{K}(M)$ are complete, within these spaces relative compactness is equivalent to total boundedness.

Lemma 2.1: *Suppose that K is a relatively compact subset of $\mathcal{K}(M)$. Then $\bigcup_{X\in K} X$ is a relatively compact subset of M.*

Proof: Since K is totally bounded, for each $\epsilon > 0$ there is a finite set $X_1, \dots, X_n$ of elements of $\mathcal{K}(M)$ such that, for any $X \in K$ there is some X_i such that $d_{\mathcal{K}(M)}(X, X_i) < \epsilon$. Let $Y = \bigcup_{i=1}^{n} X_i$ and note that $\bigcup_{X\in K} X \subseteq Y^{(\epsilon)}$. Since each X_i is compact in M, Y is also compact in M, and in particular Y is totally bounded. Hence also $\bigcup_{X\in K} X$ is totally bounded. □

Lemma 2.2: *Suppose that B is a relatively compact subset of M. Then $K = \{X \in \mathcal{K}(M)\,;\, X \subseteq B\}$ is a relatively compact subset of $\mathcal{K}(M)$.*

Proof: Since B is totally bounded, for all $\epsilon > 0$ there is a finite sequence $x_1, \dots, x_n$ such that each $x \in M$ is within distance ϵ from some x_i. For any non-empty compact subset X of B, if we set $X' = \{x_i\,;\, \exists x \in X, d_M(x, x_i) < \epsilon\}$, then $d_{\mathcal{K}(M)}(X, X') < \epsilon$. Thus, every such X is within distance ϵ in $\mathcal{K}(M)$ of one of the subsets of $\{x_1, \dots, x_n\}$. Since there are only finite many such subsets, this shows that K is totally bounded. □

2.2. *Relative compactness*

Our first objective is to characterize relative compactness in G, for which we follow a similar method to that employed in Section 3.6 of [5] for Skorohod spaces of càdlàg paths. We then transfer this characterization to Π and finally upgrade it to $\mathcal{K}(\Pi)$.

Recall that a step function is a piecewise constant function with finitely many discontinuities. Given a step function $g \in G$, let $(s_i(g))_{i=0}^{m}$ denote the (finite, ordered) sequence of times at which g is discontinuous, including also the two endpoints $s_0(g) = \sigma_g$ and $s_m(g) = 2$. With slight abuse of notation, extend $(s_i(g))$ to an infinite sequence $(s_i(g))_{i=0}^{\infty}$ by setting $s_i(g) = 2$ for all $i \geq m$.

Let $A_\delta \subseteq G$ be the set of step functions in G for which $\inf_i |s_i(g) - s_{i+1}(g)| > \delta$.

Lemma 2.3: *A_δ is a relatively compact subset of G.*

Proof: Let (g_n) be a sequence in A_δ. We will show that (g_n) has a convergent subsequence in G.

For all n, since $g_n \in A_\delta$ we have that the sequence $s_i(g_n) = 2$ for all $i \geq \delta^{-1}$. Noting that $[-1, 2]$ is compact, by diagonalization we may pass to a subsequence and assume without loss of generality that there exists $t_0, \ldots, t_{\lfloor \delta^{-1} \rfloor}, x_0, \ldots, x_{\lfloor \delta^{-1} \rfloor} \in [-1, 2]$ such that for $i \in \{0, \ldots, \lfloor \delta^{-1} \rfloor\}$

$$\lim_{n\to\infty} s_i(g_n) = t_i \qquad \text{and} \qquad \lim_{n\to\infty} g_n(s_i(g_n)) = x_i.$$

We define the function

$$g(t) = \begin{cases} x_i & \text{if } t_i \leq t < t_{i+1},\ i = 0, \ldots, \lfloor \delta^{-1} \rfloor - 1 \\ x_{\lfloor \delta^{-1} \rfloor} & \text{if } t_{\lfloor \delta^{-1} \rfloor} \leq t \leq 2. \end{cases}$$

Then $g \in G$ and it is easily seen that $d_G(g_n, g) \to 0$. □

Lemma 2.4: *For each $g \in G$, it holds that $w'(g, \delta) \to 0$ as $\delta \to 0$.*

Proof: Let $g \in G$. For any $N \in \mathbb{N}$ define a sequence of times $(T_i^N)_{i=0}^{\infty}$, inductively, by setting $T_0^N = \sigma_g$ and

$$T_i^N = 2 \wedge \inf\{t \in (T_{i-1}^N, 2] : |g(t) - g(T_{i-1}^N)| > \tfrac{1}{N}\},$$

with the convention $\inf \emptyset = +\infty$.

Since g has left limits, the increasing sequence (T_i^N) has no limits points in $[-1, 2)$, and since g is constant on $[1, 2]$ there exists $k \in \mathbb{N}$ such that $T_{k-1}^N < 2 = T_k^N$. For all $0 < \delta < \min\{T_i^N - T_{i-1}^N : i \leq k\}$ we have $w'(g, \delta) < 2/N$, which completes the proof. □

Lemma 2.5: *Let $g_n, g \in G$ such that $d_G(g_n, g) \to 0$ as $n \to \infty$. Then, for all $\delta > 0$ we have*

$$\limsup_{n\to\infty} w'(g_n, \delta) \leq w'(g, \delta).$$

Proof: We have $d_G(g_n, g) \to 0$, so for each $n \in \mathbb{N}$ there exists $\lambda_n \in \Lambda[g_n, g]$ such that $\gamma_{g_n,g}(\lambda_n) \to 0$ and $d(g_n, g, \lambda_n) \to 0$. With notation as in (1.9), for any $\delta > 0$ we have

$$\begin{aligned} w'(g_n, \delta) &\le \inf_{\{t_i\}} \max_i \sup_{s,t \in [t_{i-1}, t_i)} |g(\lambda_n(s)) - g(\lambda_n(t))| + 2d(g_n, g, \lambda_n) \\ &\le \inf_{\{\lambda_n(t_i)\}} \max_i \sup_{s,t \in [\lambda_n(t_{i-1}), \lambda_n(t_i))} |g(s) - g(t)| + 2d(g_n, g, \lambda_n) \\ &= w'(g, \delta) + 2d(g_n, g, \lambda_n). \end{aligned} \tag{2.1}$$

Here, to deduce the first line we use the triangle inequality (comparing $g(s)$ to $g(\lambda_n(s))$ and similarly for t), and to deduce the second and third lines we use the definition of $\Lambda(g_n, g)$.

The result follows by letting $n \to \infty$ in (2.1). □

Lemma 2.6: *A subset A of G is relatively compact if and only if*

$$\lim_{\delta \to 0} \sup_{g \in A} w'(g, \delta) = 0. \tag{2.2}$$

Proof: Assume, first, that A is relatively compact. Suppose that (2.2) fails: then there exists $\eta > 0$, a sequence $\delta_n \to 0$ and a sequence (g_n) in G, such that $w'(g_n, \delta_n) \ge \eta$. By relative compactness of A and completeness of G, we may pass to a subsequence and assume, without loss of generality, that there exists $g \in G$ such that $d_G(g_n, g) \to 0$. Therefore, by (2.1) we have

$$\eta \le \limsup_{n \to \infty} \; w'(g_n, 1/n) \le \limsup_{n \to \infty} w'(g, 1/n)$$

which contradicts Lemma 2.4. Hence, in fact (2.2) must hold.

It remains to prove the converse. To this end, suppose that (2.2) holds, and we now look to show that A must be relatively compact. By (2.2), for each $N \in \mathbb{N}$ there exist δ_N such that

$$\sup_{g \in A} w'(g, \delta_N) \le N^{-1}.$$

Hence, by definition of w', for each $g \in A$ there exists a δ_N-sparse partition $\sigma_g = t_0 < t_1 < \cdots < t_m = 1 < t_{m+1} = 2$ such that

$$\max_i \sup_{s,t \in [t_{i-1}, t_i)} |g(s) - g(t)| \le N^{-1}.$$

Recall the set A_δ from Lemma 2.3. We define a function $h \in A_{\delta_N}$, where $h : [\sigma_g, 2] \to [-1, 1]$, by

$$h(t) = \begin{cases} g(t_{i-1}) & \text{if } t \in [t_{i-1}, t_i) \text{ for some } i = 1, \dots, m \\ g(t_m) & \text{if } t \in [1, 2]. \end{cases}$$

Take λ to be the identity function on $[\sigma_g, 2]$ and note that $\lambda \in \Lambda[g,h]$ with $\gamma_{g,h} = 0$. By definition of h we have $d(g,h,\lambda) \le N^{-1}$, and hence $\rho(g,h) < N^{-1}$. Thus $A \subseteq A_{\delta_N}^{(N^{-1})}$ where

$$A_{\delta_N}^{(N^{-1})} = \{f \in G : \rho(f',f) < N^{-1} \text{ for some } f' \in A_{\delta_N}\}$$

is the expansion of A_{δ_N} by radius N^{-1}.

Since G is complete, within G relative compactness is equivalent to total boundedness. Hence, by Lemma 2.3, A_{δ_N} is totally bounded. Since $A \subseteq \bigcap_N A_{\delta_N}^{(N^{-1})}$, A is also totally bounded, which completes the proof. □

Lemma 2.7: *A subset B of Π is relatively compact if and only if, for all bounded $\Lambda_{L,T}$*

$$\lim_{\delta\to 0} \sup_{f\in B} w'(f,\delta,\Lambda_{L,T}) = 0. \tag{2.3}$$

Proof: From (1.21) and Lemma 2.6, we have that $B \subseteq \Pi$ is relatively compact if and only if

$$\lim_{\delta\to 0} \sup_{f\in B} w'(\bar{f},\delta) = 0. \tag{2.4}$$

Recall the re-scaling of time $\kappa_t = \tanh^{-1} t$ is a homeomorphism between $[-1,1]$ and $[-\infty,\infty]$, and note that κ_t is bi-Lipschitz on closed intervals within $(-1,1)$. It can be seen that the map (1.1) is bi-Lipschitz in a similar sense: for any bounded box $\Lambda_{L,T}$ there exists $C \in (0,\infty)$ such that for all $f \in \Pi$ and all $s,t \in [\sigma_f,\infty]$,

$$|\bar{f}(s) - \bar{f}(t)| \le |f_{L,T}(\kappa_s) - f_{L,T}(\kappa_t)| \le C|\bar{f}(s) - \bar{f}(t)|. \tag{2.5}$$

Proving the inequality (2.5) is elementary but cumbersome, and we omit the argument in the interests of brevity. However, with (2.5) in hand, it is immediate from (1.9) and (1.12) that $w'(\bar{f},\delta) \le w'(f,\delta,\Lambda_{L,T}) \le Cw'(\bar{f},\delta)$ and thus (2.3) is equivalent to (2.4). □

Lemma 2.8: *A subset C of $\mathcal{K}(\Pi)$ is relatively compact if and only if, for all bounded $\Lambda_{L,T}$,*

$$\lim_{\delta\to 0} \sup_{B\in C} \sup_{f\in B} w'(f,\delta,\Lambda_{L,T}) = 0. \tag{2.6}$$

Proof: First, assume that $C \in \mathcal{K}(\Pi)$ is relatively compact. Then, by Lemma 2.1, the union $B' = \bigcup_{B\in C} B$ is a relatively compact subset of Π. Equation (2.6) follows by applying Lemma 2.7.

It remains to prove the converse. To this end, assume that $C \in \mathcal{K}(\Pi)$ and (2.6) holds. Again, consider the union $B' = \bigcup_{B \in C} B$, and note that now Lemma 2.7 implies that B' is a relatively compact subset of Π. Hence, by Lemma 2.2, the set K' of compact subsets of B' is a relatively compact subset of $\mathcal{K}(\Pi)$. Since $C \subseteq K'$, it follows that C is relatively compact. □

2.3. *Tightness*

As in Section 2.2, if X is a subset of some metric space $\mathbb{X}$, we write $A^{(\epsilon)}$ for the expansion of the set A by radius $\epsilon > 0$.

Recall that a sequence of random variables (X_n) taking values in some metric space $\mathbb{X}$ is said to be tight if for all $\epsilon > 0$ there exists a (deterministic) compact subset K of $\mathbb{X}$ such that $\inf_n \mathbb{P}[X_n \in K] \geq 1 - \epsilon$.

In fact, by combining Lemma 2.2.1 and Theorem 2.2.2 of [5], for a complete and separable metric space $\mathbb{X}$ tightness of $(\mathcal{X}_n)$ is equivalent to asking that, for all $\epsilon > 0$ there exists a compact subset K of $\mathbb{X}$ such that

$$\liminf_{n \to \infty} \mathbb{P}\left[X_n \in K^{(\epsilon)}\right] \geq 1 - \epsilon. \tag{2.7}$$

We now provide proofs of Propositions 1.2 and 1.3.

Proof: [Of Proposition 1.2.] First, suppose that $(\mathcal{X}_n)$ is tight in $\mathcal{K}(\Pi)$. That is, for each $\epsilon > 0$ there exists a compact set K of $\mathcal{K}(\Pi)$ such that $\liminf_n \mathbb{P}[\mathcal{X}_n \in K] \geq 1 - \epsilon$. By Lemma 2.8, for any $\eta > 0$ we can choose δ_0 such that for all $\delta \in (0, \delta_0)$ we have $\sup_{B \in K} \sup_{f \in B} w'(f, \delta, \Lambda_{L,T}) < \eta$, and thus

$$\liminf_{n \to \infty} \mathbb{P}\left[\sup_{f \in \mathcal{X}_n} w'(f, \delta, \Lambda_{L,T}) \leq \eta\right] \geq 1 - \epsilon.$$

Equation (1.14) follows.

It remains to prove the converse. Suppose instead that $(\mathcal{X}_n)$ satisfies (1.14). Our plan is to show that $\mathcal{X}_n$ is likely to be close to a suitably chosen set of step functions.

For $\delta > 0$, let B_δ be the set of step functions in Π with jumps spaced at least δ apart (in time). It is easily seen, from Lemma 2.7 that B_δ is a relatively compact subset of Π. By Lemma 2.2, $C_\delta = \{B \in \mathcal{K}(\Pi)\,;\, B \subseteq \overline{B_\delta}\}$ is a relatively compact subset of $\mathcal{K}(\Pi)$.

Fix $L, T \in (0, \infty)$ and let $\eta > 0$. Consider first some f and δ such that $w'(f, \delta, \Lambda_{L,T}) \leq \eta$. We define a function $f' \in B_\delta$ as follows. Since $w'(f, \delta, \Lambda_{L,T}) \leq \eta$ there exists a δ-sparse partition (t_i) of $[(-T) \vee \sigma_f, T]$

such that $\max_i \sup_{s,t\in[t_i,t_{i+1})} |f_{L,T}(s) - f_{L,T}(t)| \leq \eta$. Set

$$f'(t) = \begin{cases} f_{L,T}(t_i) & \text{for } t \in [t_i, t_{i+1}) \\ 0 & \text{otherwise.} \end{cases}$$

Thus $\sup_{|t|\leq T} |f_{L,T}(t) - f'(t)| \leq \eta$. We have

$$\begin{aligned} \sup_{t\in[-1,2]} |\bar{f}(t) - \bar{f}'(t)| &\leq \sup_{t\in[-1,2]} |\bar{f}(t) - \overline{f_{L,T}}(t)| + \sup_{t\in[-1,2]} |\overline{f_{L,T}}(t) - \bar{f}'(t)| \\ &\leq \left(|1 - \tanh(L)| + \frac{2}{1+|T|}\right) + \left(\eta + \frac{2}{1+|T|}\right). \end{aligned}$$

Here, to deduce the second inequality, we split each sup into $|\kappa_t| \leq T$ and $|\kappa_t| > T$, and then use the left hand side of (2.5) along with (1.2). Choosing T, L sufficiently large (dependent on η but not on δ) we obtain that $\sup_t |\bar{f}(t) - \bar{f}'(t)| \leq 2\eta$, which in turn implies that $d_\Pi(f, f') = d_G(\bar{f}, \bar{f}') \leq 2\eta$. To summarise:

$$w'(f, \delta, \Lambda_{L,T}) \leq \eta \qquad \Rightarrow \qquad \exists f' \in B_\delta \text{ such that } d_\Pi(f, f') \leq 2\eta. \quad (2.8)$$

We are now in a position to complete the proof.

Let $\eta, \epsilon > 0$ and let $\Lambda_{L,T}$ be large enough that (2.8) holds. By (1.14) choose $\delta > 0$ such that

$$\liminf_{n\to\infty} \mathbb{P}\left[\sup_{f\in\mathcal{X}_n} w'(f, \delta, \Lambda_L, T) \leq \eta\right] \geq 1 - \epsilon. \quad (2.9)$$

Consider now the event that $\sup_{f\in\mathcal{X}_n} w'(f, \delta, \Lambda_{L,T}) \leq \eta$, and let $\mathcal{X}_n' = \{f' \,;\, f \in \mathcal{X}_n\}$, where f' is the function given in (2.8). Since $\mathcal{X}_n' \subseteq B_\delta$ we have $\overline{\mathcal{X}_n'} \in C_\delta$. Moreover, by (2.8) the Hausdorff distance between $\mathcal{X}_n$ and $\overline{\mathcal{X}_n'}$ is bounded above by 2η. Hence, on this event $\mathcal{X}_n \in C_\delta^{(2\eta)}$, where $C_\delta^{(2\eta)}$ denotes the expansion (in the Hausdorff metric) of C_δ by radius 2η; thus also $\mathcal{X}_n \in \overline{C_\delta}^{(2\eta)}$. Thus, from (2.9) we obtain

$$\liminf_{n\to\infty} \mathbb{P}\left[\mathcal{X}_n \in \overline{C_\delta}^{(2\eta)}\right] \geq 1 - \epsilon.$$

Take $\eta = \epsilon/2$. Noting that $\overline{C}_\delta$ is compact in $\mathcal{K}(\Pi)$, by (2.7) we have that $(\mathcal{X}_n)$ is tight. □

Proof: [Of Proposition 1.3.] Take $f \in \Pi$, and fix $\delta > 0$ and $\Lambda_{L,T}$. We aim to relate $w'(f, \delta, \Lambda_{L,T})$ to $w''(f, \delta, \Lambda_{L,T})$ and $w^+(f, \delta, \Lambda_{L,T})$. The argument is similar to the proof of Theorem 12.4 in [3].

To ease our notation, for the duration of this proof we introduce the shorthand $w' = w'(f, \delta, \Lambda_{L,T})$, similarly for w'', w^+ and also w_i from (1.13).

Let us also write $h = f_{L,T}$, and note that $\sigma_f = \sigma_h$. Without loss of generality we may assume that $\sigma_h < T$, since h is constant outside of time $[-T, T]$, and no contribution to w', w'', w^+ is made at such times.

Take α such that $w' < \alpha$. Then there exists a δ-sparse partition (s_i) of $[\sigma_h, T]$ such that, for all i, $w_i \leq \alpha$. If s, t are such that $|s-t| < \delta$ then, there exists at most one value of i such that $s_i \in [s, t)$. Therefore, if $s < u < t$, we have $s, u \in [s_i, s_{i+1})$ for some i, or $t, u \in [s_i, s_{i+1})$ for some i, which means that either $|h(s) - h(u)| < \alpha$ or $|h(u) - h(t)| < \alpha$. Thus $w'' \leq \alpha$, and since α was arbitrary we have $w'' \leq w'$.

Since the first element of the δ-sparse partition (s_i) is $s_1 = 0$, and $s_2 - s_1 > \delta$, we have that $w^+ \leq w'$. Thus,

$$w'' \vee w^+ \leq w', \tag{2.10}$$

and it remains to prove a suitable inequality in the reverse direction. The reader may wish to glance ahead at (2.12) to see where we are heading.

Our first step is to show that for all $t_1 \leq s \leq t \leq t_2$ such that $|t_2 - t_1| \leq \delta$ we have

$$|h(t_1) - h(s)| \wedge |h(t) - h(t_2)| \leq 2w''. \tag{2.11}$$

Equation (2.11) is easily seen: if $|h(t_1) - h(s)| \geq w''$ then (by definition of w'') we have both $|h(s) - h(t)| \leq w''$ and $|h(s) - h(t_2)| \leq w''$, which implies that $|h(t) - h(t_2)| \leq 2w''$.

Now take any $\alpha > w'' \vee w^+$. Suppose that h has at a pair of jumps, each of magnitude exceeding 2α, at the points $u_1 < u_2$. If $|u_2 - u_1| < \delta$ then we can find disjoint intervals $(t_1, s) \ni u_1$ and $(t, t_2) \ni u_2$ with $|t_1 - t_2| < \delta$, and this would contradict (2.11). Note also that w^+ controls movement during $[\sigma_h, \sigma_h + \delta)$. We thus have that:

(1) Any two jumps of h of magnitude exceeding 2α must occur at least δ time apart;
(2) h has no jumps of magnitude exceeding 2α during $[\sigma_h, \sigma_h + \delta)$;
(3) h is constant outside of $[-T, T]$.

Therefore, there exists a finite sequence (s_i) with $\sigma_f = s_0 < s_1 < \ldots < s_m = T$, with $s_{i+1} - s_i > \delta$ for all i, which contains all points at which h jumps more than 2α. If $s_{i+1} - s_i > \delta$ for any pair of adjacent points, enlarge the (s_i) by including their midpoint; continue doing so inductively until we have an enlarged partition satisfying, for all i,

$$\frac{\delta}{2} < s_{i+1} - s_i \leq \delta.$$

Thus, (s_i) forms a $\delta/2$-sparse partition of $[\sigma_f, T]$, and that all the jumps of h with magnitude exceeding 2α occur at some s_i.

We now look to control w', using the partition (s_i). Fix i and consider t_1, t_2 such that $s_i \leq t_1 < t_2 < s_{i+1}$, which implies $|t_1 - t_2| < \delta$. Let

$$\sigma_1 = \sup\{\sigma \in [t_1, t_2]\,;\, |h(t_1) - h(\sigma)| \leq 2\alpha\},$$
$$\sigma_2 = \inf\{\sigma \in [t_1, t_2]\,;\, |h(\sigma) - h(t_2)| \leq 2\alpha\}.$$

If $\sigma_1 < \sigma_2$ then we have s, t such that $t_1 \leq \sigma_1 < s < t < \sigma_2 \leq t_2$, which contradicts (2.11). Thus $\sigma_2 \leq \sigma_1$, which implies that $|h(t_1) - h(\sigma-)| \leq 2\alpha$ and (by right continuity of h) that $|h(\sigma) - h(t_2)| \leq 2\alpha$. Since $\sigma_1 \in [s_i, s_{i+1})$ the jump at σ_1 has magnitude at most 2α. Thus $|h(t_1) - h(t_2)| \leq 6\alpha$.

From the previous paragraph we have $w_i(f, \delta/2, \Lambda_{l,T}) \leq 6\alpha$, and since i was arbitrary we have $w'(f, \delta/2, \Lambda_{L,T}) \leq 6\alpha$. Since α was also arbitrary we have

$$w'(f, \delta/2, \Lambda_{L,T}) \leq 6\Big(w''(f, \delta, \Lambda_{L,T}) \vee w^+(f, \delta, \Lambda_{L,T})\Big). \tag{2.12}$$

Combining (2.10) and (2.12), we have that (1.14) is equivalent to (1.18), so Proposition 1.3 follows from Proposition 1.2. □

Remark 2.9: Equations (2.10) and (2.12), when applied to Lemmas 2.7 and 2.8, yield characterizations of relative compactness in Π and $\mathcal{K}(\Pi)$ based on w'' and w^+. This is left for the reader.

3. On the failure of tightness due to large jumps

In this section we will briefly outline why a system of coalescing random walkers, making suitably heavy tailed jumps, will fail to be tight in $\mathcal{K}(\Pi)$ under the space-time scaling limit that would correspond to a single random walker converging to an α-stable Lévy process. Heuristically, the presence of too many particles at once results in at least one of them making a large movement too soon after its own starting time; tightness fails as a consequence, in both $\mathcal{K}(\Pi)$ and $\mathcal{K}(\widetilde{\Pi})$.

More precisely, consider a random sequence $(\mathcal{X}_n)$ of sets of paths. If $(\mathcal{X}_n)$ consists of continuous paths and is tight in $\mathcal{K}(\widetilde{\Pi})$, then (1.6) holds and (using (1.11)) it is easily seen that for all bounded $\Lambda_{L,T}$,

$$\lim_{\delta\to 0} \limsup_{n\to\infty} \mathbb{P}\left[\sup_{f\in\mathcal{X}_n} w^+(f, \delta, \Lambda_{L,T}) > \eta\right] = 0. \tag{3.1}$$

Alternatively, if $(\mathcal{X}_n)$ consists of càdlàg paths and tightness holds in $\mathcal{K}(\Pi)$, then (1.14) holds and (3.1) is a trivial consequence. Thus, for both $\mathcal{K}(\Pi)$

and $\mathcal{K}(\widetilde{\Pi})$, tightness necessitates (3.1), which we might describe in words as uniform right continuity in probability at starting times.

Let us now restrict ourselves to the example of a collection $\mathcal{Y}$ of càdlàg paths of continuous time coalescing random walks on the space-time lattice $\mathbb{Z} \times \mathbb{R}$. More precisely, let J be a $\mathbb{Z}$ valued random variable, and specify that if a random walk is at location $x \in \mathbb{Z}$, then after an exponential time with rate 1 the random walker jumps to $x + J$. All such jumps take an independent copy of J, and all random walkers coalesce instantaneously upon meeting, leaving at most one random walker at each point of space-time. We write $f_{(y,s)} : [s, \infty) \to \mathbb{Z}$ for the càdlàg path of the random walker beginning at $(y, s) \in \mathbb{Z} \times \mathbb{R}$, and thus $\mathcal{Y} = \{f_{(y,s)} ; (y, s) \in \mathbb{Z} \times \mathbb{R}\}$.

Let $\alpha \in (0, 2]$. Following [1], which treats only the case $\alpha = 2$ of a diffusive rescaling, let $D_n(\eta, \delta)$ be the event that one of the paths within $\mathcal{Y}$ has a jump which originates within $[\eta n, \infty) \times [0, \delta n^\alpha]$ and lands within the negative half-plane $(-\infty, 0] \times [0, \delta n^\alpha]$.

Let $\mathcal{Y}_n$ denote $\mathcal{Y}$ with time sped up by a factor n^α and space compressed by a factor n. Thus, the region $[0, \eta] \times [0, \delta]$ of rescaled space-time (for $\mathcal{Y}_n$) corresponds to the unscaled region $[0, \eta n] \times [0, \delta n^\alpha]$ (for $\mathcal{Y}$).

Lemma 3.1: *If* (3.1) *holds for* $(\mathcal{Y}_n)$ *then*

$$\lim_{\delta \to 0} \limsup_{n \to \infty} \mathbb{P}\left[D_n(\eta, \delta)\right] = 0. \tag{3.2}$$

Proof: Without loss of generality, suppose that $T \geq \delta$ and $L \geq \eta$. Suppose that the event $D_n(\eta, \delta)$ occurs. We seek to show that $\sup_{f \in \mathcal{Y}_n} w^+(f, \delta, \Lambda_{L,T}) > \eta$.

We have that there is some path $f \in \mathcal{Y}$ which begins during $[0, \delta n^\alpha]$ and makes a jump originating from within $[\eta n, \infty) \times [0, \delta n^\alpha]$ that lands within $(-\infty, 0] \times [0, \delta n^\alpha]$. After rescaling f corresponds to some path in $\mathcal{Y}_n$ which has starting (rescaled) time within $[0, \delta]$ and makes a jump of (rescaled) magnitude at least η before time δ. This completes the proof. □

Remark 3.2: In the proof of Theorem 1.1 in [1], which treats the diffusive case $\alpha = 2$ and uses with linear interpolation in place of jumps, it was shown that (3.2) failed if the $3 - \epsilon$ moment of J was infinite for any $\epsilon > 0$. From this, [1] deduced that random walk 'approximations' to the Brownian web could be tight in $\mathcal{K}(\widetilde{\Pi})$ only if J had at least $3 - \epsilon$ moments.

Let us now consider $\alpha \in (0, 2)$. For clarity, we will treat only the special case in which $p(j) = \mathbb{P}[J = j] = \mathbb{P}[J = -j] \sim Cj^{-\alpha-1}$, where C is a normalizing constant. Therefore, each of our random walkers is, individually, within the domain of attraction of an α-stable Lévy process. (Note that this assumption is sufficient to ensure that our system of coalescing random walkers on $\mathbb{Z} \times \mathbb{R}$ is well defined.)

Let $F(k) = \sum_{j=k}^{\infty} p(j)$. Random walk jumps that start within $[\eta n, \infty)$ and land in $(-\infty, 0]$ occur within $\mathcal{Y}$ at rate $\sum_{k=\eta n}^{\infty} F(k)$. Thus,

$$\mathbb{P}[D_n(\eta, \delta)] = 1 - \exp\left(-\delta n^{\alpha} \sum_{k=\eta n}^{\infty} F(k)\right).$$

Hence, (3.2) holds if and only if

$$\limsup_{n\to\infty} n^{\alpha} \sum_{k=\eta n}^{\infty} F(k) < \infty. \tag{3.3}$$

Recall that $p(j) \sim Cj^{-\alpha-1}$, which means that $F(k) \sim C'k^{-\alpha}$. Hence, the argument of the lim sup on the left hand side of equation (3.3) is of order $C'n^{\alpha}(\eta n)^{-\alpha+1}\frac{1}{\alpha+1} = \mathcal{O}(n)$, and thus (3.3) fails. Hence, by Lemma 3.1 and Proposition 1.3, tightness in $\mathcal{K}(\Pi)$ fails.

Acknowledgement

The authors would like to thank the anonymous referee for their comments and corrections.

References

1. S. Belhaouari, T. Mountford, R. Sun, and G. Valle, Convergence results and sharp estimates for the voter model interfaces. *Electronic Journal of Probability*, 11:768–801, 2006.
2. N. Berestycki, C. Garban, and A. Sen, Coalescing Brownian flows: A new approach. *The Annals of Probability*, 43(6):3177–3215, 2015.
3. P. Billingsley, *Probability and Measure*. Wiley, 1995.
4. A. Etheridge, N. Freeman, and D. Straulino, Branching Brownian motion, the Brownian net and selection in the spatial Lambda-Fleming-Viot process. *Electon J. Probab.*, 22(39):1–36, 2017.
5. S. N. Ethier and T. G. Kurtz, *Markov processes: characterization and convergence*. Wiley, 1986.
6. L. R. G. Fontes, M. Isopi, C. M. Newman, and K. Ravishankar, The Brownian web: characterization and convergence. *Annal. Probab*, 32(4):2857–2883, 2004.
7. N. Freeman and S. Palau, Learning session on the Brownian web and net. Available from http://ims.nus.edu.sg/events/2017/gene/wk2.php, 2017.

8. A. Greven, R. Sun, and A. Winter, Continuum space limit of the genealogies of interacting Fleming-Viot processes on Z. *Electronic Journal of Probability*, 21, 2016.
9. T. Mountford, K. Ravishankar, and G. Valle, A construction of the stable web. *ALEA, Lat. Am. J. Probab. Math. Stat.*, 16:787–807, 2019.
10. J. Norris and A. Turner, Hastings–Levitov aggregation in the small-particle limit. *Communications in Mathematical Physics*, 316(3):809–841, 2012.
11. E. Schertzer, J. M. Swart, and R. Sun, *The Brownian web, the Brownian net, and their universality.* Advances in Disordered Systems, Random Processes and Some Applications, 270–368. Cambridge University Press, 2017.
12. B. Tóth and W. Werner, The true self-repelling motion. *Probab. Th. Rel. Fields*, 111:375–452, 1998.

On Projections of the Supercritical Contact Process: Uniform Mixing and Cutoff Phenomenon

Stein Andreas Bethuelsen

Department of Mathematics and Physics
University of Stavanger, Norway
stein.a.bethuelsen@uis.no

We consider the contact process on a countable-infinite and connected graph of bounded degree. Our focus is on its mixing properties in the supercritical regime. In particular, we prove that the projection of the stationary contact process onto a finite subset forms a process which is ϕ-mixing whenever the infection parameter is sufficiently large. The proof of this is based on large deviation estimates for the spread of an infection and general correlation inequalities. In the special case of the contact process on $\mathbb{Z}^d$, $d \geq 1$, we furthermore show the cutoff phenomenon, valid in the entire supercritical regime.

Contents

1. Introduction

1.1. *The contact process*

The contact process is a one-parameter family of interacting particle systems, introduced by Harris in [13] as a toy model for the spread of an infection in a population. In these processes an individual of the population is either *healthy* (represented by assigning it the label 0) or *infected* (represented by assigning it the label 1). With time, an individual recovers (or becomes healthy) at rate 1, irrespectively of the state of the other individuals. Additionally, an individual becomes infected at a rate λ times the number of infected individuals in its neighbourhood, where $\lambda \in (0, \infty)$ is the parameter of the model.

More formally, the contact process with parameter $\lambda \in (0, \infty)$, denoted here by $(\eta_t)_{t\geq 0}$, is a continuous-time Markov process on $\Omega := \{0,1\}^V$, where V is the vertex set of a graph $G = (V, E)$ representing the network on which the individuals live. For the process to be well defined, and to avoid certain technicalities, we assume throughout this paper that the graph G is countable, connected and of bounded degree. We denote by $dist(x, y)$ the graph distance in G between any $x, y \in V$. Then, letting $\mathcal{C}(\mathbb{R})$ denote the set of bounded and continuous functions $f \colon \Omega \to \mathbb{R}$, the contact process can be specified by its generator $L_\lambda \colon \mathcal{C}(\mathbb{R}) \mapsto \mathcal{C}(\mathbb{R})$, where, for $\omega \in \Omega$

$$L_\lambda f(\omega) := \sum_{\substack{x\in V\\ \omega(x)=1}} \left([f(\omega^{x\leftarrow 0}) - f(\omega)] + \lambda \sum_{\substack{y\in V\\ dist(y,x)=1}} [f(\omega^{y\leftarrow 1}) - f(\omega)] \right). \tag{1.1}$$

Here, for $z \in V$ and $i \in \{0, 1\}$, we denote by $\omega^{z\leftarrow i}$ the configuration where $\omega^{z\leftarrow i}(z) = i$ and $\omega^{z\leftarrow i}(x) = \omega(x)$ for any $x \neq z$.

There is also a very useful definition of the contact process via coupling, known as the *graphical construction*, which we recall in Section 3.1. As references to interacting particle systems in general and the contact process in particular we mention the books by Liggett, [17] and [19].

As can be seen immediately from the definition in (1.1), the configuration $\underline{0} \in \{0,1\}^V$ where $\underline{0}(x) = 0$ for all $x \in V$, is an absorbing state. That is, if at a certain time T all individuals are healthy then they remain healthy for all future times $t \geq T$. On the other hand, as is well known, the contact process having initially all individuals infected converges weakly towards a stationary (or invariant) distribution, called the *upper invariant measure* and denoted here by $\bar{\nu}_\lambda$. In the sequel, we denote by the *upper stationary contact process* the process $\bar{\eta} = (\bar{\eta}_t)_{t\in\mathbb{R}}$ obtained as the stationary extension of the contact process under $\bar{\nu}_\lambda$. Further, we let $\bar{\mathbb{P}}_\lambda$ denote the corresponding path-space measure on the set $D_\Omega(\mathbb{R})$ of càdlàg functions on $\mathbb{R}$ taking values in Ω, and by $\mathcal{F}$ its σ-algebra.

An important feature of the contact process is that it has a *phase transition.* That is, for G countable-infinite, there exists a unique (and critical) parameter value $\lambda_c = \lambda_c(G) \in (0, \infty)$ defined through the following properties. For all $\lambda < \lambda_c$ we have that $\bar{\nu}_\lambda = \delta_{\underline{0}}$, where $\delta_{\underline{0}}$ denotes the measure concentrating on $\underline{0}$, and for all $\lambda > \lambda_c$ it holds that $\bar{\nu}_\lambda \neq \delta_{\underline{0}}$.

The contact process is said to be *supercritical* in the regime $\lambda > \lambda_c$. This also corresponds to the notion *weak survival* in [19] and is equivalent to having that, with positive probability, the process with initially only one infected individual never reaches the state $\underline{0}$. Other critical parameter values have been considered in the literature, most notably the one defined through the notion of *strong survival.* That is, the contact process survives strongly if, with positive probability, when initiated with only one infected individual, this individual will become (re)-infected from its neighbours infinitely often.

Clearly, strong survival implies weak survival. For the contact process on the integer lattices $\mathbb{Z}^d$, $d \geq 1$, and many other graphs, these two critical parameter values coincide. Interestingly, for a large class of graphs, such as all d-regular trees with degree $d \geq 3$, there is a regime of infection parameters where the contact process survives weakly, but not strongly.

In Section 5 we introduce yet another critical parameter value for the contact process and discuss its relation to that of weak and strong survival. Besides this, for the rest of the paper, we mainly restrict our analysis to cases where λ is large and the contact process is well within the strong survival regime. Sufficient for this is the assumption that $\lambda > \lambda_c(\mathbb{N})$. This assumption allows us to state our theorems without posing strong restrictions on the underlying graph.

To this end, it should be noted that $\lambda_c(\mathbb{N}) = \lambda_c(\mathbb{Z})$, as proven in [8]. We choose to write $\lambda > \lambda_c(\mathbb{N})$ in order to emphasise that in the proofs of the following statements, we make use of the fact that, for any connected and countable-infinite graph $G = (V, E)$, to each vertex $x \in V$ there is an embedded subgraph isomorphic to the graph $\mathbb{N}$ with x as the "origin".

1.2. *Projections of the contact process onto finite subsets*

Our main objects of interest in this paper are projections of the contact process onto finite subsets of the underlying graph. That is, we only observe the evolution of a partial and finite selection of all the individuals in the population. To make this precise, denote by $\mathcal{S}$ the set of all finite subsets of V. Then, given $\Delta \in \mathcal{S}$, we consider the process $\xi^{(\Delta)} = (\xi_t^{(\Delta)})_{t\in\mathbb{R}}$ on $\Omega_\Delta := \{0,1\}^\Delta$ obtained by projecting the upper stationary contact process onto Δ, that is,

$$\xi_t^{(\Delta)}(x) := \bar{\eta}_t(x), \quad x \in \Delta, t \in \mathbb{R}. \tag{1.2}$$

Thus, the process $\xi^{(\Delta)}$ is a function of the Markov process $\bar{\eta}$ and as such it is an example of a *hidden Markov process.* By the stationarity of $\bar{\eta}$ it follows that also the process $\xi^{(\Delta)}$ is a stationary process. However, in contrast to $\bar{\eta}$, we note that $\xi^{(\Delta)}$ is not Markovian.

Our motivation for studying the processes $(\xi^{(\Delta)})_{\Delta\in\mathcal{S}}$ is manifold. Firstly, they are fairly natural processes to study in the view of the contact process as a population model. Indeed, populations are typically very large and often it is in practice impossible to study the evolution of the entire population. One therefore is essentially forced to consider observations within partial subsets of the population only.

Secondly, there has recently been a boost of work on the contact process evolving on finite graphs. Although the contact process on finite graphs can neither survive strongly nor weakly, the time until it reaches its stationary state may depend crucially on the starting configuration and the parameter value. In fact, metastable behaviour have recently been shown to hold for the contact process in great generality concerning the graph structure as soon as the infection parameter is large enough, see e.g. [22] and [28]. Motivated by the progress seen for the contact process on general finite graphs, in this paper we study the asymptotic properties of $(\xi^{(\Delta)})_{\Delta\in\mathcal{S}}$ for the contact process on general countable-infinite graphs, for which previous works have mainly been devoted its study on the particular graphs $\mathbb{Z}^d$ and $\mathbb{T}^d$, $d \geq 1$. (See, however, [24] and [25] for two notable exceptions.)

A third motivation has been to further highlight the potentials of the methods developed in [1] and [21]. These techniques rely strongly on the so-called *downward FKG* property as shown to hold for the contact process in [2]. (See Section 3.2 for a precise definition.) We have focused here on proving strong (uniform) mixing properties for $\xi^{(\Delta)}$, extending upon the mixing properties proven in [1]. These mixing properties are very useful to the mathematical analysis of the contact process, of which we provide several basic examples.

For further discussions on the current work we refer to the next section, where our main results are presented, and to Section 5, where we also phrase some in our opinion intriguing open questions.

1.3. *Outline of the paper*

In Subsection 2.1 we present our main results on the mixing properties of $\xi^{(\Delta)}$ and in Subsection 2.2 we discuss certain applications of these mixing properties. In Section 3 we introduce the graphical construction of the contact process and discuss some of its key properties. Proofs of our main results are deferred to Section 4. We end with a short discussion in Section 5.

2. Main results and applications

Throughout this section we let $\bar{\eta}$ denote the upper stationary contact process on a *countable-infinite* and *connected* graph $G = (V, E)$ of *bounded degree.* Given $\Delta \in \mathcal{S}$, we focus our study on the mixing properties of the projected process $\xi^{(\Delta)}$.

2.1. *Main results*

Projections of the contact process have previously been considered in the literature by several (e.g. [9], [12], [26] and [29]), but to our knowledge only on the particular graphs $\mathbb{Z}^d$, $d \geq 1$. In [12] and [26] two key properties of the contact process on $\mathbb{Z}$ were used in order to derive various statistics, in particular the central limit theorem, for $\xi^{(\Delta)}$ in the supercritical regime. On the one hand, the contact process is monotone (in the sense of positive association; see Section 3.2 for a precise definition) and on the other hand it has fast decay of temporal correlations.

The first property extends immediately to the contact process considered on any other graph. We next present a generalisation of the latter property. For $t \in \mathbb{R}$, we write $\mathcal{F}^{\Delta}_{\geq t}$ and $\mathcal{F}^{\Delta}_{\leq t}$ for the sub-σ-algebras of $\mathcal{F}$ generated by events in $\Delta \times [t, \infty)$ and $\Delta \times (-\infty, t)$, respectively.

Proposition 2.1: *If $\lambda > \lambda_c(\mathbb{N})$, then for any $\Delta \in \mathcal{S}$, the projected process $\xi^{(\Delta)}$ is α-mixing (or strong mixing), that is,*

$$\lim_{t\to\infty} \sup_{A_0 \in \mathcal{F}^{\Delta}_{\leq 0}} \sup_{B_t \in \mathcal{F}^{\Delta}_{\geq t}} \left| \bar{\mathbb{P}}_\lambda (B_t \cap A_0) - \bar{\mathbb{P}}_\lambda (A_0) \bar{\mathbb{P}}_\lambda (B_t) \right| = 0. \tag{2.1}$$

Moreover, there are constants $C, c \in (0,\infty)$ depending on λ only, such that, for fixed $t \in (0,\infty)$, the lefthand side of (2.1) *decays as $C|\Delta|e^{-ct}$.*

The main purpose of this subsection is to present a generalisation of the mixing property in (2.1) which, for any event $A_0 \in \mathcal{F}^{\Delta}_{\leq 0}$, compares on $\mathcal{F}^{\Delta}_{\geq t}$ the conditional measure $\bar{\mathbb{P}}_\lambda(\cdot \mid A)$ with its unconditional counterpart $\bar{\mathbb{P}}_\lambda(\cdot)$, asymptotically for large times t. For this, we introduce the (non-increasing) function $d_\Delta\colon [0,\infty) \to [0,1]$ given by

$$d_\Delta(t) := \sup_{\substack{A_0 \in \mathcal{F}^{\Delta}_{\leq 0} \\ \bar{\mathbb{P}}_\lambda(A_0)>0}} \sup_{B_t \in \mathcal{F}^{\Delta}_{\geq t}} \left| \bar{\mathbb{P}}_\lambda(B_t \mid A_0) - \bar{\mathbb{P}}_\lambda(B_t) \right|, \tag{2.2}$$

which gives a uniform bound on the distance between $\bar{\mathbb{P}}_\lambda(\cdot \mid A)$ and $\bar{\mathbb{P}}_\lambda(\cdot)$ on $\mathcal{F}^{\Delta}_{\geq t}$ with respect to the total variation distance. One of our main contributions in this paper is the following theorem.

Theorem 2.2: *If $\lambda > \lambda_c(\mathbb{N})$, then for any $\Delta \in \mathcal{S}$, the projected process $\xi^{(\Delta)}$ is ϕ-mixing, that is, $\lim_{t\to\infty} d_\Delta(t) = 0$.*

A ϕ-mixing process is also α-mixing, but the opposite is in general not true. In this sense, Theorem 2.2 is a strengthening of Proposition 2.1. As we exemplify in the next subsection, both α-mixing and ϕ-mixing are strong forms of mixing having several applications to the study of the long term behaviour of $\xi^{(\Delta)}$, and in general also to $(\eta_t)_{t\geq 0}$. For a general account on mixing properties for processes, we refer to [5].

The proof of Theorem 2.2 is based on certain modifications of a stochastic domination argument in [1] together with the following property for the contact process conditioned on survival. In order to state this property precisely let, for $\Lambda \subset V$,

$$\tau^\Lambda := \inf\left\{t \geq 0\colon \eta_t^\Lambda \equiv \underline{0}\right\} \tag{2.3}$$

be the first time $(\eta_t^\Lambda)_{t\geq 0}$ reaches the state $\underline{0}$, where $(\eta_t^\Lambda)_{t\geq 0}$ denotes the contact process with the initial state satisfying $\eta_0^\Lambda(x) = 1$ if and only if $x \in \Lambda$. We denote the corresponding path-space measure by $\mathbb{P}^\Lambda_\lambda$. If $\Lambda = \{y\}$ for some $y \in V$, we for simplicity write τ^y and $\mathbb{P}^y_\Lambda$. Further, for $y \in V$ and $\theta \in (0,\infty)$, let

$$C_\theta(y) := \{(x,s) \in V \times [0,\infty)\colon dist(y,x) \leq s\theta\} \tag{2.4}$$

be the "forward cone" of inclination θ and center at y. For $t \in [0,\infty)$, let $C_{\theta,t}(y) = C_\theta(y) \cap V \times [t,\infty)$ and denote by $\mathcal{F}_{\theta,t}(y)$ the σ-algebra generated by the events in $C_{\theta,t}(y)$.

Theorem 2.3: *Let $\lambda > \lambda_c(\mathbb{N})$. Then there are constants $\theta, C, c \in (0,\infty)$ such that, for all $y \in V$,*

$$\sup_{B_t \in \mathcal{F}_{\theta,t}(y)} \left|\mathbb{P}^y_\lambda(B_t \mid \tau^y = \infty) - \bar{\mathbb{P}}_\lambda(B_t)\right| \le Ce^{-ct}, \quad \forall t \ge 0. \tag{2.5}$$

The proof of Theorem 2.3 goes by a modification of an argument in [7], who considered the contact process on $\mathbb{Z}^d$, to more general graphs.

In general, we believe that the mixing measure $d_\Delta(t)$ in Theorem 2.2 decays exponentially in t, similarly to the bound obtained in Proposition 2.1. Further, in many cases the assumption that $\lambda > \lambda_c(\mathbb{N})$ is clearly too strong. We end this subsection with a result which shows that this indeed is the case for the contact process on $\mathbb{Z}^d$, $d \ge 1$. For this, for $n \in \mathbb{N}$, let $\Delta_n := [-n,n]^d$ and denote by $\xi^{(n)}$ the projection of $\bar{\eta}$ onto Δ_n. Further, denote by

$$t(x) := \inf\{t \ge 0 \colon \eta^o_t(x) = 1\}, \quad x \in \mathbb{Z}^d, \tag{2.6}$$

the first time the contact process with only the individual at the origin initially infected spreads its infection to the individual at x. From the proof of the *shape theorem* for the contact process (see e.g. [10], Section 5), it follows that, $\mathbb{P}^o_\lambda(\cdot \mid \tau^o = \infty)$-a.s.,

$$\lim_{n\to\infty} \frac{t(n\cdot e_1)}{n} := \beta \tag{2.7}$$

for a constant $\beta = \beta(\lambda)$ representing the asymptotic linear speed for the spread an infection in the direction of $e_1 := (1,0,\ldots,0)$, and which is strictly positive whenever $\lambda > \lambda_c(\mathbb{Z}^d)$.

Theorem 2.4: *Let $\lambda > \lambda_c(\mathbb{Z}^d)$. Then, for any $\epsilon > 0$, there are constants $C, c \in (0,\infty)$ such that for all $n \in \mathbb{N}$,*

$$d_{\Delta_n}(t) \le C|\Delta_n|e^{-ct}, \quad \forall\, t \ge \beta n(1+\epsilon). \tag{2.8}$$

Furthermore, we have that

$$\lim_{n\to\infty} d_{\Delta_n}(\beta n(1+\epsilon)) = 0 \quad \textit{and} \quad \lim_{n\to\infty} d_{\Delta_n}(\beta n(1-\epsilon)) = 1. \tag{2.9}$$

In terms of mixing times, as typically studied for Markov chains (see e.g. [16]), the second part of Theorem 2.4 says that the processes $(\xi^{(n)})_{n\in\mathbb{N}}$ has (asymptotically) a cutoff at time $\tau_n := \beta n$, called the mixing time of $(\xi^{(n)})_{n\in\mathbb{N}}$. As we discuss more thoroughly in Section 4.4 (see also Section 5), we furthermore believe that the estimate on the mixing time in (2.8) hold under minimal assumptions on the graph structure as soon as $\lambda > \lambda_c(\mathbb{N})$.

2.2. *Applications*

As a first application of the mixing properties stated in the previous subsection, we focus on large deviation estimates for $\xi^{(\Delta)}$. For $f\colon \Omega_\Delta \to \mathbb{R}$ bounded, denote by

$$Z_t^{(\Delta)}(f) := \int_0^t f(\xi_t^{(\Delta)})dt, \quad t > 0. \tag{2.10}$$

By Theorem 2.2 it follows that also $(Z_t^{(\Delta)}(f))_{t\in\mathbb{R}}$ is ϕ-mixing. Large deviation estimates for ϕ-mixing sequences have been proven in [27]. By an immediate extension of the proof in [27] to continuous-time processes, we obtain our first application of Theorem 2.2, as stated next.

Corollary 2.5: *Let $\lambda > \lambda_c(\mathbb{N})$. Then, for any $\epsilon > 0$ there are constants $C, c > 0$ such that*

$$\bar{\mathbb{P}}_\lambda\left(Z_t^{(\Delta)}(f)/t \notin (\bar{\nu}_\lambda(f) - \epsilon, \bar{\nu}_\lambda(f) + \epsilon)\right) \leq Ce^{-ct}, \quad \forall\, t \geq 0. \tag{2.11}$$

By combining the large deviation estimates in Corollary 2.5 with standard subadditivity arguments as in [15], Section 3, we furthermore obtain a large deviation principle. To state this precisely, consider the partial ordering on Ω_Δ given by $\sigma \leq \omega$ if and only if $\sigma(x) \leq \omega(x)$ for all $x \in \Delta$. We say that a function $f\colon \Omega_\Delta \to \mathbb{R}$ is *increasing* if $f(\sigma) \leq f(\omega)$ whenever $\sigma \leq \omega$.

Theorem 2.6: *Let $\lambda > \lambda_c(\mathbb{N})$ and $f\colon \Omega_\Delta \to \mathbb{R}$ bounded and increasing. Then there exists $\psi\colon \mathbb{R} \to \{-\infty\} \cup (-\infty, 0]$ concave such that $\psi(x) < 0$ for all $x \neq \bar{\nu}_\lambda(f)$ and, for all $a < b$,*

$$\lim_{t\to\infty} t^{-1} \log \bar{\mathbb{P}}_\lambda\left(\frac{Z_t^\Delta(f)}{t} \in [a, b]\right) = \sup_{a\leq x\leq b} \psi(x). \tag{2.12}$$

We remark that the large deviation estimates in Corollary 2.5 hold without the assumption that f is increasing, whereas this assumption is important to the analysis via subadditivity arguments in [15].

Other asymptotic properties of $Z_t^{(\Delta)}(f)$, such as the central limit theorem, have earlier been considered for the contact process on $\mathbb{Z}$; see [26] and [12, Section 2.4]. These works extends to the contact process on general countable-infinite graphs of bounded degree.

Corollary 2.7: *Let $\lambda > \lambda_c(\mathbb{N})$ and consider $f\colon \Omega_\Delta \to \mathbb{R}$ bounded. Then there exists $0 \leq \sigma_f^2 < \infty$ such that*

$$t^{1/2}\left[t^{-1} Z_t^{(\Delta)}(f) - \bar{\nu}_\lambda(f)\right] \to_L N(0, \sigma_f^2) \quad \textit{as } t \to \infty, \tag{2.13}$$

where $\to_L$ denotes convergence in law.

The proof of Corollary 2.7 follows by Proposition 2.1 applied to the proof strategy outlined in [26]. In fact, Corollary 2.7 is the generalisation of Lemma 1 in [26] to more general graphs and the proof of this lemma goes through after only minor modifications, replacing the last estimate with that of Proposition 2.1.

We remark that Corollary 2.7 can be further strengthened to the immediate extension of Theorem 1 in [26]. For this, the estimates in Theorem 2.3 are needed. Moreover, the proofs in [12] work in our generality as well and yield more quantitive bounds on the rate of convergence in (2.13) whenever f is increasing (for which $\sigma_f^2 > 0$; see [26]).

As a last application, we mention the complete convergence theorem, which is a consequence of Theorem 2.3.

Corollary 2.8: *Let $\lambda > \lambda_c(\mathbb{N})$. Then, for any $\Lambda \subset V$,*

$$\mathbb{P}_\lambda^\Lambda(\eta_t \in \cdot) \implies \mathbb{P}_\lambda^\Lambda(\tau^\Lambda < \infty)\delta_{\underline{0}}(\cdot) + \mathbb{P}_\lambda^\Lambda(\tau^\Lambda = \infty)\bar{\nu}_\lambda(\cdot), \quad as\ t \to \infty. \tag{2.14}$$

For a proof of Corollary 2.8 based on Theorem 2.3 we refer to [17, p. 284-287], where the proof for the case of the contact process on $\mathbb{Z}$ is given in much detail. The proof strategy therein applies after minor notational modifications only also to the contact process on general graphs as soon as the estimate in Theorem 2.3 is at hand. To the best of our knowledge, in this generality, the complete convergence theorem was first proven in [24, Theorem 5], by an elegant argument, however, not relying on any explicit mixing property.

3. Preliminaries

In this section we recall the graphical construction of the contact process and some of its basic consequences. For a more thorough description we refer to [19, Chapter 1].

3.1. *Graphical construction of the contact process*

Let $G = (V, E)$ be a connected graph having bounded degree and fix $\lambda \in (0, \infty)$. Assign Poisson processes N_x on $\mathbb{R}$ of rate 1 to each individual $x \in V$ and Poisson processes $N_{(x,y)}$ on $\mathbb{R}$ of rate λ to each ordered pair of individuals satisfying $dist(x, y) = 1$. All these processes are taken independent of each other and together yield the following space-time picture on $V \times \mathbb{R}$.

To each event t of N_x draw a *star* $\star$ at (x, t) and to each event s of $N_{(x,y)}$ draw an arrow $\rightarrow$ from (x, s) to (y, s). The stars resembles the event that an individual becomes healthy and arrows resembles the possible spreading of an infection. From these Poisson processes we can construct the contact process using terminology from percolation theory. For $x, y \in V$ and $s \leq t$, we say that (x, s) is connected to (y, t) by an *active* path, written $(x, s) \to (y, t)$, if and only if there exists a directed path in $V \times \mathbb{R}$ starting at (x, s), ending at (y, t) and going either forwards in time without hitting crosses or "sideways" following arrows in the direction of the prescribed direction. Otherwise we write $(x, s) \nrightarrow (y, t)$. In general, for $\Lambda, \Delta \subset V \times \mathbb{R}$, we write $\Delta \to \Lambda$ ($\Delta \nrightarrow \Lambda$) if there is a (there is no) active path from Λ to Δ.

Now, for $\Lambda \subset V$ and $s \in \mathbb{R}$, the contact process $(\eta_t^{(\Lambda,s)})_{t\geq s}$ on G with infection parameter λ and satisfying $\eta_s(x) = 1$ if and only if $x \in \Lambda$ can be defined by

$$\eta_t^{(\Lambda,s)}(x) = \begin{cases} 1, & \text{if } (\Lambda, s) \to (x, t); \\ 0, & \text{otherwise.} \end{cases} \tag{3.1}$$

If $s = 0$ we simply write $(\eta_t^{\Lambda})_{t\geq 0}$ for the above defined process. It is well known that the process defined by (3.1), known as the *graphical construction* of the contact process, has the same distribution as the process defined via the generator in (1.1).

In many of the proofs in the following section it will be useful to consider the contact process evolving on certain subgraphs of G. For a graph $G_1 = (V_1, E_1)$ with $V_1 \subset V$ and $E_1 \subset E$, we write $({}^{G_1}\eta_t^{(\Lambda,s)})_{t\geq s}$ for the contact process evolving on $\{0,1\}^{V_1}$ where in the graphical construction we only consider the events $(N_x)_{x\in V_1}$ and $(N_{(x,y)})_{(x,y)\in E_1}$. Further, we introduce the notation

$${}^{G_1}\tau^{(\Lambda,s)} := \inf\{t \geq s \colon \Lambda \times \{s\} \nrightarrow (y, t) \text{ in } G_1 \text{ for any } y \in V_1\}, \tag{3.2}$$

which is the time until the contact process started at time s with only the individuals inside Λ initially infected reaches the state $\underline{0}$ when restricting the graphical construction to G_1.

3.2. *Consequences of the graphical construction*

The graphical construction yields an elegant description of the upper stationary contact process:

$$\bar{\eta}_t(x) := \begin{cases} 1, & \text{if } V \times \{-\infty\} \to (x, t); \\ 0, & \text{otherwise.} \end{cases} \tag{3.3}$$

In (3.3) and later, we denote by $V \times \{-\infty\} \to (x, t)$ the event that for all $s \leq t$ there exists an active path from $V \times \{s\}$ to (x, t).

A further advantage of the graphical construction is that it yields a natural coupling of contact processes with different starting configurations or infection parameters, denoted in the following by $\widehat{\mathbb{P}}_\lambda$. Moreover, from the construction it is easy to see that the contact process is *monotone*, in the sense that, if $\Delta \subset \Lambda$, then $\widehat{\mathbb{P}}_\lambda$-a.s., $\eta_t^\Delta(x) \leq \eta_t^\Lambda(x)$ for all $x \in V$ and all $t \in [0, \infty)$. In fact, the contact process is *positively associated*. In particular, for any two increasing events $A, B \in \mathcal{F}$, we have that

$$\bar{\mathbb{P}}_\lambda(A \cap B) \geq \bar{\mathbb{P}}_\lambda(A)\bar{\mathbb{P}}_\lambda(B). \tag{3.4}$$

Here, an event A is said to be increasing if it has the property that when $(\omega_t) \in A$ then also $(\sigma_t) \in A$ whenever $\omega_t \leq \sigma_t$ for all t.

Another very useful property is that the contact process is *self-dual*, that is, for any $\Delta, \Lambda \in \mathcal{S}$,

$$\mathbb{P}_\lambda^\Delta\left(\eta_t^\Delta(x) = 1 \text{ for some } x \in \Lambda\right) = \mathbb{P}_\lambda^\Lambda\left(\eta_t^\Lambda(x) = 1 \text{ for some } x \in \Delta\right). \tag{3.5}$$

To see this from the graphical construction, first recall that $\{\eta_t^\Delta(x) = 1 \text{ for some } x \in \Lambda\} = \{(\Delta, 0) \to (\Lambda, t)\}$. This event again equals $\{(\Delta, 0) \leftarrow (\Lambda, t)\}$, where we write $\{(\Delta, 0) \leftarrow (\Lambda, t)\}$ if there exists a directed *backwards*-path in $V \times \mathbb{R}$ starting at Δ, ending at Λ and going either *backwards* in time without hitting crosses or "sideways" following arrows in the *opposite* direction of the prescribed direction. The latter event defines the *dual process* $(\tilde{\eta}_t^{(\Lambda,s)})_{t\geq 0}$ given by

$$\tilde{\eta}_t^{(\Lambda,s)}(x) := \begin{cases} 1, & \text{if } (x, s-t) \leftarrow (\Lambda, s); \\ 0, & \text{otherwise,} \end{cases} \tag{3.6}$$

It is clear by the construction that $(\tilde{\eta}_t^{(\Lambda,s)})$ has the same distribution as $(\eta_t^{(\Lambda,0)})$ and hence that the contact process is self-dual.

The last property of the graphical construction we want to discuss is that of *downward FKG* (abbreviated by dFKG), which is an even stronger correlation inequality than positive association introduced above. In particular, for each $\Lambda \in \mathcal{S}$ and any interval $[s_1, s_2] \in \mathbb{R}$, it says that

$$\bar{\mathbb{P}}_\lambda\left(\cdot \mid \bar{\eta}_t(x) = 0 \,\forall\, (x,t) \in \Lambda \times [s_1, s_2]\right) \tag{3.7}$$

is positively associated. This property was first proven in [2] for the upper invariant measures and in [1] it was noted that the proof in [2] extends, again by use of the graphical construction, to the upper stationary contact process.

The dFKG property is very useful when considering conditional measures since it implies that

$$\bar{\mathbb{P}}_\lambda\left(B \mid \bar{\eta}_t(x) = 0 \,\forall\, (x,t) \in \Lambda \times [s_1, s_2]\right) \leq \bar{\mathbb{P}}_\lambda\left(B \mid A\right), \tag{3.8}$$

where $B \in \mathcal{F}$ is any increasing event depending only on the contact process outside $\Lambda \times [s_1, s_2]$ and $A \in \mathcal{F}$ is any event depending only on the contact process within $\Lambda \times [s_1, s_2]$. In other words, the measure $\bar{\mathbb{P}}_\lambda(\cdot \mid A)$ *stochastically dominates* the measure $\bar{\mathbb{P}}_\lambda(\cdot \mid \bar{\eta}_t(x) = 0 \, \forall \, (x,t) \in \Lambda \times [s_1, s_2])$. For a proof of this fact, see [1].

To this end, we should emphasise that positive association may fail for other conditional measures of the contact process than that on the lefthand side of (3.7). For instance, [18] proved that this is the case for the upper invariant measure of the contact process on $\mathbb{Z}$ conditioned on having an infected individual at the origin. On the other hand, the dFKG property is known to hold for a wide range of percolation-like models as shown in [2]. (See also [20] and [3] for further discussions on correlation inequalities for the contact process.)

4. Proofs

In this section we present the proofs of the statements given in Section 2.1. Detailed proofs of the statements in Section 2.2 are left to the reader. As standard, throughout this section, the numbers $C, c \in (0, \infty)$ represent constants whose value might change from line to line in the calculations, but which remain strictly bounded away from 0 and ∞.

4.1. *Proof of Proposition 2.1*

Let $G = (V, E)$ be a countable-infinite and connected graph of degree bounded by $D \in \mathbb{N}$. Further, to each individual $y \in V$ we assign a self-avoiding nearest neighbour path $\gamma^{(y)} = (\gamma_0, \gamma_1, \gamma_2, \ldots)$ in V. That is, the path $\gamma^{(y)}$ satisfies $\gamma_0 = y$, $dist(\gamma_i, \gamma_{i+1}) = 1$ for each $i \geq 0$, and $\gamma_i \neq \gamma_j$ for $i \neq j$. We denote by $\Gamma^{(y)}$ the corresponding subgraph with vertex set $V_1 = \gamma^{(y)}$ and edge set $E_1 := \{e \in E \colon e = (\gamma_{i-1}, \gamma_i), i \in \mathbb{N}\}$.

The following lemma is a very useful generalisation of Proposition 4 in [7], important to the proofs of Proposition 2.1 and Theorem 2.3.

Lemma 4.1: *Let $\lambda > \lambda_c(\mathbb{N})$. Then there are constants $C, c \in (0, \infty)$ such that, for all $y \in V$,*

$$\mathbb{P}_\lambda^y\left(t < \tau^y < \infty\right) \leq Ce^{-ct}, \quad \forall \, t \geq 0. \tag{4.1}$$

Proof: This follows by exactly the same argument as in the proof of Proposition 4 in [7] via a so-called restart argument. For completeness, we present the key ideas of the proof. Fix $t \in (0, \infty)$ and $y \in V$. Let $v_0 = 0$, and define iteratively $(v_k)_{k\geq 1}$ as follows. If $\eta^y_{v_k} = \emptyset$, then we set $v_{k+1} = v_k$. Otherwise, if $\eta^y_{v_k} \neq \emptyset$, let x_k be a vertex with the property that $\eta^y_{v_k}(x_k) = 1$, chosen in an arbitrary manner. Let $(\eta^{(k)}_{v_k+t})$ denote the truncated contact process started at time v_k with only individual x_k infected and using the graphical construction only within the subgraph $\Gamma^{(x_k)}$. Consequently, we have that $\eta^{(k)}_{v_k+t} \leq \eta^y_{v_k+t}$ for all $t \geq 0$ a.s. Now, set $v_{k+1} = \inf\{s \geq v_k \colon \eta^{(k)}_{v_k+s} = \emptyset\}$ and let

$$k_0 := \inf\{k \geq 0 \colon v_k = v_{k+1} \text{ or } v_{k+1} = \infty\}. \tag{4.2}$$

Thus, since $\lambda > \lambda_c(\mathbb{N})$, on each trial time k there is a positive probability that $v_k = \infty$. In particular, k_0 is majorised by a geometrically distributed random variable and, moreover, we have that

$$\{t < \tau^y < \infty\} = \{t < v_{k_0} < \infty\}. \tag{4.3}$$

Further, from the graphical construction we have that, given $k_0 = k$, the times $v_j - v_{j-1}$ for $1 \leq j \leq k$ are i.i.d. Moreover, they have exponential tails since (4.1) holds for the contact process on $\Gamma^{(y)}$ uniformly in y. Indeed, the contact process on $\Gamma^{(y)}$ is isomorphic to the contact process on $\mathbb{N}$ for which this property was proven in [8, Theorem 5]. (The paper [8] treated the contact process on $\mathbb{Z}$, but their methods can easily be adapted to the process on $\mathbb{N}$, see the discussion on p. 6 in [8].) By combining these two properties, we conclude the proof of the lemma. □

Corollary 4.2: *Let $\Delta \in \mathcal{S}$. Then there are constants $C, c \in (0,\infty)$ such that,*

$$\widehat{\mathbb{P}}_\lambda(A_{\Delta,0,t}) \leq C|\Delta|e^{-ct}, \quad \forall\, t \geq 0, \tag{4.4}$$

where $A_{\Delta,0,t} := \{(V\times\{0\}) \leftarrow (\Delta\times[t,\infty))\} \cap \{(V\times\{-\infty]) \not\leftarrow (\Delta\times[t,\infty))\}$.

Proof: Together with the backwards-paths started from $\Delta\times\{t\}$, it is sufficient to control the paths starting at the (randomly distributed) space-time points

$$B_k := \{(x, N_{x,y}) \in \Delta\times[t+k, t+k+1) \colon dist(x,y) = 1\}, \quad k \geq 0. \tag{4.5}$$

Hence, by a union bound estimate, using the self-duality property and the independence structure of the graphical construction, we have that

$$\widehat{\mathbb{P}}_\lambda(A_{\Delta,0,t}) \leq Ce^{-ct}(|\Delta| + \sum_{k\geq 0} e^{-ck}\sum_{l\geq 0} l\widehat{\mathbb{P}}_\lambda(B_k = l)) \leq C'|\Delta|e^{-c't} \tag{4.6}$$

for some constants $C', c' \in (0,\infty)$, since $\sum_{l\geq 0} l\mathbb{P}(B_k = l) < \lambda|\Delta|D$ for any k, and since $\sum_{k\geq 0} e^{-ck} < \infty$. □

Proof: [Proof of Proposition 2.1] The proof follows by essentially the same argument as that for the proof of Lemma 2 in [26], where a slightly weaker property was shown. Indeed, [26] proved the inequality (2.1) for the contact process on $\mathbb{Z}$ in the particular case where the events $A \in \mathcal{F}^{(\Delta)}_{\leq 0}$ and $B \in \mathcal{F}^{(\Delta)}_{\geq t}$ only depend on the contact process at time 0 and t, respectively. For our extension, it is sufficient to consider events of the form

$$A = \{\eta_s(x) = \omega_s(x), (x,s) \in \Delta \times (-r_1, 0]\}; \tag{4.7}$$
$$B = \{\eta_s(x) = \omega_s(x), (x,s) \in \Delta \times [t, t + r_2)\}, \tag{4.8}$$

where $r_1, r_2 \in (0,\infty)$ and $(\omega_s)_{s\in\mathbb{R}} \in D_\Omega(\mathbb{R})$. In particular, we allow for any events A and B having finitely many "jumps" within $\Delta \times (-r_1, 0]$ and $\Delta \times [t, t + r_2)$ respectively. Then, letting

$$\Delta_B := \{(x,s) \in \Delta \times [t, t + r_2) \colon \omega_s(x) = 1\}, \tag{4.9}$$

and by replacing the definitions of B' and E' in the proof of Lemma 2 in [26] by their natural generalisations, namely

$$B' := \{V \times \{-\infty\} \leftarrow \Delta_B\} \text{ and } E := \{V \times \{0\} \leftarrow \Delta_B\}, \tag{4.10}$$

we attain, by the same proof as that of Lemma 2 in [26], the inequality

$$|\bar{\mathbb{P}}_\lambda(A \cap B) - \bar{\mathbb{P}}_\lambda(A)\bar{\mathbb{P}}_\lambda(B)| \leq \widehat{\mathbb{P}}_\lambda(E \cap (B')^c) \tag{4.11}$$
$$\leq \widehat{\mathbb{P}}_\lambda(A_{\Delta,0,t}), \tag{4.12}$$

from which we conclude the proof by applying Corollary 4.2. □

4.2. *Proof of Theorem 2.3*

For the proof of Theorem 2.3 we follow the idea of the proofs of Propositions 5 and 6 in [7]. These propositions yield slightly stronger statements for the contact process on $\mathbb{Z}^d$, $d \geq 1$, under the same assumption that $\lambda > \lambda_c(\mathbb{N})$. The proofs in [7] use geometrical properties of $\mathbb{Z}^d$, $d \geq 1$, and the argument therefore needs to be adapted in order to work for general graphs of bounded degree. (In fact, in [7] only a detailed proof of the case $d = 2$ is given.)

One important estimate for the proofs in [7], and for the proof of Theorem 2.3 below, is the following estimate for the contact process on $\mathbb{N}$.

Lemma 4.3: *Consider the contact process on* $\mathbb{N}$ *with* $\lambda > \lambda_c(\mathbb{N})$. *Then there are constants* $C, c, \alpha \in (0, \infty)$ *such that,*

$$\mathbb{P}_\lambda^0\left(t(x) > t \mid \tau^0 = \infty\right) \le Ce^{-ct}, \quad \forall\, t > 0, \tag{4.13}$$

whenever $dist(0, x) \le \alpha t$.

Proof: In [7] it is referred to [8] for a proof. That paper, however, concerns the contact process on $\mathbb{Z}$ for which they prove the analog of Lemma 4.3, see Theorem 4 therein. To be precise, Theorem 4 in [8] states that there are constants $C, c, \alpha \in (0, \infty)$ such that

$$\mathbb{P}_\lambda^{(-\infty,0]}(r_t < \alpha t) \le Ce^{-ct}, \quad t \ge 0. \tag{4.14}$$

Here r_t denotes the position of the rightmost infected individual at time t. Letting $x \le \alpha t$, we have that

$$\mathbb{P}_\lambda^0\left(t(x) > t \mid {}^{\mathbb{N}}\tau^0 = \infty\right) = \mathbb{P}_\lambda^0\left(r_t < \alpha t \mid {}^{\mathbb{N}}\tau^0 = \infty\right) \tag{4.15}$$

$$\le \mathbb{P}_\lambda^{(-\infty,0]}\left(r_t < \alpha t \mid {}^{\mathbb{N}}\tau^0 = \infty\right) \tag{4.16}$$

$$\le \mathbb{P}_\lambda^{(-\infty,0]}(r_t < \alpha t), \tag{4.17}$$

where in the last line we have used that $\mathbb{P}_\lambda^{(-\infty,0]}$ is positively associated, noting that $\{{}^{\mathbb{N}}\tau^0 = \infty\}$ and $\{r_t < \alpha t\}^c$ are increasing events. This derivation is formally with respect to the contact process on $\mathbb{Z}$, however, by the graphical construction coupling the estimate immediately transfers to the contact process on $\mathbb{N}$. □

From Lemma 4.1 and Lemma 4.3 we derive the following proposition for the contact process on a general graph.

Proposition 4.4: *Let* $\lambda > \lambda_c(\mathbb{N})$. *Then there are constants* $\alpha, C, c \in (0, \infty)$ *such that,*

$$\hat{\mathbb{P}}_\lambda\left(\eta_t^x(y) \ne \eta_t^V(y) \mid \tau^x = \infty\right) \le Ce^{-ct} \tag{4.18}$$

for any $x, y \in V$ *satisfying* $dist(x, y) \le \alpha t$.

Proof: Let $x, y \in V$ satisfy $dist(x, y) \le \alpha_0 t$ for some $\alpha_0 \in (0, \infty)$ to be determined. Note that the event $A := \{\eta_t^x(y) \ne \eta_t^V(y), \tau^x = \infty\}$ equals, by the graphical construction in Section 3.1, the intersection of the following three events:

$$\{\exists \text{ active path from } V \times \{0\} \text{ to } (y, t)\} \tag{4.19}$$

$$\{\exists \text{ active path from } (x, 0) \text{ to } V \times \{s\}\, \forall\, s > 0\} \tag{4.20}$$

$$\{\nexists \text{ active path from } (x, 0) \text{ to } (y, t)\}. \tag{4.21}$$

From this we see that A is contained in the union of A_1 and A_2;

$$A_t^{(1)} := \left\{\hat{\eta}_t^{(y,t)} \neq \underline{0}, \hat{\eta}_s^{(y,t)} = \underline{0} \text{ for some } s > t\right\} \tag{4.22}$$

$$A_t^{(2)} := \left\{\tau^x = \infty, \hat{\eta}_s^{(y,t)} \neq \underline{0}\, \forall\, s > 0, \eta_s^x \hat{\eta}_{t-s}^{(y,t)} = \underline{0}\, \forall\, s \in [0,t]\right\}, \tag{4.23}$$

where we recall that $(\hat{\eta}_s^{(y,t)})_{s\geq 0}$ denotes the dual process started from time t with only the individual y initially infected, and $\eta_s^x \hat{\eta}_{t-s}^{(y,t)}$ denotes the configuration obtain by multiplying the corresponding values of the individuals, i.e. $\eta_s^x \hat{\eta}_{t-s}^{(y,t)}(z) = \eta_s^x(z)\hat{\eta}_{t-s}^{(y,t)}(z)$, $z \in V$.

By Lemma 4.1 and the self-duality property of the contact process, the probability of $A_t^{(1)}$ is bounded by Ce^{-ct}. The rest of the proof is devoted to showing that a similar bound holds for $A_t^{(2)}$. For this, choose the self-avoiding paths $(\gamma^{(z)})_{z\in V}$ introduced in the previous subsection to be the concatenation of a shortest path from z to the graph $\Gamma^{(y)}$, say with endpoint $y_l \in \gamma^{(y)}$, and the path $(y_l, y_{l+1}, \dots)$. As in the proof of Lemma 4.1, it follows that $\mathbb{P}_\lambda^x(T > \epsilon t \mid \tau^x = \infty)$ decays exponentially in t, where $\epsilon \in (0,1)$ and

$$T := \inf\left\{t \geq 0\colon \eta_t^x(z) = 1 \text{ and } {}^{\Gamma^{(z)}}\tau_t^z = \infty \text{ for some } z \in V\right\} \tag{4.24}$$

denotes the first time that the contact process started from only x infected spreads its infection within one of the subgraphs $(\Gamma^{(z)})_{z\in V}$. By a comparison with a continuous-time branching process with branching rate $D\lambda$ it furthermore follows that, for δ sufficiently large, the probability

$$\mathbb{P}_\lambda^x\left(\exists\, z \in V, dist(z,x) \geq \delta t\colon \eta_t(z) = 1\right) \tag{4.25}$$

decays exponentially in t. Hence, we may assume that there is a $z \in V$ within distance $\delta\epsilon t$ of x satisfying $\eta_T(z) = 1$ and ${}^{\Gamma^{(z)}}\tau_T^z = \infty$.

By self-duality, the above argumentation also applies to the dual process $(\hat{\eta}_s^{(y,t)})$ and yields that, with a probability exponentially close to 1 in t, at a time $\hat{T} \leq \epsilon t$, there is a vertex w at distance at most $\delta\epsilon t$ from y and satisfying that $(y,t) \to (z, t-\hat{T})$ and that this infection is spread (backwards in time) within the subgraph $\Gamma^{(w)}$.

Finally, in order to conclude the proof, it is sufficient to control the event that the infection paths from (z,T) (forward in time) intersects with the (backwards-)paths from $(w, t-\hat{T})$ in the time-interval $[T, t-\hat{T}]$. After a bit of thought, it is not difficult to see that this happens at a probability bounded from below by $1 - Ce^{-ct}$ for some constants $C, c \in (0,\infty)$. Indeed, due to the particular construction of the paths $(\gamma^{(z)})_{z\in V}$, we may apply the estimate in (4.15) to control the spread of infections from (z,T) and

$(w, t-\hat{T})$. Moreover, we can control the probability of intersecting infection paths by noting that, under the above analysis, we have $dist(z, y) \leq \alpha_0 t + \delta \epsilon t$ and $dist(w, y) \leq \delta \epsilon t$ and by taking the constants ϵ and α_0 sufficiently small. From this we conclude the proof of the proposition. □

Proof: [Proof of Theorem 2.3] From the graphical construction, and by self-duality of the contact process, we have that, for any $y \in V$,

$$\widehat{\mathbb{P}}_\lambda\left(\eta_t^V(y) \neq \bar{\eta}_t(y)\right) = \mathbb{P}_\lambda^y\left(t < \tau^y < \infty\right) \leq Ce^{-ct}, \quad \forall t > 0, \tag{4.26}$$

where the inequality follows by Lemma 4.1. Hence, by combining this with Proposition 4.4, we obtain that, for any $x \in V$,

$$\widehat{\mathbb{P}}_\lambda\left(\eta_t^x(y) \neq \bar{\eta}_t(y) \mid \tau^x = \infty\right) \leq Ce^{-ct}, \quad \forall t > 0, \tag{4.27}$$

whenever $dist(y, x) \leq \alpha t$. In almost exactly the same manner as in the proof of Corollary 4.2, we conclude from this that in fact

$$\widehat{\mathbb{P}}_\lambda\left(\eta_s^x(y) \neq \bar{\eta}_s(y) \text{ for some } s \in [t, t+1) \mid \tau^x = \infty\right) \leq Ce^{-ct}, \tag{4.28}$$

since, by the graphical construction, it is sufficient to consider the times in the time interval $[t, t+1)$ at which there is an arrow event towards y. Now, let $\alpha_0 \in (0, \alpha)$ and consider an event $B \in \mathcal{F}_{\alpha_0, t}(x)$. We have that

$$\begin{aligned} &\left|\mathbb{P}_\lambda^x(B \mid \tau^x = \infty) - \bar{\mathbb{P}}_\lambda(B)\right| && (4.29)\\ &\leq \widehat{\mathbb{P}}\left(\eta_s^x(y) \neq \bar{\eta}_s(y) \text{ for some } (y, s) \in C_{\alpha_0 t}(x) \mid \tau^x = \infty\right) && (4.30)\\ &\leq \sum_{l=\lfloor t \rfloor}^{\infty} \sum_{y \in \Delta_{l,\alpha_0}(x)} \widehat{\mathbb{P}}_\lambda\left(\eta_s^x(y) \neq \bar{\eta}_s(y), s \in [l, l+1) \mid \tau^x = \infty\right), && (4.31)\end{aligned}$$

where $\Delta_{l,\alpha_0}(x) := \{y \colon dist(y, x) \leq \alpha_0(l+1)\}$. In particular, by (4.28) the latter sum is bounded by $\sum_{l=\lfloor t \rfloor}^{\infty} D^{\alpha_0(l+1)} Ce^{-cl}$, which, for α_0 close to 0, decays exponentially in t. This completes the proof of the theorem. □

4.3. *Proof of Theorem 2.2*

From Lemma 4.1 above, and by use of similar arguments as in the proof of [1, Theorem 1.3], we derive the following lemma.

Lemma 4.5: *Let $\Delta \in \mathcal{S}$. Then there is $\delta > 0$ and $x \in V$ such that*

$$\inf_{A_0 \in \mathcal{F}_{\leq 0}^{\Delta}} \bar{\mathbb{P}}_\lambda\left(\bar{\eta}_0(x) = 1 \mid A_0\right) \geq \delta. \tag{4.32}$$

Proof: Let $X = (X_i)_{i\in\mathbb{Z}}$ be the (discrete-time) process on $\{0,1\}$ given by

$$X_i := \max\{\bar{\eta}_t(x) \colon x \in \Delta, t \in [i, i+1)\}. \tag{4.33}$$

Since $\bar{\mathbb{P}}_\lambda$ is dFKG and the dFKG property is preserved under taking maximum, it follows that also the process (X_i) is dFKG. Moreover, by Lemma 4.1 applied to [1, Lemma 2.5], we have that, for some $\delta > 0$,

$$\bar{\mathbb{P}}_\lambda (X_i = 0, i \in [0, m-1]) \leq \bar{\mathbb{P}}_\lambda (\bar{\eta}_t(x) = 0 \text{ for all } (x,t) \in \Delta \times [0,m]) \leq (1-\delta)^m. \tag{4.34}$$

Since (X_i) is translation invariant with respect to time shifts, we can thus apply Theorem 1.2 in [21], which says that the estimate in (4.34) is equivalent to

$$\lim_{T\uparrow\infty} \bar{\mathbb{P}}_\lambda (X_0 = 1 \mid X_i = 0, -T < i < 0) \geq \delta. \tag{4.35}$$

Note that, by the dFKG property, the probability on the left hand side of (4.35) is decreasing in T and hence the limit is well defined.

Now, consider the function $h \colon V \setminus \Delta \to [0,1]$ given by

$$h(x) := \lim_{T\to\infty} \bar{\mathbb{P}}_\lambda (\eta_0(x) = 1 \mid X_s = 0, -T < s < 0), \tag{4.36}$$

which, again by the dFKG property, is well defined. We claim that $h(x) > 0$ for some $x \in V \setminus \Delta$. Indeed, otherwise the measure $\bar{\mathbb{P}}_\lambda (\cdot \mid X_s = 0, s < 0)$ equals $\delta_{\underline{0}}$ on $\mathcal{F}_0 \subset \mathcal{F}$, the σ-algebra generated by events in $V \times \{0\}$. By the Markov property of the contact process, this readily contradicts (4.35). We thus conclude the proof of the lemma since, as a consequence of (3.8), we have that the righthand side of (4.32) can be taken to equal the δ in (4.34). □

Corollary 4.6: *There exists an $\epsilon > 0$ such that, for all t large, we have that*

$$\sup_{B_t \in \mathcal{F}^\Delta_{\geq t}} \sup_{A_0 \in \mathcal{F}^\Delta_{\leq 0}} \left| \bar{\mathbb{P}}_\lambda(B_t \mid A_0) - \bar{\mathbb{P}}_\lambda(B_t) \right| \leq 1 - \epsilon. \tag{4.37}$$

Proof: This follows immediately by combining Lemma 4.5 with Theorem 2.3 together with the Markov property of the contact process. □

Proof: [Proof of Theorem 2.2] Firstly, the contact process on G is mixing (in the ergodic-theoretic sense), that is,

$$\lim_{s\to\infty} \bar{\mathbb{P}}_\lambda(A \cap T_{-s}B) = \bar{\mathbb{P}}_\lambda(A)\bar{\mathbb{P}}_\lambda(B), \quad \forall\, A, B \in \mathcal{F}. \tag{4.38}$$

This follows as a consequence of Proposition 2.1. In particular, (4.38) holds for the process $\xi^{(\Delta)}$, which we also recall is a stationary process.

For a stationary process satisfying (4.38), the mixing measure $d_\Delta(t)$ either converges towards 0 or it is equal to 1 for all t. A detailed proof of this fact is given in [5], Theorem 22.3. ([5] considers discrete-time processes, however, the argument extends immediately to continuous-time processes.) From this and the two observations above, we thus conclude the proof of Theorem 2.2 by noting that, by Corollary 4.6, we have $d_\Delta(t) < 1$ for all sufficiently large t. □

4.4. *Proof of Theorem 2.4*

Important to the proof of Theorem 2.4 is the following lemma, which can be seen as an extension of Lemma 4.5 for the particular case of the contact process on $\mathbb{Z}^d, d \geq 1$.

Lemma 4.7: *Consider the upper stationary contact process on $\mathbb{Z}^d$, $d \geq 1$, with $\lambda > \lambda_c(\mathbb{Z}^d)$. Then there exists $l \in \mathbb{N}$ and $\delta > 0$ such that, for all $n \in \mathbb{N}$ and $T < 0$,*

$$\bar{\mathbb{P}}_\lambda\left(\bar{\eta}_0((n+l)\cdot e_1) = 1 \mid \bar{\eta}_t(x) = 0, x \in \Delta_n, t \leq [T, 0]\right) \geq \delta, \tag{4.39}$$

where $e_1 = (1, 0, \ldots, 0)$ is a unit vector.

Proof: We first consider the case of the contact process on $\mathbb{Z}$ with $\lambda > \lambda_c(\mathbb{Z})$. For $n \in \mathbb{N}$, let $(Y_i^{(n)})_{i\in\mathbb{Z}}$ be the (discrete-time) process on $\{0,1\}$ defined by

$$Y_i^{(n)} := \max\{\bar{\eta}_t(y) \colon y = j = -n, \ldots, -1 \text{ and } t \in [i, i+1)\}, \quad i \in \mathbb{Z}. \tag{4.40}$$

By the same argument as in the proof of Lemma 4.5 we conclude that the process $(Y_i^{(n)})$ is dFKG and, moreover, that for some $\delta > 0$,

$$\lim_{T\uparrow\infty} \bar{\mathbb{P}}_\lambda\left(Y_0^{(n)} = 1 \mid Y_i^{(n)} = 0, -T < i < 0\right) \geq \delta. \tag{4.41}$$

Observe that this estimate holds irrespectively of n.

Further, consider the function $f \colon \mathbb{N} \to [0,1]$ given by

$$f(x) := \lim_{n\to\infty} \lim_{T\uparrow\infty} \bar{\mathbb{P}}_\lambda\left(\bar{\eta}_0(x) = 1 \mid Y_i^{(n)} = 0, -T < i < 0\right) \tag{4.42}$$

which, again by the dFKG property, is well defined. We claim that $f(x) > 0$ for some $x \in \mathbb{N}$. For a contradiction, assume that this is not the case. Hence, for any fixed $M \in \mathbb{N}$ and $\epsilon > 0$, we may chose N so large so that, for all $n \geq N$, we have that

$$\lim_{T\uparrow\infty} \bar{\mathbb{P}}_\lambda\left(\bar{\eta}_0(x) = 1 \mid Y_i^{(n)} = 0, -T < i < 0\right) < \epsilon \tag{4.43}$$

for all $x \in [1, M] \cup [-(M+n), n]$, where we also make use of the translation invariance of the contact process. The estimate (4.43), however, is easily seen to contradict (4.41) by choosing ϵ and M appropriately. We thus conclude that $f(x) > 0$ for some $x \in \mathbb{N}$ and from this, by translation invariance and the dFKG property, we conclude (4.39) for the case $\mathbb{Z}$ and $\lambda > \lambda_c(\mathbb{Z})$.

We next consider the case of the contact process on $\mathbb{Z}^d$ with $\lambda > \lambda_c(\mathbb{Z}^d)$ for which we need to adapt the above proof slightly. For $n \in \mathbb{N}$, let $(Z_i^{(n)})$ be the (discrete-time) process on $\{0,1\}^{\mathbb{Z}^{d-1}}$ defined by

$$Z_i^{(n)}(x) := \max\left\{\bar{\eta}_t(y) \colon y = (j, x) \in \mathbb{Z}^d, j \in [-n, 0], t \in [i, i+1)\right\}. \quad (4.44)$$

Again this process is dFKG. Moreover, it holds that, for some $\delta > 0$,

$$\bar{\mathbb{P}}_\lambda\left(Z_i^{(n)}(x) = 0 \text{ for all } (x, i) \in [0, m]^{d-1} \times [0, m]\right) \leq (1-\delta)^{m^d}, \quad (4.45)$$

as shown in [1, Theorem 1.5]. Thus, by applying Theorem 4.1 in [21] (see also Lemma 2.4 in [1]), we conclude that

$$\bar{\mathbb{P}}_\lambda\left(Z_0^{(n)}(o) = 1 \mid Z_i^{(n)}(x) = 0 \text{ for all } x \in \mathbb{Z}^{d-1} \text{ and } i < 0\right) \geq \delta, \quad (4.46)$$

where $o \in \mathbb{Z}^{d-1}$ denotes the origin. Next, let $g \colon \mathbb{N} \to [0,1]$ be given by

$$g(m) := \lim_{n\to\infty} \lim_{T\uparrow\infty} \bar{\mathbb{P}}_\lambda\left(\exists x \in \Delta_m \colon \bar{\eta}_0(x) = 1 \mid Z_i^{(n)}(x) = 0, i \in (-T, 0), x \in \mathbb{Z}^{d-1}\right) \quad (4.47)$$

which, again by the dFKG property, is well defined. We claim that $g(m) > 0$ for some $m \in \mathbb{N}$. Indeed, the opposite contradicts (4.46), as can be seen by arguing similarly as in the $d = 1$ case. By this claim, and using translation invariant and the dFKG property, we conclude (4.39) also in this case and hence the proof of Lemma 4.7 is complete. □

Remark 4.8: The above proof strongly relies on symmetries of the contact process on $\mathbb{Z}^d$ and cannot immediately be extended to general graphs. These problems, however, can be circumvented for discrete-time analogs of the contact process for which a version of (4.39) holds, replacing Δ_n by any simply connected set Δ. Indeed, by using that infections spread only to neighbouring individuals and that it is a process in discrete-time, by following the argument for the contact process on $\mathbb{Z}$ above, it is not difficult to see that

$$\bar{\mathbb{P}}_\lambda\left(\bar{\eta}_0(x) = 1, x \in \partial\Delta \mid \bar{\eta}_t(x) = 0, x \in \Delta, t \leq [T, 0] \cap \mathbb{Z}\right) \geq \delta, \quad (4.48)$$

where $\partial\Delta = \{x \notin \Delta \colon dist(x, y) = 1 \text{ for some } y \in \Delta\}$ is the outer boundary of Δ.

Before presenting the proof of Theorem 2.4, we first recall a couple of large deviation estimates from [11], see Theorems 1 and 4 therein, for the contact process on $\mathbb{Z}^d$. ([11] in fact considered a generalisation of the contact process, evolving in a *random environment.*) For this, denote by

$$t'(x) := \inf\{T \geq 0 \colon \forall t \geq T, \eta^o_t(x) = \eta^{\mathbb{Z}^d}_t(x)\}, \quad x \in \mathbb{Z}^d, \tag{4.49}$$

and recall the definitions of $t(x)$ and β; see Equations (2.6) and (2.7).

Lemma 4.9: *Consider the contact process on $\mathbb{Z}^d$, $d \geq 1$, with $\lambda > \lambda_c(\mathbb{Z}^d)$. Then there are constant $C, c \in (0,\infty)$ such that for any $\epsilon > 0$ we have that*

$$\mathbb{P}^o_\lambda\left(t'(n \cdot e_1) \geq \beta n(1+\epsilon) \mid \tau^o = \infty\right) \leq Ce^{-cn} \tag{4.50}$$

$$\mathbb{P}^o_\lambda\left(t(x) \leq \beta n(1-\epsilon), dist(x,o) \geq n \mid \tau^o = \infty\right) \leq Ce^{-cn}. \tag{4.51}$$

Proof: [Proof of Theorem 2.4] Let $\bar{\eta}$ be the upper stationary contact process on $\mathbb{Z}^d$, $d \geq 1$, with $\lambda > \lambda_c(\mathbb{Z}^d)$. Then there are constants $C, c \in (0,\infty)$ such that, for any $B \in \mathcal{F}^{\Delta_n}_{\geq t}$,

$$\left|\bar{\mathbb{P}}_\lambda(B) - \mathbb{P}^{\mathbb{Z}^d}_\lambda(B)\right| \leq Cn^d e^{-ct}, \quad \forall\, t \geq 0. \tag{4.52}$$

This follows by self-duality and a standard union bound together with the fact that the contact process started from all individuals infected converges exponentially fast to $\bar{\nu}_\lambda$, that is, Theorem 1.2.30 in [19].

In order to conclude the first statement of the theorem, by the dFKG property (in particular, (3.8)), it is thus sufficient to show that for each $\epsilon > 0$ there are constant $C, c \in (0,\infty)$ such that, for every $B \in \mathcal{F}^{\Delta_n}_{\geq t}$ we have that

$$|\bar{\mathbb{P}}_\lambda(B) - \bar{\mathbb{P}}_\lambda(B \mid A_0)| \leq Cn^d e^{-ct}, \quad \forall\, t \geq \beta n(1+\epsilon), \tag{4.53}$$

where A_0 here denotes the event $\{\bar{\eta}_t(x) = 0\, \forall\, (x,t) \in \Lambda \times (-\infty, 0]\}$.

To this end, denote by $\widehat{\mathbb{P}}_\lambda$ the coupling of $(\eta_t)_{t\geq 0}$ and $(\eta^\mu_t)_{t\geq 0}$ via the graphical construction, where $\mu(\cdot) = \mathbb{P}(\eta_0 \in \cdot \mid A_0)$ is a measure on $(\Omega, \mathcal{F}_0)$ and η^μ_t is the contact process started with a configuration drawn according to μ at time 0. We have that

$$\left|\bar{\mathbb{P}}_\lambda(B) - \mathbb{P}^\mu_\lambda(B)\right| \leq \sum_{x \in \Delta_n} \widehat{\mathbb{P}}_\lambda\left(\bar{\eta} \neq \eta^\mu \text{ on } \{x\} \times [t,\infty)\right). \tag{4.54}$$

Let $\epsilon > 0$ and consider for each $x \in \Delta_n$ the event

$$M_x := \inf\{dist(y,x) \colon \eta^\mu_0(y) = 1 \text{ and } \tau^y = \infty\}. \tag{4.55}$$

In order to control the terms within the sum of (4.54), we use that

$$\widehat{\mathbb{P}}_\lambda\left(\bar{\eta} \neq \eta^\mu \text{ on } \{x\} \times [t, \infty)\right) \tag{4.56}$$

$$\leq \widehat{\mathbb{P}}_\lambda\left(\{M_x > \epsilon t\}\right) + \widehat{\mathbb{P}}_\lambda\left(\{\bar{\eta} \neq \eta^\mu \text{ on } \{x\} \times [t, \infty)\} \cap \{M_x \leq \epsilon t\}\right). \tag{4.57}$$

The first term on the righthand side of (4.57) decays exponentially in δt. Indeed, by Lemma 4.7 above there is an $l \in \mathbb{N}$ and $\delta > 0$ such that the measure μ stochastically dominates a Bernoulli product measure on the subset $\{y \in \mathbb{Z}^d : y = x + e_x \cdot (n \pm rl), r \in \mathbb{N}\}$ with density δ. Hence, by standard large deviation estimates for Bernoulli measures combined with Theorem 1.2.30 in [19], we conclude the exponential decay of $\mathbb{P}^\mu_\lambda\left(M_x > \epsilon t\right)$ in $t \geq 0$. Next, the second term on the righthand side of (4.57) decays exponentially whenever $t \geq \beta n(1+\epsilon)$, as a consequence of the first inequality in Lemma 4.9. From these two bounds, which hold uniformly in $x \in \Delta_n$, we conclude that (4.57) decays exponentially in t. In particular, we have that

$$(4.54) \leq \sum_{x \in \Delta_n} Ce^{-ct} \leq Cn^d e^{-ct}, \quad \forall\, t \geq \beta n(1+\epsilon), \tag{4.58}$$

which together with (4.52) yields the statement of (2.8).

For the second part, that is (2.9), we first note that (2.8) implies that, for any $\epsilon > 0$, we have $d_{\Delta_n}(\beta n(1+\epsilon)) \to 0$ as $n \to \infty$. In order to prove a lower bound, denote by $(\eta_t^{\Delta_n^c})$ the contact process started from the configuration where all individual inside Δ_n are healthy and all the other individuals are infected. Further, let $B_r := \{\eta_0(x) = 1 \text{ for some } x \in \Delta_r\}$. Then, for any $r \in \mathbb{N}$ and $\epsilon > 0$,

$$\mathbb{P}_\lambda^{\Delta_n^c}(\eta_{\beta n(1-\epsilon)}^{\Delta_n^c} \in B_r) \leq Cr^d e^{-c(n-r)}, \quad \forall\, n \geq 0, \tag{4.59}$$

which for fixed r goes to 0 as $n \to \infty$. Indeed, this bound follows as a consequence of the second inequality in Lemma 4.9 above and by the self-duality of the contact process. From this we obtain the second limit in (2.9) since $\bar{\mathbb{P}}_\lambda(B_r) \to 1$ as $r \to \infty$, and by this we conclude the proof. □

Remark 4.10: Some of the arguments in the proof of Theorem 2.4 can surely be extended to other graphs. In particular, assuming that $\lambda > \lambda_c(\mathbb{N})$, one can instead of Lemma 4.9 use the estimate on the spread of an infection obtained in Theorem 2.3. An important step in the proof of (2.8) above is, however, the use of Lemma 4.7. On general graphs, we do not know how to obtain such an uniform estimate.

Interestingly, these problems can be completely circumvented for discrete-time analogs of the contact process. By applying (4.48) in place of Lemma 4.7, the first part of the proof of Theorem 2.4 goes through without difficulty, using a discrete-time version of Theorem 2.3 for the large deviation estimates in the final estimate.

5. Discussion and open questions

In this last section we mention some potential extensions of the theory so far presented and discuss some in our opinion intriguing open questions.

(1) In Section 1.1 we introduced the notions weak and strong survival. Other critical parameter values have also been considered in the literature, see for instance [23]. We next define yet another critical parameter value which to our knowledge has not appeared in the literature before. Given a connected and countable-infinite graph $G = (V, E)$, define

$$\lambda_T(G) := \sup\{\lambda > 0 \colon \rho(\lambda, x) = 0 \text{ for all } x \in V\}, \tag{5.1}$$

where, for $x \in V$, we set

$$\rho_\lambda(x) := \lim_{t\to\infty} \frac{1}{t} \log\left(\bar{\mathbb{P}}_\lambda(\eta_s(x) = 0, s \in [0, t))\right), \quad x \in V. \tag{5.2}$$

Note that both $\lambda_T(G)$ and $\rho_\lambda(x)$ are well defined due to monotonicity of the contact process (both in G and λ) and by the large deviation principle in Theorem 2.6.

It is not difficult to see that $\lambda_T(G) \in (0, \infty)$. A lower bound on $\lambda_T(G)$ holds since its value is larger than that of weak survival. Indeed, $\bar{\nu}_\lambda$ has to be non-trivial in order for $\rho(\lambda)$ to be positive. Further, $\lambda_c(\mathbb{N})$ yields an upper bound on $\lambda_T(G)$ as follows from Lemma 4.1 applied to Lemma 2.5 in [1]. In fact, for many graphs, such as $\mathbb{Z}^d$, $d > 1$, the upper and lower bound can be shown to match and hence $\lambda_T(G) = \lambda_c(G)$ in these cases.

Question 1: How does $\lambda_T(G)$ relate to the critical values defined through the notions weak survival and strong survival? In particular, does there exist graphs G for which $\lambda_T(G) > \lambda_c(G)$?

Question 2a: Does the contact process on a graph G always converge exponentially fast (in the weak sense) to its equilibrium state when started from all individuals infected when $\lambda > \lambda_c(G)$?

Question 2b: Equivalently (by duality), for the contact process on a graph G, does there exists constants $C, c > 0$ such that, for all $x \in V$, we have that $\mathbb{P}(s \leq \tau^x < \infty) \leq Ce^{-cs}$ for all $s > 0$ whenever $\lambda > \lambda_c(G)$?

Remark 5.1: Any values of λ fulfilling the requirements of Question 2 also satisfy that $\rho(\lambda) > 0$ as can been seen by a classical restart argument. Hence, a positive answer to Question 2 would yield a negative answer to Question 1.

Based on Corollary 2.5 a natural followup question is whether large deviation properties hold as soon as $\lambda > \lambda_T$.

Question 3: Does the contact process on a graph G projected onto a finite subset always satisfy large deviation estimates similar to those seen in Corollary 2.5, when $\lambda > \lambda_c(G)$ or $\lambda > \lambda_T(G)$?

(2) As discussed in Section 4.4, the estimates on the mixing times in Theorem 2.4 can also be proven for certain other graphs and in more generality for discrete-time analogs of the contact process. Motivated by this the following question seems natural.

Question 4: For each $\lambda > \lambda_c(\mathbb{N})$ and $\epsilon > 0$, does there exist constants $\alpha, C, c \in (0, \infty)$ such that the contact process on a graph G projected onto any $\Delta \in \mathcal{S}$ satisfies that $d_\Delta(t) \leq C|\Delta|e^{-ct}$ whenever $t \geq \alpha\text{diam}(\Delta)(1 + \epsilon)$? If yes, does it also hold for all $\lambda > \lambda_c(G)$ or $\lambda > \lambda_T(G)$?

A natural followup questions to Question 4 is whether the contact process in the regimes considered therein also exhibits the cutoff phenomena, as obtained in Theorem 2.4 only for the special case of the contact process on $\mathbb{Z}^d$. It would also be interesting to study the dependency on $\Delta \in \mathcal{S}$ for other statistics of (ξ^Δ). One case treated in the literature to date is the asymptotic behaviour of the occurrence times of rare events, which has been studied for the contact process on $\mathbb{Z}$ in [14] and [9]. We postulate that these works can be extended to general countable-infinite graphs by similar methods to those developed in this paper.

(3) We find it somewhat surprising that projections of the contact process satisfy such a strong "loss of memory"-property as seen by Theorems 2.2 and 2.4, and it would be interesting to check whether other types of interacting particle systems may satisfy the same kind of mixing properties. The main technical tools used in this paper are the dFKG property, self-duality and estimates on the rate of convergences towards the equilibrium state. Presumably, many of our results can be extended to a larger class of attractive spin-flip systems for which such properties are known to hold, see [4] and [20]. Another interesting model to consider is the voter model as studied e.g. in [6].

Question 5: Is the occupation time process of the Voter model on $\mathbb{Z}^d$, as considered in [6], a ϕ-mixing process for any dimension $d \geq 3$?

Acknowledgement

The author thanks the IMS and the organisers of the program *Genealogies of Interacting Particle Systems*, July-August 2017, Institute for Mathematical Sciences, Singapore, where this work was initiated. He thanks Daniel Valesin for stimulating discussions at an early stage of this project and the referee for valuable comments. Financial support from the DFG, project GA582/7-2, is greatly acknowledged.

References

1. J. van den Berg and S. A. Bethuelsen. Stochastic domination in space-time for the contact process. *Random Structures Algorithms*, 53(2), 221–237, 2018.
2. J. van den Berg, O. Häggström, and J. Kahn. Some conditional correlation inequalities for percolation and related processes. *Random Structures Algorithms*, 29(4), 417–435, 2006.
3. J. van den Berg, O. Häggström, and J. Kahn. Proof of a conjecture of N. Konno for the 1D contact process. *Dynamics & stochastics*, IMS Lecture Notes, 48, 16–23, 2006.
4. C. Bezuidenhout and L. Gray. Critical attractive spin systems. *Ann. Probab.*, 22(3): 1160–1194, 1994.
5. R. C. Bradley. Introduction to strong mixing conditions. Vol. 2. Kendrick Press, 2007.
6. M. Bramson, J. T. Cox and D. Griffeath. Occupation time large deviations of the voter model. *Probab. Theory Related Fields,* 77(3), 401–413, 1988.
7. R. Durrett and D. Griffeath. Contact processes in several dimensions. *Z. Wahrsch. Verw. Gebiete*, 4(59), 535–552, 1982.
8. R. Durrett and D. Griffeath. Supercritical contact processes on $\mathbb{Z}$. *Ann. Probab.*, 11(1), 1–15, 1983.
9. A. Galves, F. Martinelli and E. Olivieri. Large-density fluctuations for the one-dimensional supercritical contact process. *J. Statist. Phys.*, 55(3-4), 639–648, 1989.
10. O. Garet and R. Marchand. Asymptotic shape for the contact process in random environment. *Ann. Probab.*, 22(4), 1362–1410, 2012.
11. O. Garet and R. Marchand. Large deviations for the contact process in random environment. *Ann. Probab.*, 42(4), 1438–1479, 2014.
12. L. Goldstein and N. Wiroonsri. Stein's method for positively associated random variables with applications to the Ising and voter models, bond percolation, and contact process. *Ann. Inst. H. Poincaré Probab. Statist.*, 54(1), 385–421, 2018.
13. T. E. Harris. Contact interactions on a lattice. *Ann. Probab.*, 2, 969–988, 1974.

14. J. L. Lebowitz and R. H. Schonmann. On the asymptotics of occurrence times of rare events for stochastic spin systems. *J. Statist. Phys.*, 48(3-4), 727–751, 1987.
15. J. L. Lebowitz and R. H. Schonmann. Pseudo-free energies and large deviations for non-Gibbsian FKG measures. *Probab. Theory Related Fields*, 77(1), 49–64, 1988.
16. D. A. Levin, Y. Peres and E. L. Wilmer. Markov chains and mixing times. *American Mathematical Society*, Providence, RI, 2017.
17. T. M. Liggett. *Interacting particle systems*, Springer, New York, 1985.
18. T. M. Liggett. Survival and coexistence in interacting particle systems, *Probability and phase transition*, Kluwer Academics, Dordrecht, 1994.
19. T. M. Liggett. *Stochastic interacting systems: contact, voter and exclusion processes*, Springer, Berlin, 1999.
20. T. M. Liggett. Conditional association and spin systems, *ALEA Lat. Am. J. Probab. Math. Stat.*, 1, 1–19, 2006.
21. T. M. Liggett and J. E. Steif. Stochastic domination: the contact process, Ising models and FKG measures. *Ann. Inst. H. Poincaré Probab. Statist.*, 42 (2), 223–243, 2006.
22. T. Mountford, J.-C. Mourrat, D. Valesin and Q. Yao. Exponential extinction time of the contact process on finite graphs. *Stochastic Process. Appl.*, 127(7), 1974–2013, 2016.
23. R. Pemantle. The contact process on trees. *Ann. of Probab.*, 20(4), 2089–2116, 1992.
24. M. Salzano and R. H. Schonmann. The second lowest extremal invariant measure of the contact process. *Ann. of Probab.*, 25(4), 1846–1871, 1997.
25. M. Salzano and R. H. Schonmann. The second lowest extremal invariant measure of the contact process. II. *Ann. of Probab.*, 27(2), 845–875, 1999.
26. R. H. Schonmann. Central limit theorem for the contact process. *Ann. Probab.*, 14(4), 1291–1295, 1986.
27. R. H. Schonmann. Exponential convergence under mixing. *Probab. Theory Related Fields*, 81, 235–238, 1989.
28. B. Shapira and D. Valesin. Extinction time for the contact process on general graphs. *Probab. Theory Related Fields*, 169, 871–899, 2017.
29. A. Simonis. Filling the hypercube in the supercritical contact process in equilibrium. *Markov Process. Related Fields*, 4(1), 113–130, 1998.

Phase Transition for the Contact Process in a Random Environment on $\mathbb{Z}^d \times \mathbb{Z}^+$

Qiang Yao
School of Statistics
East China Normal University
No. 500 Dongchuan Road
Shanghai 200241, P. R. China
qyao@sfs.ecnu.edu.cn

We review the results in Chen and Yao [5], [6] which concerns the contact process in a static random environment on the half space $\mathbb{Z}^d \times \mathbb{Z}^+$ and make some additions to them. Furthermore, we explain why our methods cannot apply to the whole space case and compare our results with some related works.

1. Introduction

1.1. *Basic definitions of the contact process*

The basic contact process, which will be denoted by "contact process" in the following, was introduced in Harris [12]. It is a model to describe the spread of diseases. The process is defined as follows. Given a graph $G = (V, E)$, where V denotes the vertex set of G, and E denotes the edge set of G, the contact process $(\xi_t : t \geq 0)$ is a continuous-time Markov process, whose state space is $\{A : A \subseteq V\}$. At each t, each vertex is either healthy or infected. Denote by ξ_t the collection of infected vertices at time t. The transition rates are as follows:

$$\begin{cases} \xi_t \to \xi_t \setminus \{x\} \text{ for } x \in \xi_t \text{ at rate } 1, \\ \xi_t \to \xi_t \cup \{x\} \text{ for } x \notin \xi_t \text{ at rate } \lambda \cdot |\{y \in \xi_t : x \sim y\}|, \end{cases} \tag{1.1}$$

where $\lambda > 0$ is a positive constant, $|\cdot|$ denotes the cardinality of a set, and "$x \sim y$" denotes that the vertices x and y are neighbors. The intuitive interpretation of the above transition rates is that an infected vertex becomes healthy at fixed rate 1, while a healthy vertex becomes infected at rate proportional to the number of its infected neighbors. The proportional coefficient λ is the parameter of the contact process. Readers can refer to the standard references Liggett [14] and Durrett [7] for how the above rates determine a Markov process in a rigorous way. Often we use the notation $(\xi_t^A : t \geq 0)$ to denote the contact process with initial state A, that is, at time 0 all vertices in A are infected, while all vertices outside of A are healthy. There is another viewpoint for the contact process which treats infected vertex as "1" while treats healthy vertex as "0". Under this viewpoint, the contact process is a Markov process with state space $\{0,1\}^V$. Therefore, the contact process is a special example of "spin system" (see Liggett [14] for rigorous definition).

The main problem in studying the contact process is its asymptotic behavior. For the process ξ_t^O with a single infected vertex $O \in V$ at time 0, we say that the process survives if $\mathbf{P}(\xi_t^O \neq \emptyset \text{ for any } t \geq 0) > 0$, otherwise we say that the process dies out. Furthermore, we say that the process survives strongly if $\mathbf{P}(\forall T \geq 0,\ \exists t > T, \text{ such that } O \in \xi_t^O) > 0$. And we say that the process survives weakly if it survives but not survives strongly. By the monotonicity (or attractiveness) of the contact process (which implies that the process is inclined to survive with larger infection parameter λ), we can define two critical values as follows:

$$\begin{cases} \lambda_1 := \inf\{\lambda : \ \xi_t^O \text{ survives}\}, \\ \lambda_2 := \inf\{\lambda : \ \xi_t^O \text{ survives strongly}\}. \end{cases}$$

If G is a connected graph, then the value of λ_1 and λ_2 do not depend on the choice of the vertex O. Since strong survival implies survival, it can be easily seen that $\lambda_1 \leq \lambda_2$.

1.2. *Known results for the contact process on* $\mathbb{Z}^d$

The contact process was firstly studied on the straight line $\mathbb{Z}^1$. Liggett [14] and Durrett [7] contain the main results for the one-dimensional case. The seminal work of Bezuidenhout & Grimmett [2] used different geometric constructions to get the results for the high dimensional case, including:

(a) $\lambda_1 = \lambda_2$ (denote by λ_c the common value);
(b) the process with parameter λ_c dies out;
(c) the complete convergence theorem holds for all $\lambda > 0$, that is, for any $A \subseteq \mathbb{Z}^d$,

$$\xi_t^A \Rightarrow \overline{\nu} \cdot \mathbf{P}(\xi_t^A \neq \emptyset \text{ for any } t \geq 0) + \delta_\emptyset \cdot \mathbf{P}(\xi_t^A = \emptyset \text{ for some } t \geq 0)$$

as t tends to infinity, where $\overline{\nu}$ denotes the upper invariant measure (that is, the weak limit of the distribution of $\xi_t^{\mathbb{Z}^d}$ as $t \to \infty$), $\delta_\emptyset$ denotes the measure putting mass one on the empty set, and "$\Rightarrow$" stands for weak convergence;

(d) the shape theorem holds, that is, there exists a convex subset $U \subseteq \mathbb{R}^d$, such that for any $\varepsilon > 0$,

$$(1-\varepsilon)U \subseteq \frac{1}{t}H_t^0 \subseteq (1+\varepsilon)U \quad \text{eventually}$$

almost surely on the event that $\xi_t^0 \neq \emptyset$ for any $t \geq 0$, where 0 denotes the origin of $\mathbb{Z}^d$, and $H_t^0 = \bigcup_{0 \leq s \leq t} \xi_s^0$ denotes the set of vertices that have ever been infected before time t.

Remark. (1) In the following, we use **(a)**, **(b)**, **(c)** and **(d)** to denote the above four results for short.

(2) In Bezuidenhout and Grimmett [2], the notations of the two critical values λ_1 and λ_2 have not been mentioned (they first appeared in Pemantle [19]). But Theorem 3 in [2] implies this conclusion.

(3) Bezuidenhout and Grimmett [2] contains the proof of **(a)** and **(b)**. They didn't give the formal proof of **(c)** and **(d)**. The detailed proof of these results are provided in Liggett [17].

1.3. *Contact processes in random environments*

Liggett [15] gives a general setting for the contact process in random environment. That is, the transition rates in (1.1) are modified by

$$\begin{cases} \xi_t \to \xi_t \setminus \{x\} \text{ for } x \in \xi_t \text{ at rate } \delta_x, \\ \xi_t \to \xi_t \cup \{x\} \text{ for } x \notin \xi_t \text{ at rate } \sum\limits_{y \in \xi_t,\ y \sim x} \lambda_{(y,x)}, \end{cases} \tag{1.2}$$

where $\{\delta_x : x \in V\}$ and $\{\lambda_e : e \in E\}$ are random variables chosen in a stationary ergodic manner. That means, the recovery rates and infection rates become random.

The contact process in random environment was first studied on $\mathbb{Z}^1$ (see Bramson, Durrett & Schonmann [3], Liggett [15], [16], Klein [13], Newman & Volchan [18], etc), focusing on the conditions for survival (extinction). The high-dimensional case is more challenging. Chen & Yao [5], [6] settled **(c)** in the half space case when $\delta_x \equiv 1$ and λ_e's are independent and identically distributed. Garet & Marchand [8], [10] settled **(d)** when $\delta_x \equiv 1$ and λ_e's are stationary, ergodic and properly bounded.

All the above models belong to contact processes in static random environments, that is, the environment does not change as time goes. There are some models concerning contact processes in dynamic random environments; see, for example, Broman [4], Remenik [20], Steif & Warfheimer [21], etc. The main difficulty in studying the processes in static random environments is that the process is not Markovian under the annealed (or averaged) law.

1.4. *Organization of this article*

In Section 2, we will consider **(a)**, **(b)** and **(c)** in the half space case. **(c)** has been proved in Chen & Yao [6], so we only state the proof sketch heuristically. **(a)** and **(b)** must be posed in a "parameterized version", and will be proved using the idea of Grimmett and Mastrand [11] (they considered the percolation model). A special case is an addition to Chen & Yao [5], since we did not prove that the critical process dies out for the half space percolation cluster case there. In Section 3, we will compare our results with some related works and explain why our methods cannot apply to the whole space case.

2. Contact Process in a Random Environment on $\mathbb{Z}^d \times \mathbb{Z}^+$

The graph we are considering is $(\mathbb{H}, \mathbb{E})$, where $\mathbb{H} = \mathbb{Z}^d \times \mathbb{Z}^+$ $(d \geq 1)$, with $\mathbb{Z} = \{0, \pm 1, \pm 2, \ldots\}$ and $\mathbb{Z}^+ = \{0, 1, 2, \ldots\}$; and $\mathbb{E} = \{(x, y) : \ x, y \in \mathbb{H}, \ \|x - y\| = 1\}$, with $\|\cdot\|$ denoting the Euclidean norm. The graph is treated as unoriented; that is, (x, y) and (y, x) denote the same edge for all $x, y \in \mathbb{H}$ satisfying $\|x - y\| = 1$. The environment is defined via (1.2) with $\delta_x \equiv 1$ and λ_e's being i.i.d.$\sim \mu$, where $\mu([0, +\infty)) = 1$.

2.1. *The complete convergence theorem (c)*

The complete convergence theorem **(c)** was proved in Chen & Yao [5] for the half space percolation cluster case (where μ follows the Bernoulli distribution), and was proved later in Chen & Yao [6] for the general half space case. In [5] and [6] we only proved the half plane case (when $d = 1$). The higher dimensional case (when $d \geq 2$) can be settled with no substantial difficulty.

Step 1: The block conditions

The "building block" of the proof procedure is the setup of the "block conditions" (denoted by **(BC)** in the following) for the survival of the process (Proposition 3.2 in Chen & Yao [6]). Intuitively, suppose the process survives, then for any $\varepsilon > 0$ sufficiently small, we can construct two kinds of boxes whose sizes depend on ε but with almost fixed shape, such that with probability greater than $1 - \varepsilon$, a horizontal seed (i.e. an interval with all infected vertices) on the bottom of each box can give birth to another vertical seed with the same length on the right side of the box, with infection path being entirely contained in the interior of the box. The two kinds of boxes are called by "S-box" and "L-box" respectively, where "S" stands for "short" and "L" stands for "long". See Figure 1 for illustration. Note that when $d \geq 2$, only one kind of boxes are needed by using the skew lines.

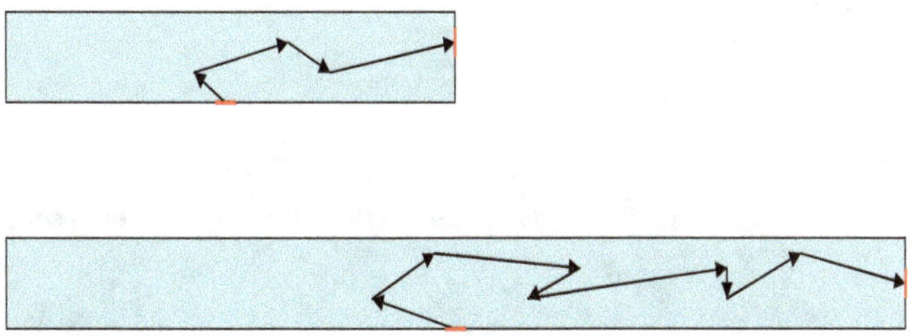

Fig. 1. Construction of blocks

The proof of Proposition 3.2 in Chen & Yao [6] (which is the main contribution of that paper) was divided into three cases (with three totally different proofs), one of which covers the proof of Lemma 3.4 in Chen & Yao [5] as a special case. Note that there is a similar disjunction in Garet & Marchand [9].

Step 2: The dynamic renormalization construction

Then we will use the S-boxes and L-boxes to construct a route, such that with probability greater than $1 - \varepsilon$, a seed in a fixed square is joined through this route to some seeds in the other two fixed squares depending on ε and having the same size (one above, the other on the right). See Figure 2 for illustration.

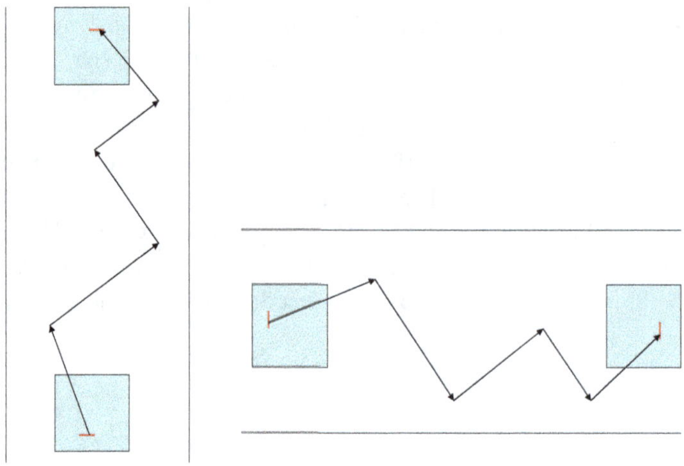

Fig. 2. Producing new seeds in the north square and the east square

Next, we fix $\varepsilon > 0$ sufficiently small and iterate the above procedure several times in both directions (to the "east" and to the "north"), then treat the graph in a larger scale (called the "dynamic renormalization" procedure). Figure 3 gives an illustration for the case that the "large scale length" of the "large scale square" is 3. Furthermore, denote by $T(n, \varepsilon)$ the time span that the seed in the "southwest" generate the seed in the "northeast" in the "large scale square" with "large scale length" n.

The next proposition is Proposition 4.1 in Chen & Yao [6], which is the main result for the dynamic renormalization. It tells us that as n tends to infinity, $T(n, \varepsilon)$ follows the "almost linear growth" property asymptotically. The detailed proof can be found in Appendix 2 of Chen & Yao [5].

Proposition 2.1: *Suppose that (BC) holds. Then there exists* $\overline{W} > 0$, *such that*

$$\lim_{\varepsilon \to 0+} \liminf_{n \to \infty} \mathbf{P}\left(\frac{7\overline{W}}{6} n < T(n, \varepsilon) < \frac{11\overline{W}}{6} n\right) = 1.$$

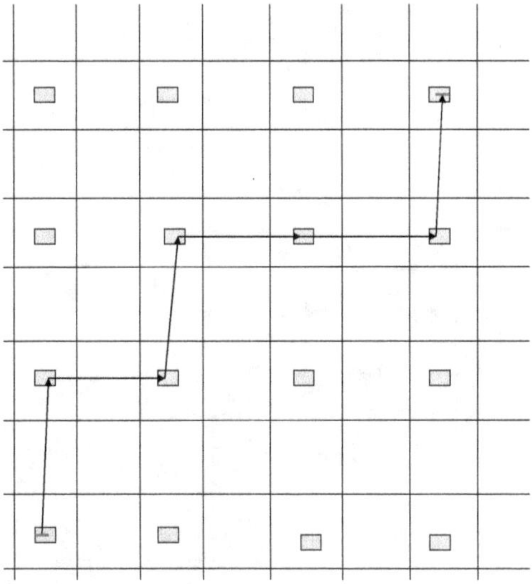

Fig. 3. Dynamic renormalization

Step 3: Checking the equivalent conditions for (c)

Having made the above preparations, we can prove **(c)** by checking the following two assertions (Theorem 1.12 of Liggett [17]):

(c1) $\mathbf{P}\left(x \in \limsup_{t\to\infty} \xi_t^A\right) = \mathbf{P}(\xi_t^A \neq \emptyset \text{ for any } t \geq 0)$ for all $x \in \mathbb{H}$ and $A \subset \mathbb{H}$.

(c2) $\lim_{M\to\infty} \liminf_{t\to\infty} \mathbf{P}(\xi_t^{B_x(M)} \cap B_x(M) \neq \emptyset) = 1$ for all $x \in \mathbb{H}$, where $B_x(M)$ is defined to be the "ball" centered at x and with radius M (but restricted on $\mathbb{H}$).

The rigorous proof of **(c1)** and **(c2)** can be found in Subsections 5.1 and 5.2 of Chen & Yao [6], respectively. The intuitive idea of **(c1)** is to iterate the construction posed in Proposition 2.1 four times, as shown in Figure 4 intuitively. **(c2)** can be obtained by the above observation together with some extra tricks to prove the assertion that every remote site cannot be infected in a short time.

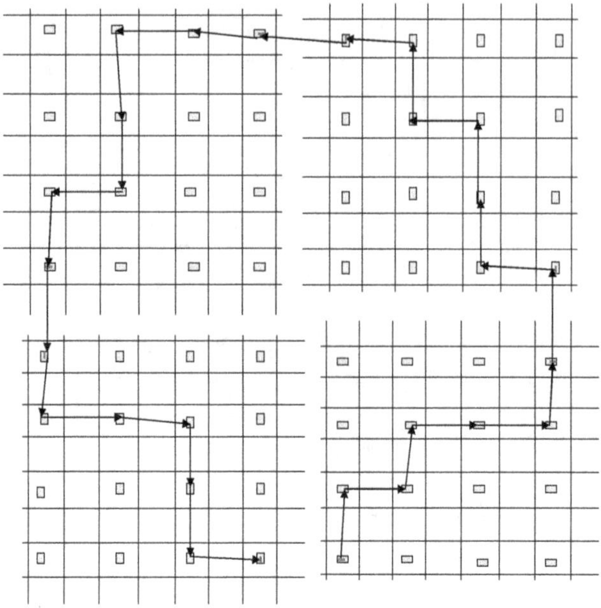

Fig. 4. Description of **(c1)**

2.2. *(a) and (b) in the parameterized version*

In order to consider **(a)** and **(b)**, we need to parameterize the model by changing the infection rate λ_e by $\lambda \cdot \lambda_e$, where the λ_e's are still i.i.d. $\sim \mu$, and $\lambda > 0$ is a free parameter. By monotonicity, we can still define λ_1 and λ_2, which are almost surely constants by translation invariance.

Therefore, there is no difference between quenched law and annealed law when considering **(a)** and **(b)**. For simplicity, we use $\mathbf{P}$ for the measure, and use $\mathbf{P}_\lambda$ if we want to stress that the parameter is λ.

It is easy to see that **(c)** also holds for the parameterized version. Therefore, **(a)** holds trivially by **(c1)**. So we can denote by λ_c the common value of λ_1 and λ_2.

Remark. Whether $\lambda_c \in (0, +\infty)$ or not depends on the distribution μ. When there exists $M \in (0, +\infty)$ such that $\mu([0, M]) = 1$, then $\lambda_c > 0$. And when there exists $b \in (0, +\infty)$ such that $\mu([b, +\infty)) = 1$, then $\lambda_c < +\infty$. It will be interesting to consider the case when the support of μ is $(0, +\infty)$.

To prove **(b)**, that is, the critical process dies out in the parameterized version (which includes the half space percolation cluster case in Chen & Yao [5]), we need the following lemma. The idea comes from Grimmett & Mastrand [11], where they considered the percolation model.

Lemma 2.2: *If* $\mathbf{P}_\lambda(\forall t > 0,\ \xi_t^0 \neq \emptyset) > 0$*, then there exists* $\delta > 0$*, such that*

$$\mathbf{P}_{\lambda-\delta}(\forall t > 0,\ \xi_t^0 \neq \emptyset) > 0.$$

Proof. Since $\mathbf{P}_\lambda(\forall t > 0,\ \xi_t^0 \neq \emptyset) > 0$, the block conditions **(BC)** hold. Fix $\varepsilon > 0$ sufficiently small as well as the variables (including the sizes of the two kinds of boxes, the length of the seeds, and the time span) which guarantee **(BC)** to hold. Since all these variables have upper bounds depending only on the above ε, it follows from the continuity of the finite-time process in the parameter λ that there exists $0 < \delta < \lambda$, such that **(BC)** also hold with parameter $\lambda - \delta$ under the same ε and the above variables. By Proposition 2.1, we have

$$\lim_{\varepsilon \to 0+} \liminf_{n \to \infty} \mathbf{P}_{\lambda-\delta}(T(n, \varepsilon) < \infty) = 1,$$

where $T(n, \varepsilon)$ is defined in Subsection 2.1. So we can choose $\varepsilon' > 0$ sufficiently small, such that

$$\liminf_{n \to \infty} \mathbf{P}_{\lambda-\delta}(T(n, \varepsilon') < \infty) > \frac{1}{2}.$$

Therefore, we have

$$\mathbf{P}_{\lambda-\delta}(T(n, \varepsilon') < \infty \quad i.o.) \geq \liminf_{n \to \infty} \mathbf{P}_{\lambda-\delta}(T(n, \varepsilon') < \infty) > \frac{1}{2}.$$

Furthermore, since $T(n, \varepsilon') < \infty$ *i.o.* implies that $\xi^{[-r,r]}$ can infect infinitely many sites, and therefore, the infection will persist forever. Here $r = r(\varepsilon')$ denotes half of the length of the initial seed in **(BC)** corresponding to the above ε'. See Figure 4 for intuition. This implies

$$\mathbf{P}_{\lambda-\delta}(\forall t > 0,\ \xi_t^{[-r,r]} \neq \emptyset) > \frac{1}{2}.$$

Together with the trivial fact $\mathbf{P}_{\lambda-\delta}(\xi_1^0 = [-r, r]) > 0$, we obtain

$$\mathbf{P}_{\lambda-\delta}(\forall t > 0,\ \xi_t^0 \neq \emptyset) > 0,$$

as desired. $\square$

Proof of (b). Suppose $\mathbf{P}_{\lambda_c}(\forall t > 0,\ \xi_t^0 \neq \emptyset) > 0$. Then by Lemma 2.2, there exists $0 < \delta < \lambda_c$, such that $\mathbf{P}_{\lambda_c-\delta}(\forall t > 0,\ \xi_t^0 \neq \emptyset) > 0$, contradicting with the definition of λ_c. □

3. Concluding Remarks and Discussions

The main idea of the proof procedure of **(c)** (then **(a)** and **(b)** in the parameterized version) is enlightened by Bezuidenhout & Grimmett [2], that is, using the "dynamic renormalization" argument. The argument first appeared in Grimmett & Mastrand [11] and Barsky, Grimmett & Newman [1], where the authors considered the percolation model. But there are some big differences in our model. The "block conditions" in Bezuidenhout & Grimmett [2] contain their Lemma 7 (which deals with "space" by using the fact that events depending on disjoint subgraphs are relatively independent) and Lemma 18 (which deals with "time" by using the Markov property of the process). However, In order to make good use of some symmetric properties, we need to consider the annealed law first (Step 1 and Step 2 in the proof procedure), then go back to the quenched law to get the desired result (Step 3 in the proof procedure). Under the annealed law, the fact that events depending on disjoint subgraphs are relatively independent still holds, but the Markov property does not hold any more. In consequence, if we consider the whole space case, we can only get a result similar to Lemma 7 in [2] and cannot get the result similar to Lemma 18 in [2]. And furthermore, we cannot get the desired result in the whole space case. On the other hand, the "space block" in $\mathbb{Z}^d \times \mathbb{Z}^+$ constructed by Proposition 3.2 in Chen & Yao [6] has similar function as the "space–time block" in $\mathbb{Z}^d \times \mathbb{R}^+$ constructed by Lemmas 7 and 18 in Bezuidenhout & Grimmett [2]. That is why we can get the results in the half space case.

We believe that the results are true for the whole space case. There may be some possible ways to prove the whole space case. The first possible idea is to prove directly. It is not easy. Even in the percolation cluster case, it is of the same difficulty as the long-existing problem that whether there is percolation at the critical point in the whole space case, which is clear in the half space case. The second possible idea is to prove that the critical value in the whole space case is the same as it in the half space case, which is clear for the percolation case as well as the contact process case, but it is not known for the contact process on the percolation cluster case. We will think about it in future research.

We mention at the end of this article that Garet & Marchand [8], [10] deal with the shape theorem **(d)** in the whole space case under the assumption that $\delta_x \equiv 1$ and λ_e's are stationary, ergodic, and take value in $[\lambda_{\min}, \lambda_{\max}]$, where $\lambda_{\min} > \lambda_c(\mathbb{Z}^d)$, and $\lambda_{\max} < +\infty$. Getting rid of the boundedness assumption of λ_e's may be a similar challenge as our model.

Acknowledgment

We would like to thank Matthias Birkner, Rongfeng Sun, Jan Swart, as well as an anonymous referee for their useful suggestions on an earlier version of the manuscript. Partial supports from the Institute for Mathematical Sciences at the National University of Singapore and a grant from the National Science Foundation of China (No.11671145) are gratefully acknowledged.

References

1. D. J. Barsky, G. R. Grimmett and C. M. Newman: Percolation in half-spaces: equality of critical densities and continuity of the percolation probability, *Probab. Th. Rel. Fields* **90** 111-148 (1991).
2. C. Bezuidenhout and G. R. Grimmett: The critical contact process dies out, *Ann. Probab.* **18** 1462-1482 (1990).
3. M. Bramson, R. Durrett and R. H. Schonmann: The contact process in a random environment, *Ann. Probab.* **19** 960-983 (1991).
4. E. I. Broman: Stochastic domination for a hidden Markov chain with applications to the contact process in a randomly evolving environment, *Ann. Probab.* **35** 2263-2293 (2007).
5. X. X. Chen and Q. Yao: The complete convergence theorem holds for contact processes on open clusters of $\mathbb{Z}^d \times \mathbb{Z}^+$. *J. Statist. Phys.* **135** 651-680 (2009).
6. X. X. Chen and Q. Yao: The complete convergence theorem holds for contact processes in a random environment on $\mathbb{Z}^d \times \mathbb{Z}^+$. *Stoch. Proc. Appl.* **122** 3066-3099 (2012).
7. R. Durrett: *Lecture Notes on Particle Systems and Percolation.* Wadsworth, Pacific Grove, CA, 1988.
8. O. Garet and R. Marchand: Asymptotic shape for the contact process in random environment, *Ann. Appl. Probab.* **22** 1362-1410 (2012).
9. O. Garet and R. Marchand: The critical branching random walk in a random environment dies out, *Electron. Commun. Probab.* **18** Article 9 (2013).
10. O. Garet and R. Marchand: Large deviations for the contact process in random environment, *Ann. Probab.* **42** 1438-1479 (2014).
11. G. R. Grimmett and J. M. Mastrand: The supercritical phase of percolation is well behaved, *Proc. R. Soc. Lond. Set. A* **430** 439-457 (1990).
12. T. E. Harris: Contact interactions on a lattice, *Ann. Probab.* **2** 969-988 (1974).
13. A. Klein: Extinction of contact and percolation processes in a random environment, *Ann. Probab.* **22** 1227-1251 (1994).

14. T. M. Liggett: *Interacting Particle Systems.* New York: Springer-Verlag, 1985.
15. T. M. Liggett: Spatially inhomogeneous contact processes. In *Spatial Stochastic Processes. A Festschrift in honor of the Seventieth Birthday of Ted Harris.* Birkhäuser, Boston, 1991.
16. T. M. Liggett: The survival of one dimensional contact processes in random environments, *Ann. Probab.* **20** 696-723 (1992).
17. T. M. Liggett: *Stochastic Interacting Systems: Contact, Voter and Exclusion Processes.* Springer, Berlin, Heidelberg, 1999.
18. C. M. Newman and S. B. Volchan: Persistent survival of one-dimensional contact processes in random environments, *Ann. Probab.* **24** 411-421 (1996).
19. R. Pemantle: The contact process on trees, *Ann. Probab.* **20** 2089-2116 (1992).
20. D. Remenik: The contact process in a dynamic random environment, *Ann. Appl. Probab.* **18** 2392-2420 (2008).
21. J. E. Steif and M. Warfheimer: The critical contact process in a randomly evolving environment dies out, *ALEA* **4** 337-357 (2008).

www.ingramcontent.com/pod-product-compliance
Lightning Source LLC
LaVergne TN
LVHW010330220926
848484LV00004BB/871